Planetary Systems from the Ancient Greeks to Kepler

Planetary Systems from the Ancient Greeks to Kepler

Theodor S. Jacobsen

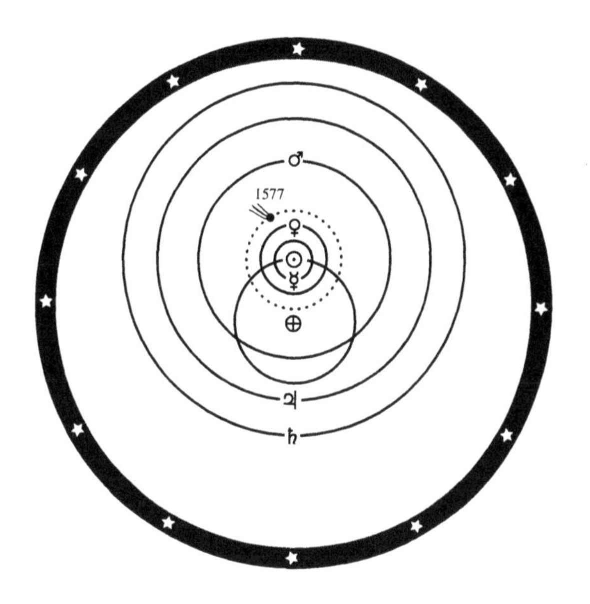

Department of Astronomy, University of Washington
in association with
The University of Washington Press

Copyright ©1999 by the Astronomy Department, University of Washington

All rights reserved. No part of this publication may be reproduced or transmitted in any form or by any means, electronic or mechanical, including photocopy, recording, or any information storage or retrieval system, without permission in writing from the publisher.

Library of Congress Cataloging-in-Publication Data
Jacobsen, Theodor S.
Planetary systems from the ancient Greeks to Kepler / Theodor S. Jacobsen.
p. cm.
Includes bibliographical references.
ISBN 0-295-97821-X (alk. paper)
1. Planetary theory—History. I. Title.
QB361.J33 1999
521′.3—dc21

99-14005
CIP

Contents

Figures

Professor Jacobsen measuring radial velocities in his office at the University of Washington in 1985.

Photograph by Harlan H. Chinn©, Seattle, Wash.

Foreword

Placing the sheet music for Grieg's piano concerto in front of him, Professor Jacobsen squints up from the keyboard of his beloved baby grand and matter-of-factly intones, "To play well, you must know the details." Although certainly true of the demanding passage that began to fill the room, he was actually speaking metaphorically of his personal motivation for writing this book. He originally set out only to educate himself regarding the substance of the contributions of the great astronomers of antiquity – those who brought the quest for cosmic mechanism from the bold, impious questions of the pre-Socratics to Kepler's relentless pursuit of answers tightly constrained by observation. For Theodor Jacobsen, telling the story of the profound intellectual struggle at the fertile interface between ideas and observation of the natural world is an undertaking that lives almost wholly within the details. In his recounting, those details are revealed within the 82 diagrams that form the backbone of this monograph and illuminate the history of a grand model-building enterprise that sought to capture celestial motions in a web of geometry.

Theodor Siegumfeldt Jacobsen was born to a middle-class family in the small border town of Nyborg in southern Denmark in 1901. During this period the socialist movement was making life difficult for capitalists, and the Jacobsens consequently relocated not far from Palo Alto, California, when Theodor was 16. Not surprisingly, he later attended Leland Stanford Jr. University and found himself the only astronomy major in his graduating class. But this was no idle career choice, as Tycho Brahe remained a hero to many Danes of his generation. Theodor went on to Berkeley as a Lick Observatory fellow and received his Ph.D. in 1926, working with his advisor, J. H. Moore, as well as with W. W. Campbell and Joel Stebbins. Following a brief but productive appointment to the staff of Mt. Wilson Observatory, where he made hundreds of spectroscopic observations with both the 100-inch and the 60-inch telescope, Dr. Jacobsen returned to Lick to continue his work with Moore, Stebbins, and Campbell. The University of Washington hired him in 1928, and he remained the university's sole astronomer until 1965. It was only after his retirement in 1971 that he could find the time to pursue his lifelong interest in the astronomy of antiquity.

We discovered the manuscript for this book more than a decade after its completion, neatly stacked on a table in his home where he had brought it out to recapture some forgotten detail. What a delightful surprise for us. We determined that this manuscript should be more widely appreciated, and over the subsequent three years we have polished the text and figures and secured its publication. In reviewing the major historical planetary systems, it finds a niche between the popular books of Dreyer, Caspar, and Koyré and the technical treatises of Schmeidler, Neugebauer, and Swerdlow and Neugebauer, which are accessible only to specialists. But again, the truly distinguishing feature is the abundance of detail in the geometrical diagrams, through which Professor Jacobsen offers a clear exposition of the intricacies of ancient astronomy and the developing art of geometrizing celestial motion.

Despite his desire to have the figures bound separately, enabling the reader to juxtapose a chosen diagram with any page of text, that idea was not practical. For the foreseeable future, however, high-resolution images of all figures will be available for downloading from the University of Washington Department of Astronomy Web site.

In bringing this work to publication, we editors have left the text largely as it was received, only occasionally restructuring or reordering passages to enhance clarity. We gratefully acknowledge the contributions of Professor James Evans of the University of Puget Sound, who lent his expert advice to both the selection and seamless joining of sections in several chapters. We have not attempted to homogenize the mathematical notation within or among chapters, fearing we would introduce confusing disparities, especially between text and figures. Certainly we have made every effort to faithfully reproduce Professor Jacobsen's remarkable diagrams. With great patience and skill, early electronic drafts of the figures and text were prepared by Karen Fisher. The text was elegantly typeset in LaTeX by John Deeter. In the end, copy editor Pamela Bruton, working with the University of Washington Press, brought it all together with grace and style.

From our vantage at the close of the 20th century, we see this work as a distinctive chronicle of an epic quest for cosmic order. Copernican heliocentrism emerged from this process of discovery and, through Kepler, the subsequent formulation of Newton's universal gravitation. But implications extend beyond the 17th-century codification of empirical science. As Louis Dupré observes, on a grander scale this quest was a key element in our passage to modernity. We are gratified to have been parties to the publication of Theodor Jacobsen's bridge of ideas between past and present.

Paul Boynton
Donald Brownlee
Woodruff Sullivan III

Seattle, June 1999

Planetary Systems from the Ancient Greeks to Kepler

Introduction

Among specialists there is now and has always been a deplorable lack of knowledge regarding the historical basis of their subjects. The present book has resulted from the author's desire to broaden his own education in astronomy beyond the few specialties mastered during his specific training in the observational astronomy and astrophysics of the early part of this century. Its inception was triggered by an invitation in 1973 to give a public appreciation of "Copernicus as an Astronomer" – which ought to have been a delightful task for anyone who had posed as an astronomer for nearly fifty years! Yet I had to forgo discussing this magnificent subject because of my insufficient firsthand knowledge of the professional quality of the great man's *astronomical* work. The same restrictions also applied to the works of Ptolemy and Kepler. I knew that they were great names in astronomy – but exactly why?

In 1973, the year of the semimillennium of Copernicus's birth, one could read a score of pamphlets and appreciations extolling the enormity of the influence of his heliocentric viewpoint, not only on astronomy but on the advancement of science in general. However, so far as the substance of his work was concerned, one learned almost nothing. It would seem obvious that he strongly advocated "placing the Sun at the center of the planetary orbits" (which is not strictly true) and that "he was a great man" (which undoubtedly is true). But what one did not find in all this eulogy was a basis for judging him *as an astronomer.*

Similarly, Kepler is popularly thought to have hit upon his first law, that of elliptic motion, by merely plotting a few of Tycho's observations of Mars and then immediately noticing that the orbit was an ellipse with the Sun at one focus. "And why bother to study the work of the earlier astronomers," said one of my colleagues, "now that we know the work of Newton, which shows not only *how* but also *why* the planets move as they must?" Such a viewpoint, apparently due to admiration for Newton's genius (which, it is true, can scarcely be exaggerated), entirely misses the point. In an age before Newtonian mechanics one cannot judge astronomical excellence by the later success of Newton's work.

One aim of the present study is to supply a bird's-eye view of the *astronomical* nature of the work of some of the illustrious predecessors of Newton. Throughout history these astronomers were endlessly computing and figuring details to improve their mechanisms of prediction for the positions of heavenly bodies. For example, the systems of Eudoxus and Ptolemy were explicitly considered by their authors to be mere mathematical devices for predicting positions. Kepler may be said to have spent a lifetime in curve fitting. An astronomically minded reader should not be satisfied by a mere assertion of these facts but should wish to examine just what the various systems involved and how they were used. Histories of science contain paragraph after paragraph on the influence of great ideas on "this, that, and everything" but rarely the details of their development. In this work, however, the details *will* be shown.

This book, then, is not another history of astronomy, still less a history of astronomers. Nor is it a history of pre-Newtonian astronomy – for this would require detailed discussions of the numerous approaches to gravitational astronomy that took place, especially in the seventeenth century. Instead, it is primarily an attempt to describe the processes whereby real astronomers prior to Newton derived a detailed knowledge of the cosmos by observing the heavens and actually trying out detailed models to account for their observations. We will not include among astronomers those figures such as Thales and Plato whose work was primarily in philosophy or cosmogony. We will concentrate on purely astronomical principles useful in their time and only briefly comment on cosmological systems, however brilliant.

Concerning the style of the exposition, I have chosen to make it as elementary and geometrical as possible, because it is my predilection and because that was the original style of the ancients. There will of course be some analytical results quoted where geometry may be inadequate or too cumbersome. However, since geometrical visualization is essential to any typically astronomical thinking, I have sometimes chosen a geometrical demonstration, although perhaps not quite rigorous, in preference to an analytical one. Although there is some discussion of the basic principles of astronomy, a good background in astronomy will be necessary to follow the bulk of the arguments herein. Many of the mathematical details were suggested by N. Herz's *Introduction to Astronomy* (1897). The clarity of his expositions first called my attention to the fact that the present project was possible, and that nearly all of ancient and Renaissance astronomy could be explained using no more than elementary mathematics.

Ideally, one should study the *original* writings of all the great astronomers to fully appreciate their efforts. Such a procedure, however, would be beyond any but the most energetic specialists, and would bring no closer the goal of elucidating in a perspicuous manner the astronomical verities buried in the mountains of details. Fortunately, excellent condensations of the great books by Ptolemy, Copernicus, and Kepler already exist. By using three or four of these, condensing yet further remaining details, supplying diagrams, and incorporating occasional simplifications from my own lecture notes on general and spherical astronomy, I believe I have reached the goal of this study. The book represents what I myself would like to have known about historical astronomy before taking up my advanced special education. It is designed to be a useful contribution with an approach between that of popular yet exact astronomical information, and that of formal, fully referenced, scholarly investigation.

A special feature of the book is the comprehensive nature of some of the diagrams, in which complete orbital mechanisms for a celestial body are represented in a single picture. This device permits the astronomically trained intuition to visualize the positions of a body that derive from its nonuniform motion. As an example, consider a good elementary diagram of the approximate progress of the so-called evection and variation of the Moon's apparent motion, as defined in Tycho's system. Such an illustration enables one to answer, in a far more direct way than would be possible from years of study of advanced celestial mechanics, the shamefully simple but intensely *astronomical* question: "How in this system does the Moon *actually appear* to move on the sky at any given moment?" In this way the present elementary study of these historical astronomical systems may also prove valuable for visualizing the salient facts of general astronomy.

As to the contents of the book, Chapter I contains the general body of observational astronomy known to the ancients and forming the subject of their explanatory endeavors. Each of the following chapters describes the work of one or more great astronomers. Each chapter has three parts: a short biographical introduction, an exposition of the astronomical system, and a brief evaluation of the astronomer's greatness.

CHAPTER I

Astronomical Knowledge of the Ancient Greeks

However deficient by modern standards was the astronomical knowledge of most early philosophers, one never ceases to wonder at the completeness and precision of some of their results, derived as they were from inaccurate observations made with the naked eye or with crude instruments. Whether one considers the early Babylonians or the Greeks, the main results of astronomy that *could* be discovered by the naked eye actually *were* discovered by the best ancient observers. What is more, if all astronomical knowledge were destroyed, and a primitive civilization again undertook to replace it by naked-eye observation, the same kind of results would eventually be sorted out and would survive as being relevant to the description of the appearances and movements in the sky.

The question of how the thinkers of this civilization would try to *explain* the astronomical appearances is, however, completely unanswerable. An infinite variety of myth and fable would probably again encumber any effort of rational description. How rich and anthropocentric were the imaginations of even the greatest thinkers in the dawn of our present civilization can be garnered from the descriptions that follow at the end of this chapter. First we will summarize, as briefly as is consistent with clarity, the observed facts that gradually became common knowledge among the ancient savants. Then very briefly we will state some of their interpretations, in order to see how much mental dross had to be cleared away before they could develop even primitive planetary systems. That some of these interpretations seem preposterous today does not in the least alter the fact that because of much less stringent observational tests then possible, several systems were fully scientific in their day, even by modern standards.

Knowledge of the Sun's, the Moon's, and the Planets' Apparent Motions

We will first record the main facts, visible to the naked eye, of the motions on the sky of the Sun, the Moon, and the planets, to serve as a basis for judging the completeness or inadequacy of the observational knowledge of each natural philosopher. We shall find that there were enormous differences in their astronomical knowledge, which varied from the ludicrously insufficient to the highly sophisticated. Differences as great can be found in the representations of the observed motions by their planetary systems. Yet the best of these astronomers probably knew more than they succeeded in having their systems explain.

In the following summary of observed facts approximate modern values are given. Some of the ancient values differ somewhat from these. When appreciable and/or relevant, these differences will be noted.

1. The **rising, setting,** and **diurnal motions** of the Sun, Moon, stars, planets, and comets. The main concepts of practical astronomy – the celestial poles, the celestial equator, the ecliptic, altitudes, azimuths, hour angles, celestial (ecliptic) latitudes and longitudes, right ascensions, and declinations – were well known to the best-informed ancients. Refraction was imperfectly known to a few (Cleomedes, Ptolemy).
2. The **Sun's diurnal motion**, rising, culmination, setting. Also known were seasonally unequal or seasonal diurnal arcs (time interval from rising to setting) and unequal or seasonal hours (the hour was defined as 1/12 of the diurnal arc).
3. The **motions of the planets** among the fixed stars are very irregular, mostly direct (eastward), often retrograde (westward), but always within the zodiac, an 18°-wide strip of sky centered on the ecliptic and running for 360° around the sky. Also well known were the aspects of the planets: conjunctions, oppositions, and quadratures. The stations (stopping and turning points) of the planets were known. The sidereal periods of the planets were known to range from about 88 days for Mercury to about $29\frac{1}{2}$ years for Saturn.
4. **Eclipses** of the Sun and the Moon are repeated in a period called the *saros* (meaning "repetition"). The approximate length of the saros is 18 years, $11\frac{1}{3}$ days, i.e., 223 synodic lunar months.
5. The **equinoxes** (points of intersection of the celestial equator with the ecliptic) slowly move westward on the ecliptic. This motion, called the "precession of the equinoxes," gives rise to a gradual change in the positions of the reference circles used for measuring right ascension, declination, and celestial longitude and latitude. In the course of centuries, the altered declinations of the fixed stars appreciably change their rising and setting points on the horizon and the lengths of their diurnal arcs.
6. The **Sun's apparent annual motion** on the ecliptic is always eastward, but not perfectly uniform. Each month the Sun travels approximately one sign of the zodiac (30° along the ecliptic). The length of the tropical year is $365\frac{1}{4}$ mean solar days. The mean solar day is approximately $3^{m}56^{s}$ longer than the sidereal day (the diurnal period of the fixed stars).
   a. Except for very minor perturbations, the annual apparent path of the Sun is along the **ecliptic**, a great circle inclined 23°27′ to the celestial equator. The Sun's declination δ thus changes continually, oscillating between limits of ±23°27′ in a period of one year. The Sun's right ascension α increases continuously, though not uniformly, gaining 360° in a sidereal year. The Sun's celestial (or ecliptic) longitude λ also increases continuously, though not uniformly, gaining 360° in a sidereal year. The Sun's celestial (or ecliptic) latitude β is zero at all times, except for very minor perturbations.
   b. The **Sun's apparent angular speed** eastward on the ecliptic is a maximum about 1 January (61′/mean solar day), and a minimum about 4 July (57′/day). As a consequence of this, the Sun on 5 April is about 2° ahead in celestial longitude, and on 5 October 2° behind, with respect to a fictitious reference body moving uniformly eastward on the ecliptic at a rate of 59′/day.

c. These continuously varying daily increments of the Sun's annual eastward motion on the ecliptic are symmetrically distributed as regards size with respect to a line joining the positions of the Sun on approximately 1 January and 4 July. This line, called **the line of apsides** of the Sun's orbit, extends in the direction from the constellation Gemini to Sagittarius. The Sun's perigee lies in Sagittarius at celestial longitude 282°, while the apogee and the center of its apparent orbit lie in Gemini at celestial longitude 102°.

d. The **apparent diameter** of the Sun varies continuously with a period of one year. It is a maximum (about 32.5′) on 1 January, and a minimum (about 31.5′) on 4 July. In a modern view the Sun is nearest the Earth (147 million km) on 1 January, and farthest from it (152 million km) on 4 July. The distance from the Sun of the Earth's orbital center in Sagittarius is about 2.5 million km.

e. The time taken by the Sun to move the 180° of the ecliptic from vernal equinox through summer solstice to autumnal equinox (the length of the summer **"half-year"**) is 186 days, while the time from autumnal equinox through winter solstice to vernal equinox (the length of the winter "halfyear") is 179 days.

7. The Moon:

a. The **Moon's apparent monthly motion** is always eastward, continuous but nonuniform, varying from about 11° to 15° per day with an average of 13° per day. The motion is along an orbit inclined 5°8′ to the ecliptic. The Moon is thus always within the 18°-wide zodiac. The average length of the Moon's sidereal period is about $27\frac{1}{3}$ days, and the average length of the synodic period is about $29\frac{1}{2}$ days. Thus the Moon's declination changes monthly between limits continually varying from ±18°19′ to ±28°35′ nine years later. Its right ascension and celestial longitude increase continuously, though not uniformly. Its celestial latitude changes continually between limits ±5°8′ (neglecting perturbations of ±10′).

b. **Phases** of new Moon, first quarter, full Moon, and last quarter refer to the moments when the celestial longitude differences between the Moon and the Sun are, respectively, 0°, 90°, 180°, and 270°.

c. The **nodes** of the Moon's orbit (crossing points on the ecliptic) **regress** (move westward) on the ecliptic, continuously though not uniformly, by 360° in 18.60 years. The **line of apsides** of the Moon's orbit **advances** (moves eastward) in the plane of the lunar orbit by 360° in 8.85 years.

d. Large perturbations continually modify the Moon's orbit. The size of the largest one, the **evection**, was first estimated by Hipparchus. The four largest ones can be detected by naked-eye measurements and were all recognized by Tycho Brahe.

The Aspects, Stations, and Retrograde Motions of the Planets

The **apparent motions** of the planets on the sky are not so simple as those of the Sun or Moon. As noted above, the Sun seems to move with nearly constant angular speed always toward the east on the great circle called the ecliptic, which is practically fixed among the stars. The Moon's apparent motion is also eastward at all times; much faster, more variable in speed, and more complicated than the Sun's motion; and always close to a great circle inclined about 5° to the ecliptic but slightly variable in inclination. The apparent paths of the planets are still more complicated but always within the 18°-wide zodiac belt centered on the ecliptic. The motion of a planet is alternately eastward and westward, generally either speeding up or slowing down, and often zigzag shaped or looped. The long-lasting eastward motion is called **progression**, or **direct** motion, and the shorter-duration westward motion is called **retrogression**, or **retrograde** motion. The positions and times when direct motion changes into retrograde, or vice versa, are called **stationary points**, or simply **stations**, of the planet. At these points, occupied only for an instant, the motion of the planet in celestial longitude momentarily stops, then reverses direction. The orbital motions of all planets outside Earth's orbit (i.e., the **superior** planets) are slower than that of Earth. Hence, as seen from Earth they move westward in the sky relative to the Sun. The planets inside Earth's orbit (the **inferior** planets) move faster than Earth. They move rapidly west of the Sun during their retrograde motion, then slowly overtake the Sun after their motion becomes direct. They appear as "morning stars" when on the west side of the Sun and as "evening stars" when on the east.

The angle between the lines of sight from the Earth to a body and to the Sun is called the **elongation** of the body from the Sun, or simply the elongation (see Fig. I.1). The name **aspect** has been given to a few relative positions specified by certain standard values of the elongations. Thus the values of 0, 90°, and 180° define the aspects **conjunction, quadrature,** and **opposition** (see Fig. I.1). Generally the differences are of geocentric celestial longitude, unless the term *absolute* is prefixed, in which case they are geocentric great-circle distances. Mercury and Venus never reach opposition, or even quadrature; their greatest possible elongations are, respectively, 28° and 47°. Their conjunctions reached during retrograde motion are named **inferior**, whereas **superior** conjunctions occur during direct motion for all planets. The superior planets have no inferior conjunctions, and the Moon has only inferior conjunctions; hence in these cases the term *conjunction* with no qualifying term is used. Opposition of an outer planet is reached somewhere near the middle of its retrograde arc. About this time its distance from the Earth is the smallest, and its brightness the greatest. An inner planet reaches its smallest and greatest distances from the Earth near the positions of inferior and superior conjunction, respectively. Its greatest brilliancy, however, occurs at a point intermediate between elongation and inferior conjunction.

The concept of aspects and elongations is often generalized to involve any three bodies. For example, when Venus is at greatest eastern elongation as seen from the Earth, the Earth is at western quadrature as seen from Venus. Aspects of bodies other than the Sun can also be described, for instance, "quadrature of Jupiter and Saturn, as seen from Mars." Unless there are modifiers, however, these terms always refer to phenomena with respect to the Sun as seen from the Earth.

The Periods of the Planets

The following two definitions will often be used:

1. The **sidereal period** of an orbiting body is the time of its revolution about a central body, using as reference a fixed direction as seen from the central body.

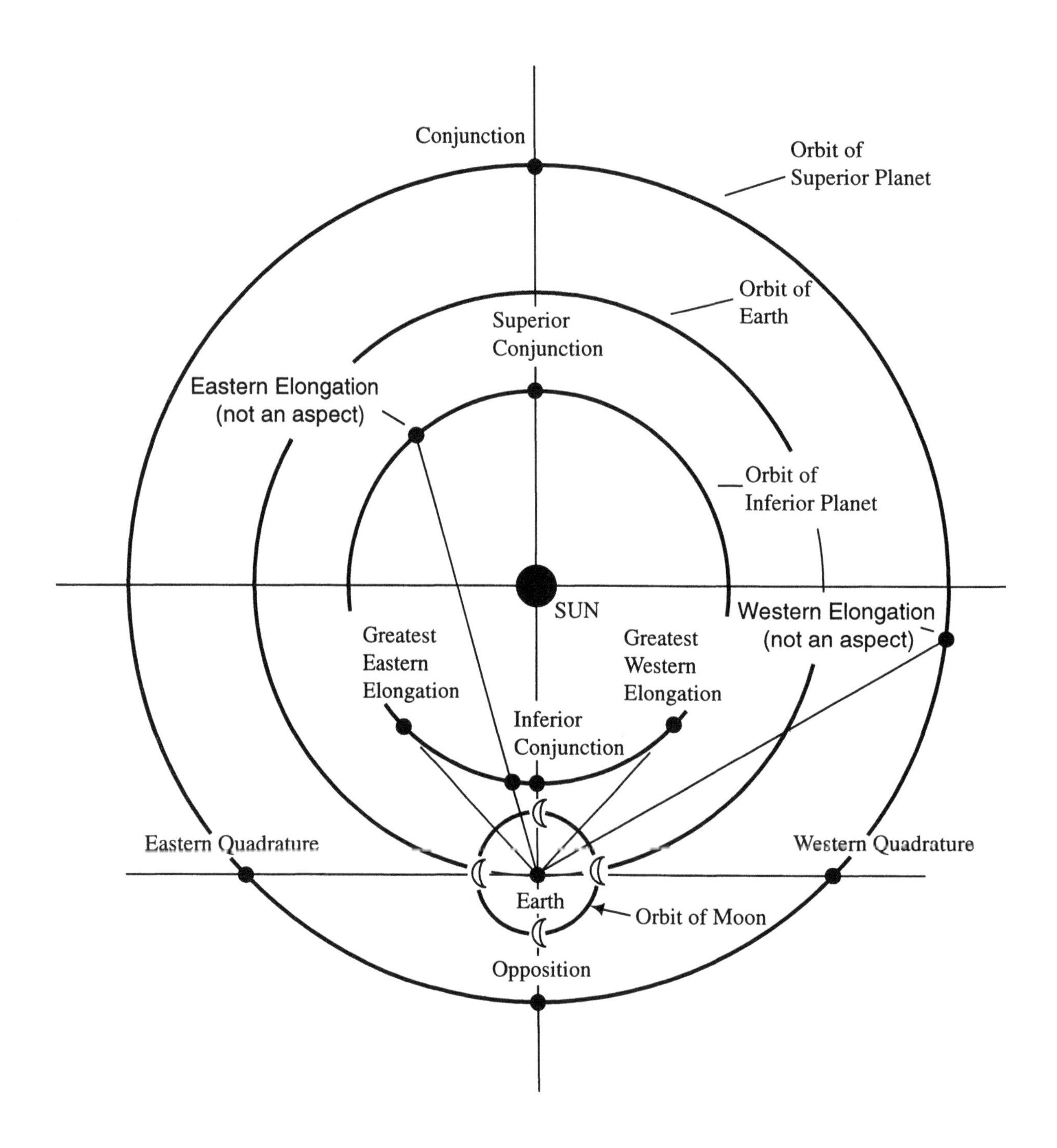

Figure I.1 — The aspects of the planets (configuration as seen from the north side of the ecliptic).

2. The **synodic period** of an orbiting body is the time between two successive similar aspects (usually conjunctions or oppositions) of the body as seen from the central body for which the aspects are defined.

A planet's **sidereal period** can be determined as follows. When a planet is passing its node, it is in the plane of the ecliptic. Since the Earth is also always in that plane (neglecting perturbations amounting at most to 0.8″), the planet's celestial latitude, both geocentric and heliocentric, will be zero, no matter where the Earth is in its orbit. At any other point of the planet's orbit except the node, its apparent latitude is not independent of the Earth's place. If, then, we observe the planet at two successive passages of the same node, the interval between the moments when the latitude becomes zero defines the planet's sidereal period – exactly if the node and the perihelion are stationary, and very approximately if the node and the perihelion are not absolutely stationary, as is actually the case. This method is often referred to as "Kepler's method of finding the sidereal period by observation."

A planet's **synodic period** is the interval between two successive oppositions or conjunctions, the opposition being the moment when the planet's longitude differs from that of the Sun by 180°. This angle between the planet and the Sun cannot be measured directly, but one can make (with a meridian circle) a series of observations of both the planet's and the Sun's right ascensions and declinations for several days before and after the date of opposition and reduce the observations to celestial longitude and latitude. The Sun will have to be observed at noon and the planet near midnight, but from the sequence of solar observations one can deduce the longitude of the Sun corresponding to any exact moment when the planet is observed. From these one finds the difference of longitude between the planet and the Sun at the time of each planetary observation and can deduce the moment of opposition when the difference was 180°. This can be ascertained with an accuracy of a few seconds of time. Since the orbits are not exactly circular, the interval between any two particular successive observations will be only an approximation to the desired quantity, which is the long-term *mean* synodic period. But repeated determinations in all parts of the orbit can yield a fair mean value. This can be checked, or perhaps improved, by comparing results from oppositions taken many years apart and dividing the elapsed time interval by the number of synodic periods (which is easily determined when the approximate value of the period is known). The accuracy of such a determination is especially great if the two oppositions occur at about the same time of year.

By inspection of Figure I.2, the simple relations between the sidereal and synodic periods of a planet may be visualized. In Figure I.2(a), let the Sun S, Earth E, and planet P_1 be in line with some fixed point at a particular moment. After a certain interval of time, say one mean solar day, let the Earth be at E_2, and the planet at P_2. Let

S = synodic period of planet (in mean solar days),

P = sidereal period of planet (in mean solar days),

E = sidereal period of Earth (in mean solar days).

Then $\frac{360}{P}$ is the angular rate of motion of the planet about the Sun in degrees per day. Similarly, $\frac{360}{E}$ is the angular rate of motion of the Earth about the Sun, and $\frac{360}{S}$ is the angular rate of gain of the Earth on the planet as they orbit about the Sun.

Inspection of the figure shows that $\angle P_1SE_2$ is the sum of $\angle P_1SP_2$ and $\angle P_2SE_2$. Therefore $\frac{360}{E} = \frac{360}{P} + \frac{360}{S}$, or $\frac{1}{P} = \frac{1}{E} - \frac{1}{S}$, which is the relation sought (for an outer planet). For an inner

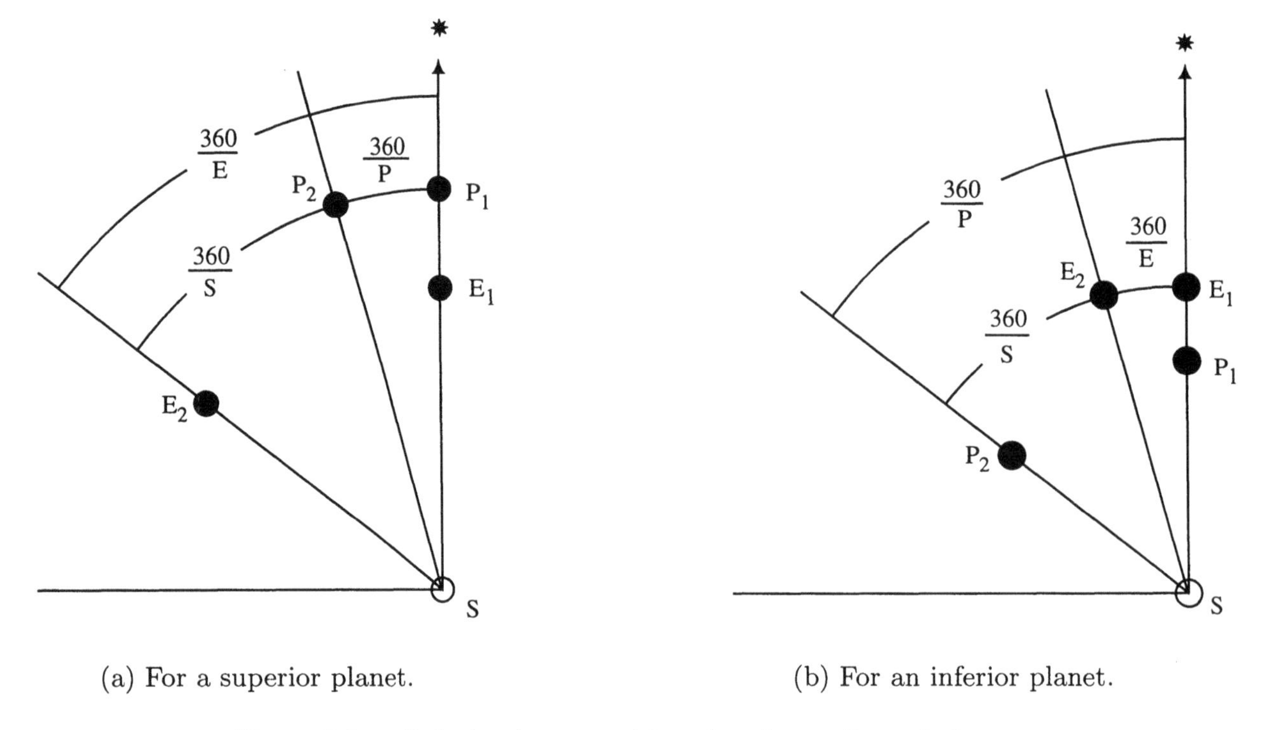

(a) For a superior planet.

(b) For an inferior planet.

Figure I.2 — Relation between sidereal and synodic periods.

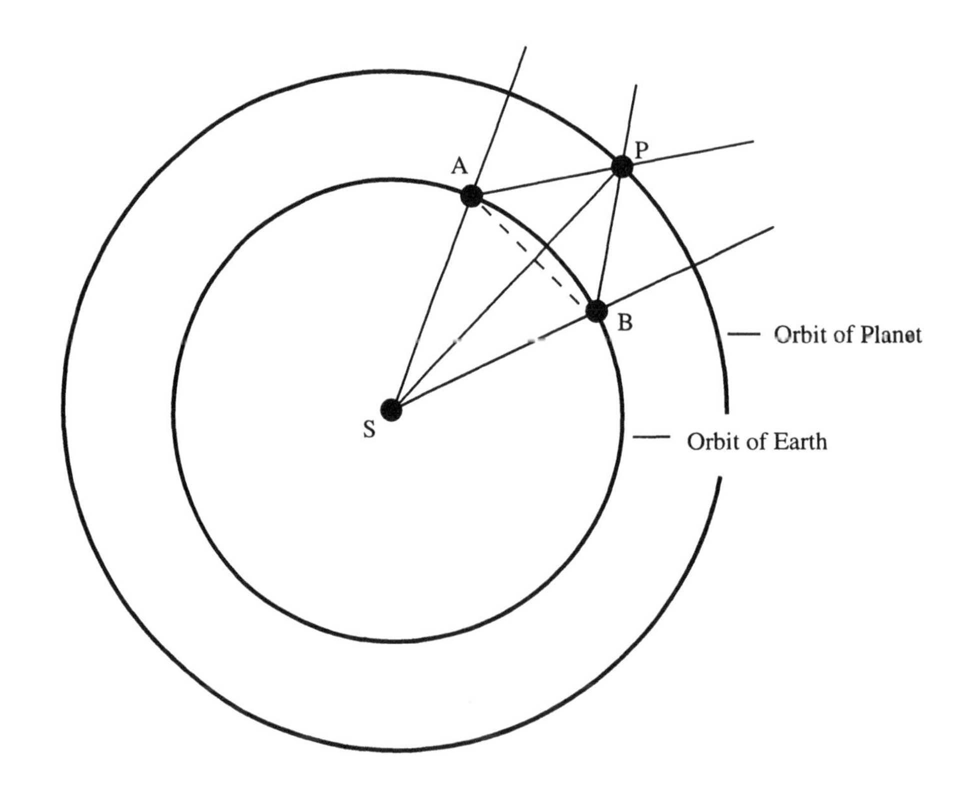

Figure I.2(c) — Kepler's method of triangulating the distance of a planet from the Sun

planet a similar derivation may be made by inspection of Figure I.2(b), giving $\frac{1}{P} = \frac{1}{E} + \frac{1}{S}$.

Having found the sidereal period of planet P (Fig. I.2(c)), one can determine by triangulation SP, the planet's distance from the Sun at different points of its orbit, using as baseline a chord of the Earth's orbit. In Figure I.2(c), let S represent the Sun, and P a point of Mars's orbit occupied by the planet at two times separated by one sidereal period (687 days). At these two times let the Earth in its orbit about the Sun be first at A, later at B, after having completed somewhat less than two yearly revolutions. Let the elongations of P from the Sun, i.e., $\angle SAP$ and $\angle SBP$, be measured. On average $\angle ASB = (2 \times 365 - 687)$ days $\times$ 0.99°/day = 42.6°, and in specific cases it can be found to the degree of accuracy to which the Earth's orbit and motion are known. Since SA and SB are also known, three parts of ΔASB are thus supplied. One can now compute the chord AB, and since $\angle SAB$ and $\angle SBA$ are already known or easily computed, one can find $\angle BAP = \angle SAP - \angle SAB$, and similarly $\angle ABP$. With these angles and the chord AB known, one may compute AP, the distance of Mars from Earth. Finally SP, the distance of Mars from the Sun, follows from ΔSAP, in which two sides and their included angle are now known. This process is often referred to as Kepler's method of triangulation. Originally Kepler obtained the elongations of Mars from the Sun by using Tycho Brahe's extensive journal of observations, interpolating the exact positions where necessary. From many such pairs of observations separated in time by 687 days, he found the distance of Mars from the Sun at many points. He thus determined the size and shape of the orbit and the speed with which the planet moved over various parts of its orbit.

Apparent Planetary Loops and Zigzags

Some typical shapes of apparent retrograde arcs of an outer planet, as seen from the Earth, may be visualized by contemplating a few special cases with the help of Figure I.3. When the planet is in the line of nodes nn', its heliocentric and geocentric latitudes are both zero, regardless of the Earth's location. The heliocentric longitude is then equal to ☊ at n or ☊ + 180° at n' (where ☊ is the longitude of the ascending node), but the geocentric longitude may have a wide range of values depending upon the position of the Earth in its orbit at that moment.

Case 1 (Figs. I.4(a) and I.4(c)). Suppose an opposition takes place near the line of nodes, with the Earth at a and the planet at n. The retrograde motion sets in sometime before, at the first station with the Earth at a_1 and the planet at n_1, and ends at the second station with the Earth at a_2 and the planet at n_2. While both heliocentric longitudes increase (that of the Earth faster than that of the planet), the geocentric longitude of the planet decreases due to the inner body overtaking the outer one. For an *ascending* passage the heliocentric latitude of the planet will steadily increase, being negative before node passage, zero at node passage, and positive after it. As seen from the Earth, i.e., from a point in the plane of the ecliptic closer to the planet than the Sun, the geocentric latitude of the planet will increase even faster than as seen from the Sun, reaching the value zero at node passage. If the latitude increase sets in slightly before the first station and lasts until after the second, the resulting apparent path of the planet as seen from the Earth will be akin to a zigzag motion as in Figure I.4(a). For an opposition occurring near the *descending* node, the shape of the path will be approximately as indicated in Figure I.4(c).

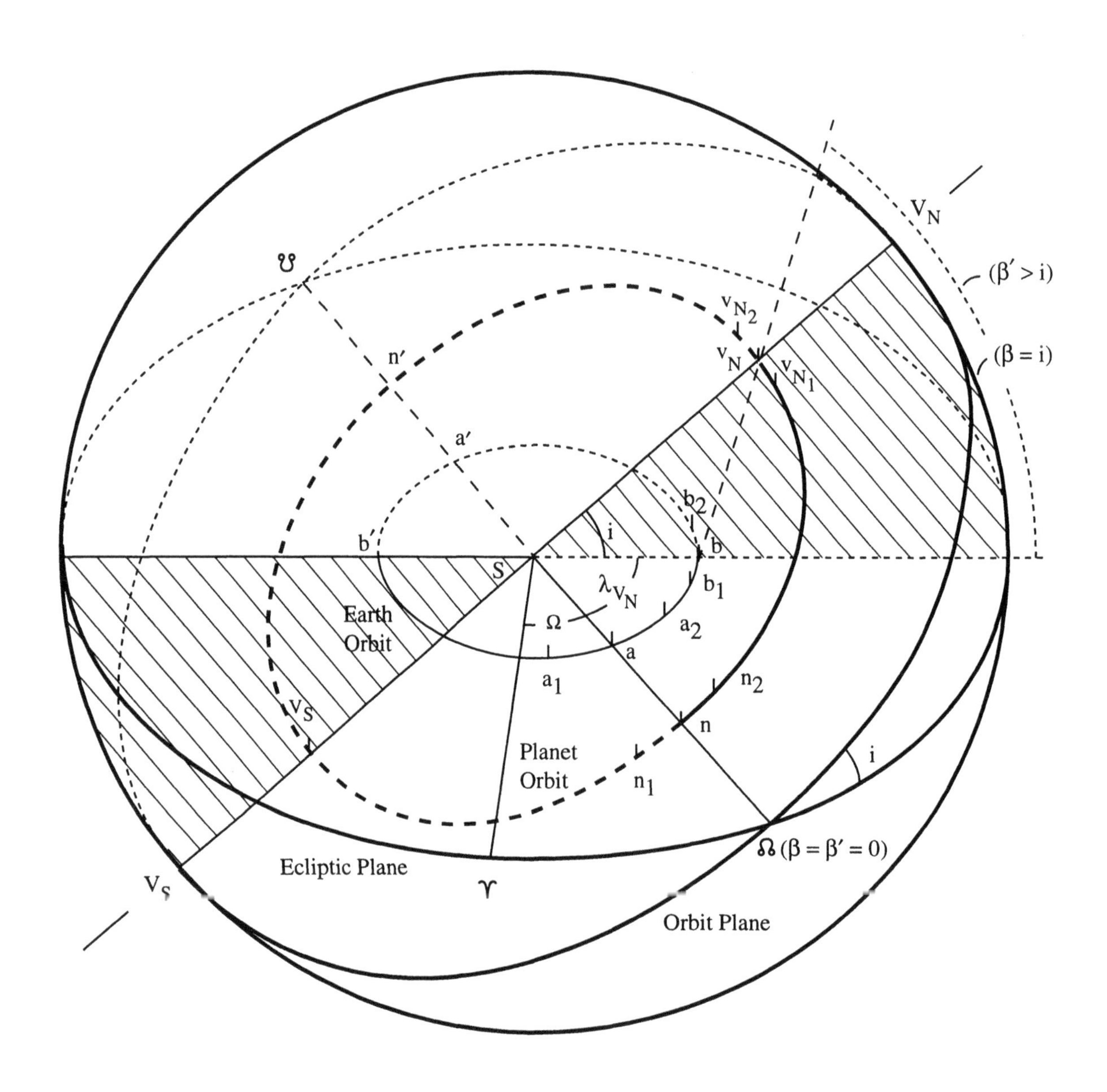

Figure I.3 — Orbits of the Earth and a superior planet, showing the projections of the orbit planes on the celestial sphere.

Case 2 (Figs. I.4(b) and I.4(d)). Suppose an opposition takes place near a point at right angles to the line of nodes, as when the Earth is at b and the planet at v_N. The retrograde motion sets in at the first station defined by the direction of $b_1v_{N_1}$ and lasts until the second station in direction $b_2v_{N_2}$. As seen from the Sun, the planet at opposition is north of the ecliptic near its maximum heliocentric latitude $EV_N = i$. As seen from the Earth it will be at an even greater geocentric latitude, defined by the direction bv_N. Since the Earth is closer to the planet than the Sun, the geocentric latitude will exceed the heliocentric ($\beta' > \beta$); and since the curvature of the Earth's orbit exceeds that of the planet's orbit, the increment of the geocentric latitude will be positive before and negative after opposition; i.e., the resulting geocentric latitude will increase from v_{N_1} to v_N and decrease from v_N to v_{N_2}. Hence the resulting apparent path of the planet as seen from the Earth will be akin to the loop pictured in Figure I.4(b). For an opposition occurring near the direction $b'v_S$, the path is as depicted in Figure I.4(d). Loops or zigzags of different degrees of symmetry (as in Figs. I.4(e) and I.4(f)) will occur at intermediate locations of the opposition points.

The term *great conjunction* is commonly applied to conjunctions of Jupiter with Saturn. These are separated by approximately 19.85 years and can occur in any part of the zodiac. If the oppositions of the two planets occur on not too different dates, two or even three conjunctions (hence the adjective *great*) may occur within a year's time, as shown schematically in Figure I.4(g). A famous example discussed by Kepler is the great conjunction of the year 7 B.C. It happened in Pisces, which was about halfway between the descending and ascending nodes of both planets. The paths in the sky were therefore both south of the ecliptic and all celestial latitudes were negative. A rough check of Ahnert's (1971) tables for the situation in 7 B.C. shows the progress of the phenomenon to have been as indicated in Figure I.5. The dates of opposition were 14 September for Saturn and 16 September for Jupiter, and the first conjunction happened on 29 May. Both planets were then moving eastward, approaching their first stations. Jupiter overtook Saturn in celestial longitude and produced the first conjunction. About 1 July both planets reached their first stationary points and started retrograding, coming to oppositions near the middle of their retrograde arcs about 15 September. On 3 October both planets were now retrograding, and Jupiter again raced past Saturn, producing the second conjunction. The third conjunction happened on 4 December, when Saturn was still retrograding and slowing down near its second station, while Jupiter, having already started its eastward motion, again passed the longitude of Saturn. Due to the slow motions of each planet near its second station, their optical propinquity would have been striking all through November and December. Kepler advanced this conjunction as an explanation of the "Star of Bethlehem."

The Elements of a Planetary Orbit

The elements of a planetary orbit are seven numbers, of which the first two, a and e, describe the size and shape of the orbit. The next three, i, Ω, and ω, describe its tilt and orientation in space. The last two, P and T, give the orbital position of the planet at a given time. Figure I.6 illustrates the elements of an elliptical orbit $Pe-A-P_1$ with major axis $2a$, minor axis $2b$, focus at S, and center at C. The planet is situated at P_1 at time t, at perihelion Pe at time T, and at aphelion A at time $T + P/2$. The planes of the ecliptic $EKLM$ and of the orbit $ORBT_1$, intersecting in the line of nodes NN', are both depicted from the north side.

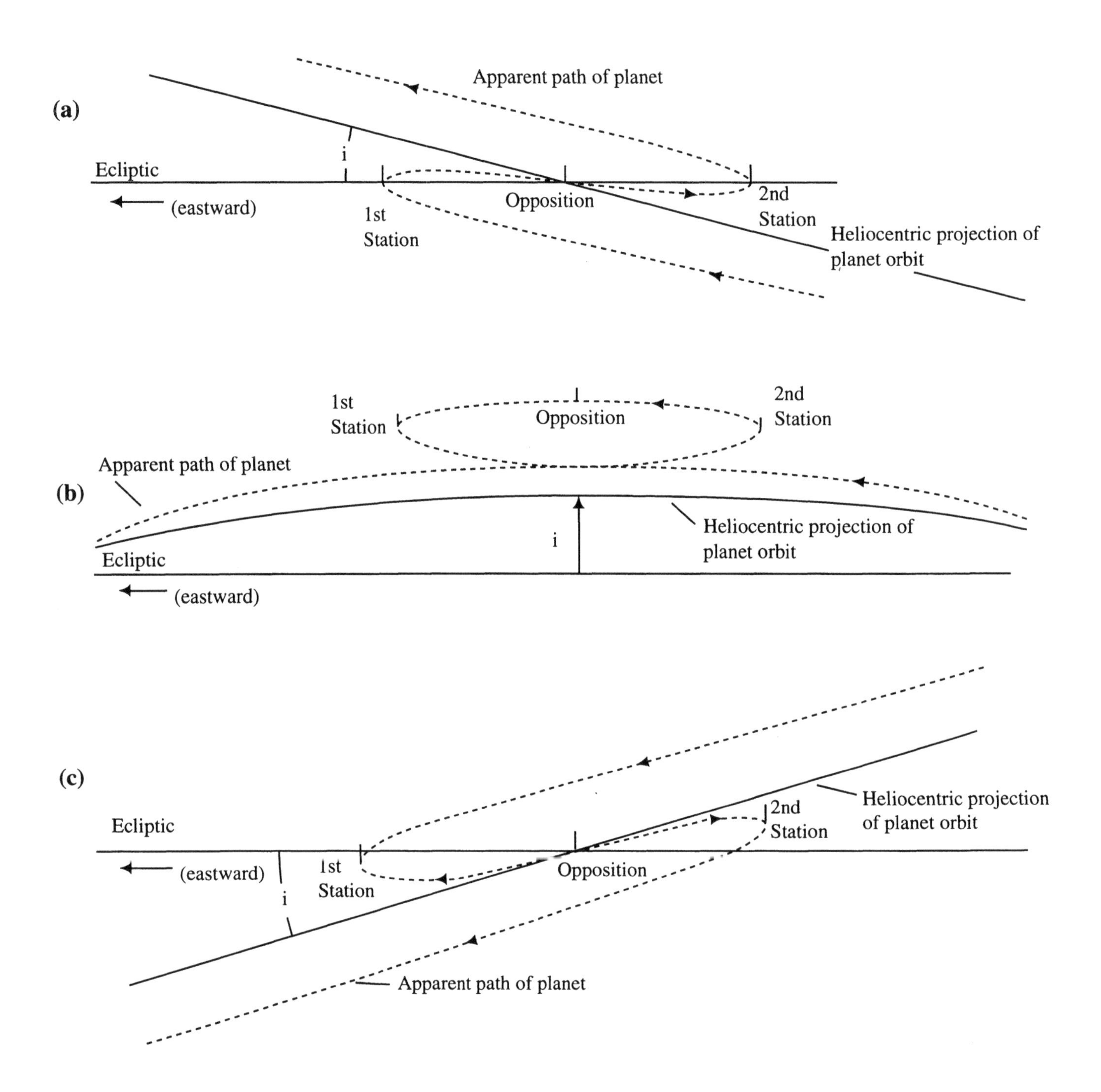

Figure I.4 — Schematic representations of typical apparent celestial motions of a superior planet near the time of opposition. In all diagrams the celestial sphere is seen from the "inside" by an observer looking toward the position of opposition, i.e., toward a point near the middle of the retrograde arc. (a) Retrograde motion occurring near the ascending node. (b) Retrograde loop occurring about 90° east of the ascending node. (c) Retrograde motion occurring near the descending node.

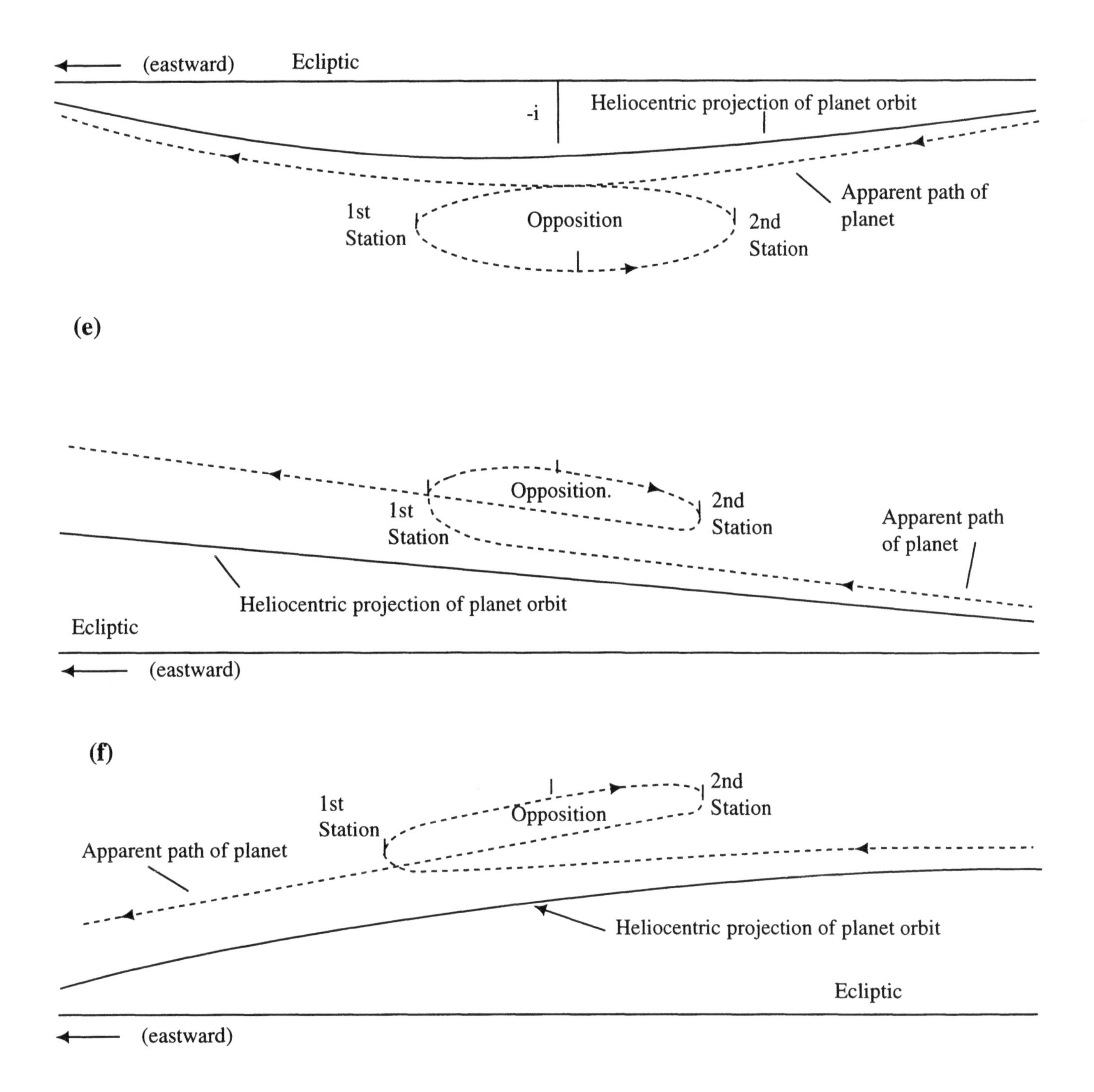

Figure I.4, cont. — (d) Retrograde loop occurring about 90° west of the ascending node. (e) Partial loop occurring about 45° east of the ascending node. (f) Partial loop occurring about 135° east of the ascending node.

Planet	Retrograde Arc	Inclination	Long. of Desc. Node	Time of Node Passage and Opposition
P_1	10°, or 120^d	1.3°	$\lambda + 2°$	$T + 2^d$
P_2	6.5°, or 141^d	2.5°	λ	T

Apparent position and path of P_1, (Jupiter)
Apparent position and path of P_2, (Saturn)

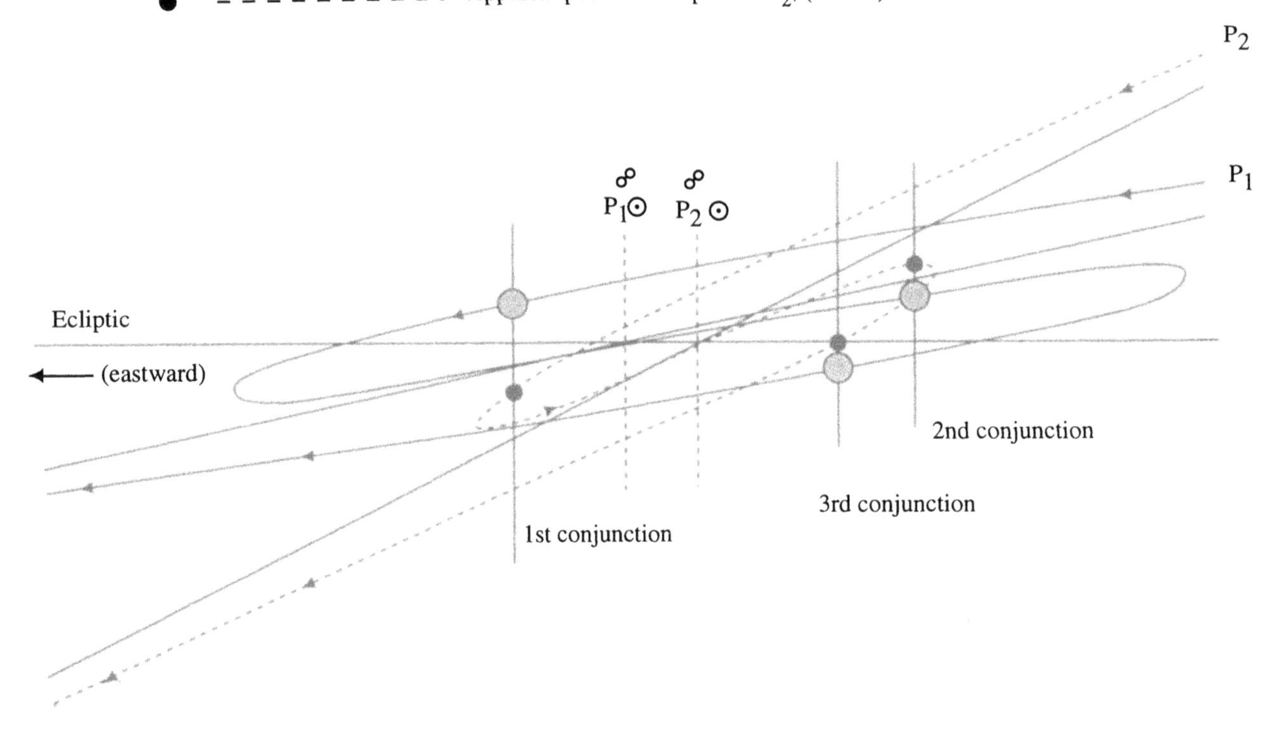

Apparent Motion at Moment of Conjunction

Conjunction	P_1	P_2
1st	direct	direct
2nd	retrograde	retrograde
3rd	direct	direct

Figure I.4(g) — A possible triple conjunction of two superior planets (P_1 and P_2) having nearly the same descending nodes (☋$_1$ and ☋$_2$) when their node passages (oppositions with the Sun) occur within a few days.

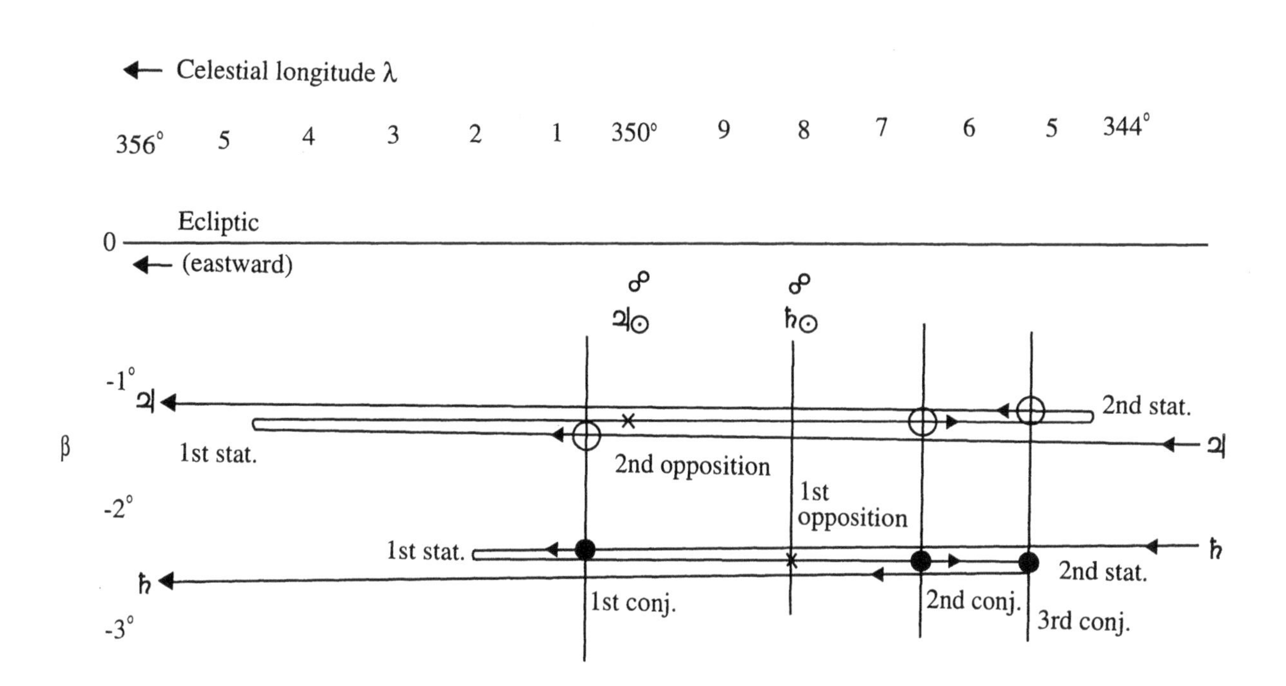

Apparent Motion of Moment of Conjunction

Conjunction	Jupiter	Saturn
1st	direct	direct
2nd	retrograde	retrograde
3rd	direct	retrograde

Approximate Dates of Configurations

Configuration	Jupiter	Saturn
1st station	July 1	July 1
1st conj.	May 29	May 29
Oppos. to ☉	September 16	September 14
2nd conj.	October 3	October 3
2nd station	December 1	December 7
3rd conj.	December 4	December 4

Figure I.5 — Triple conjunction of Jupiter and Saturn in Pisces, 7 B.C.

Table I.1 — Modern Values of the Elements for the Ancient "Planets"

	a (A.U.)	e	i	Ω	ω	T (UT)	P (trop. yr)
Sun	1.000	0.017	0.00°	0.00°	282.95°	1 Jan 1970	1.000
Moon	0.00257	0.055	5.15				0.0748
Mercury	0.387	0.206	7.00	48.33	77.46	25 Dec 1970	0.241
Venus	0.723	0.007	3.39	76.68	131.53	21 May 1970	0.615
Mars	1.524	0.093	1.85	49.58	336.04	21 Oct 1969	1.881
Jupiter	5.203	0.048	1.31	100.56	14.75	26 Sept 1963	11.862
Saturn	9.537	0.054	2.48	113.72	92.43	8 Sept 1944	29.458

Note: All values are for the epoch J2000. Times of perihelia are those before 1971. Values for the "Sun" are derived from the modern Earth's orbit.

In more detail, the elements are as follows. Table I.1 gives modern values of the elements for the seven "planets" of the ancients.

1. a, the **semimajor axis** of the ellipse, defines the size of the orbit.

2. e, the **eccentricity** of the ellipse, defines the particular form of the orbit. If the semiminor axis is denoted by b, then $e^2 = \frac{a^2 - b^2}{a^2}$. The distance c between center and focus is given by $c = SC = ae$. The "flattening" of the ellipse, or its **oblateness** o, is given by $o = \frac{a-b}{a}$. In many cases no eccentricity is given, but instead the value of the **eccentric angle** φ is given: $\varphi = \sin^{-1} e$.

3. i is the **inclination**, or angle between the plane of the ecliptic $EKLM$ and the plane of the orbit $ORBT_1$.

4. Ω is the **longitude of the ascending node** of the orbit plane on the plane of the ecliptic. The elements Ω and i completely determine the position of the orbit plane in space. This plane passes through the Sun's center (in a two-body problem with negligible planetary mass). The angle Ω is measured eastward along the ecliptic from the **vernal equinox** Υ to the **ascending node** N.

5. ω is the **perihelion distance**, or the angle in the plane of the orbit measured eastward from the ascending node to the perihelion. It describes the orientation of the major axis of the ellipse in the plane of the orbit. (The obsolete element π, called the **longitude of perihelion**, was defined as $\pi = \Omega + \omega$, but it is not strictly a longitude at all, since it is the sum of two angles, one measured in the ecliptic and the other in the orbit plane. It may be thought of as being measured all in the orbit plane from a point Υ' that is back from N by the distance Ω.)

6. T is the **time of perihelion passage**. This element is necessary to find the instantaneous position of the planet at any time t. Any epoch at which the planet occupies a defined point of the orbit may serve as well for reference, but since M (the **mean anomaly**), E (the **eccentric anomaly**), and v (the **true anomaly**) are all measured from perihelion, and since $M = n(t - T)$, where $n =$ the **mean daily motion**, the

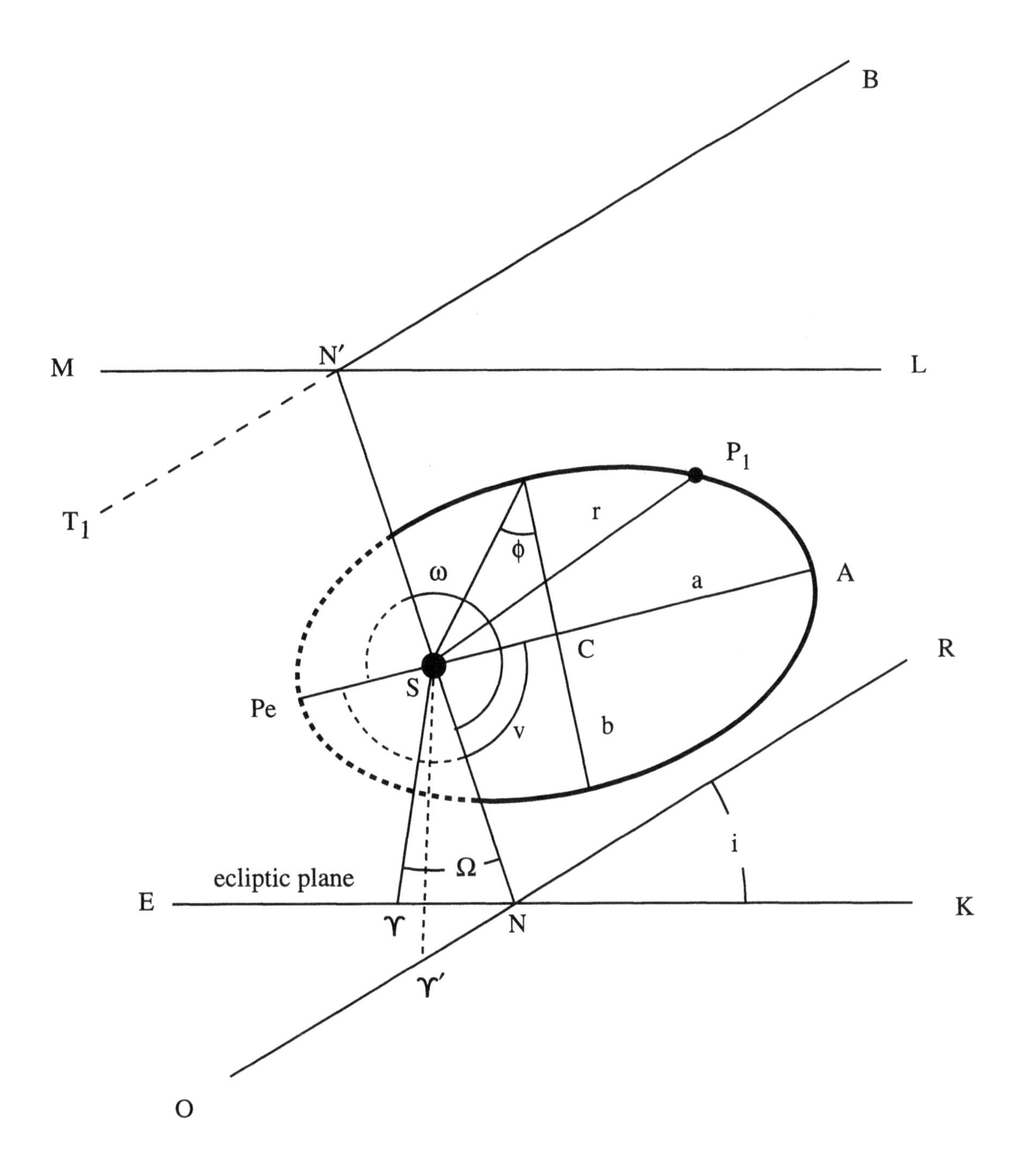

Figure I.6 — The elements of an elliptical orbit (as seen from the north side of the ecliptic).

perihelion point Pe has the advantage of simplicity.

7. P is the **period** of the planet's orbit about the Sun. This is usually given in mean solar time units. The **mean daily motion**, $n = \frac{2\pi}{P}$ (in radians per day) or $\frac{360^\circ}{P}$ (in degrees per day), is often given instead of P.

The reason for quoting both a and P, in spite of the known relation between them (Kepler's third law), is that the planetary mass correction, which enters in the exact expression for the relation, is not derivable until both a and P have been observationally determined.

The position P_1 of the planet at some definite moment t is usually specified by giving the **true anomaly** v and the **radius vector** r at that moment. A planet's radius vector r is its distance from a focus of the orbit (usually the center of the Sun). Its true anomaly v is the angle measured in the plane of the orbit from the perihelion point Pe to the current position P_1. Values for r and v are easily found if elliptical elements are given. In most work by early astronomers, v was measured from the aphelion point of the orbit.

Orbit Computation

Neglecting at first the mass of the planet, a defines P through Kepler's third law. Thus only six elements are necessary to define an orbit, and at least six constants must be supplied in order to find these elements. The six constants are commonly taken to be the observed right ascensions and declinations for three positions of the planet at three known times, although other, less convenient choices of data are possible. By an **orbit method** is meant the mathematical transformations necessary to derive the elements from the observations. This involves, first, practical astronomy to correct positions of reference stars for refraction, stellar aberration, nutation, and precession, and to correct the planetary observations for refraction, stellar aberration, and, provisionally, parallax (since the distance is not known at the beginning of the computation). From these data directional differences are then formed to give angular velocities and accelerations in right ascension and declination. The formulas of spherical trigonometry are then applied to transform coordinates into celestial longitudes and latitudes, as in the older methods, or into heliocentric equatorial rectangular coordinates, as in some later methods. The "triangle equation," which expresses the fact that the Sun, Earth, and planet always lie at the corners of a plane triangle, generally enters the formulation, and a sixth- or seventh-degree equation for the distance ρ from the Earth to the planet has to be solved by trials, graphically, or by the use of specially constructed tables. (The equation for the distance is of the sixth degree for a parabolic orbit, but of the seventh degree for an elliptical orbit. Because hyperbolic orbits in solar system astronomy generally have eccentricities close to unity, they are usually computed by introducing certain modifications in a previously computed, approximate, parabolic orbit.) After the distance has been found, a complete correction for parallax can be performed. Then the problem of finding the elements and of computing from them the right ascensions and declinations of the original sightings (for checking and improvement) and future positions (to serve as an ephemeris) becomes a question of spherical trigonometry and dynamics.

A great number of orbit methods have been invented since the time of Newton. Some of these, such as the various ways of computing the position of a body in a conic orbit at a given time, as well as Lambert's and Euler's theorems, involve principles derived from the two-body problem. Perturbations usually need not enter into a first approximation for a planet's orbit. However, they must be immediately taken into account in the equations for satellite orbits about Earth. A technical discussion of how to perform the actual computations of elements from observations is beyond the scope of the present work.

Concepts of Some Early Greek Philosophers

Although we are mainly concerned with the precise astronomical systems of a few early thinkers who qualify as "real" astronomers, it may not be amiss to outline concepts of other early thinkers. This will show the depths of the gulfs to be overcome before rational explanations could be reached. It will also allow us to take stock of the primitive knowledge, or lack of knowledge, that existed as a background for the early astronomical explanations. Even erroneous explanations may focus the mind on the body of facts that has to be explained.

Thales of Miletus (c. 624–546 B.C.)

Thales is generally considered the founder of Greek science, but he is not reckoned much as an astronomer. He studied in Egypt, the fountain of knowledge, and returned to found the Ionian school of philosophy. It is said that he invented geometry and taught it to the Egyptians, but Antoniadi (1934), in his thorough book *L'Astronomie Egyptienne,* holds that he learned geometry from the Egyptian priests, who were certainly familiar with the 3-4-5 triangle. Yet Thales may himself have developed the subject to apply to other triangles. Thales was credited by the Greeks as being the first to measure the length of the year. He was said to have predicted the solar eclipse of 585 B.C., which gave him an enormous reputation. Perhaps he knew of the eclipse in Babylon of 603 B.C. and merely added 18 years, the length of the eclipse cycle called the saros and known to the Babylonians. However, while lunar eclipses can be predicted very reliably by means of the saros, solar eclipses visible at a particular locality cannot. Thus, most modern scholars put little stock in the ancient claims for Thales's skill as an eclipse predictor. Aristotle describes Thales's cosmogony as follows:

> The Earth is considered by Thales to be a circular disk, floating like a piece of wood or something of that kind, on the ocean. Water is the principle or origin of everything, and from its evaporation the air is formed. As to the quantity and form of this principle there is a difference of opinion, but Thales, the founder of this sort of philosophy (he was the first to explain the gods), says that it is water, getting the idea, I suppose, because he saw that the nourishment of all beings is moist, and that heat itself is generated from moisture, and getting the idea also from the fact that the germs of all beings are of a moist nature. The celestial vault limits the world above, while nothing is said about the lower limit or the support of the ocean; but apparently the water, being the first principle of everything, does not require anything to support it and was possibly regarded as infinite.

Anaximander (c. 611–545 B.C.)

Anaximander was a follower of Thales and taught that the Sun was a bright spot, the size of the Earth, on the inside of a hollow, invisible ring, filled with fire, concentric with the Earth, probably in the plane of the ecliptic, and 28 times the size of the Earth in circumference. This is the famous Anaximander's wheel, common in poetry, sculpture, and painting but so completely misunderstood through the ages. In old statues of the Sun god the face is correctly represented by the disk of the Sun, but the wheel of Anaximander is erroneously shown around his face as a sort of halo with several spokes, and outside the rim more rays project radially like flames, as on the Statue of Liberty. According to this view, the wheel represents the radiance surrounding the Sun, which is wrong, and the plane of the wheel is tangent to the surface of the sky at the Sun, which is also wrong. The real Anaximander's wheel is an invisible hollow tube encircling the Earth, lying in the plane of the ecliptic, and concentric with the Earth, so that the invisible spokes (if any) would everywhere pierce the sky if prolonged. It is like a bicycle tire in the sky, covering the ecliptic. The "tire" is filled with a fiery substance, also invisible except at one spot. This spot is the Sun, a hole the size of the Earth on the inside of the tube. Only in this one spot do the fiery contents of the tube become luminous. As the wheel skews around, turning westward on an invisible oblique axis, the Sun rises in the east, crosses the sky during the daytime, and sets in the west. The details of the system were never worked out. It is not even known whether the celestial bodies passed under the Earth at night or merely passed behind its elevated north side.

Similar rings with fiery vents were invoked to explain the stars. The Moon was considered to be a hole in another ring, 19 times the Earth in diameter, concentric with the Sun's ring. The ring for the Moon, like that for the Sun, was placed obliquely. This shows some familiarity with the sky, about as much as an attentive observer could derive in a few months of naked-eye observation. The phases of the Moon and eclipses were explained as regular and irregular, partial or complete, cloggings of the flue. Occultations of bright fixed stars by the Moon must have been unknown. The celestial bodies were spaced from the Earth in four layers. Closest to the Earth were the planets and stars, then the Moon, and finally the Sun. The heavens were a fiery shell, a sort of diffused blue fire in the daytime, together with a dark, concentric shell studded with concentrations of fire that were the stars at night. The planets were small holes punched on the inside of this shell. The Earth was a cylinder, which was in equilibrium in the center of the universe since, being in the center, there was no reason why it should fall in any particular direction. Everything on the Earth's surface was on the flat, circular (upper) end of the cylinder.

Anaximenes of Miletus (c. 585–528 B.C.)

Anaximenes, the third philosopher of the Ionian school, taught that the first principle of all things was air. The solid Earth was condensed air. When vaporized, air passed into fire, which in turn could condense back into air. In his cosmogony the Earth was a flat table, supported on air. Over the Earth was a hemisphere made of some crystalline material. The stars were attached to this like nails. It turned around the Earth like a hat around the head, but somewhat askew. After setting, the stars passed, not below the Earth, but merely behind the high part of it. Most of the Greeks entertained the view of a world mountain north of Greece, i.e., that the ground gradually sloped upward toward the north. Anaximenes maintained that the Sun, Moon, planets, and stars consisted of fire that had risen aloft from the Earth. He imagined the Sun to be a sort of flat leaf of fire that floated in the air because of its thinness – rather like a glider or an airplane. Like the Earth itself, the

Sun was supported by air, at a height determined by the density of the air. The Moon and planets were something of the same sort, but the stars were of an entirely different nature, being more like fiery studs nailed into the crystalline hemisphere. As there was nothing in all this to make eclipses, he supposed the sky to contain also dark bodies "of an earthy nature." Although he did not say so, this is near enough to the time of the Pythagoreans for us to surmise that he held the view that the dark bodies produced solar eclipses by coming between us and the Sun.

Anaxagoras of Clazomenae (c. 500–428 B.C.)

The world was made up of an infinite number of elements, one essence for each kind of substance; no absolutely pure substance existed. One could find gold in water or any other substance, for there was always a trace of every substance, be it ever so slight, in each object. The infinite number of elements also existed outside the world. A great meteorite that fell around 467 B.C. is associated with his name. He is even said to have predicted its fall! It fell in the daytime, and he therefore supposed it to have fallen from the Sun, which he took to be a red-hot stone, not very distant and "larger than the Peloponnese." The Moon was partly fiery, partly of the same nature as the Earth, the inequalities of its face being due to this mixture of materials. There were plains and valleys on the Moon. He correctly explained the phases of the Moon as being a consequence of the fact that the Moon was a spherical body which received its light from the Sun and circled around the Earth. His ordering of the planets was the same as was current in the time of Plato, namely: Moon, Sun, Venus, Mercury, Mars, Jupiter, and Saturn. Lunar eclipses were correctly explained as being due to the Earth's shadow, and solar eclipses as being due to the Moon's shadow falling upon the Earth. With Anaximenes he assumed the existence of other, cold, dark bodies to produce some of the effects.

Anaxagoras thought that the stars were stony particles torn away from the circumference of the flat Earth, like sparks from a grindstone, and prevented from falling by the rapidly revolving fiery ether. When the stars set, they pursued their courses under the Earth, which was supported in the center of the celestial sphere by the air underneath. The inclination of the axis of the heavens to the vertical was caused by a tilting of the Earth toward the south after the appearance of living creatures, in order that there might be inhabitable and uninhabitable regions. Another idea was that the Sun upon reaching a solstice was driven back in declination by the air, which had been thickened by its heat. A similar cause kept the Moon in its course.

Pythagoras of Samos (c. 580–500 B.C.)

The Pythagoreans were another early group of philosophers standing apart from both the Ionians and the Eleatics. Their system, as artificial though perhaps not quite as fantastic as the others, had an immense influence upon the later development of cosmological systems. They had the order of the outer planets correctly represented. Pythagoras, the founder of the school, was famous as a philosopher of number relations and as a mathematician. He left no writings from which to judge his excellence as an astronomer. It is said that he and Parmenides were the first to assert that the Earth is a sphere, and that he was aware that the Sun's annual motion in the sky was eastward along an oblique path, the ecliptic.

The details of the planetary system bearing his name were probably worked out by his followers, many of whom were anxious to credit the founder of the school with discoveries or ideas of their own. The system most admired was that of Philolaus, one of the teachers of Democritus.

Philolaus, the Pythagorean (c. 470–400 B.C.)

Philolaus assumed that the center of the Universe was occupied, not by an "impure" body like the Earth, but by a pure fire called the "central fire" or "Hearth of the Universe." The Earth, the Moon, the Sun, and the planets all revolved eastward in circular orbits of different diameters and inclinations, and in different periods, about the central fire. The Earth was closest to the central fire, with the other distances in order of the lengths of the bodies' sidereal periods of revolution. The Earth revolved about the central fire every 24 hours, producing an apparent 24-hour rotation of all the other bodies and the firmament about the Earth and accounting for the sky's diurnal motion parallel to the celestial equator. The Earth also rotated once every 24 hours, which forever kept the inhabited hemisphere facing outward from the central fire and ensured climate stability. In addition to this diurnal motion, the Sun, for instance, revolved eastward about the central fire in a slanting orbit coincident with the ecliptic in a period of one year, thus nicely explaining its annual apparent motion. The Moon and the planets moved eastward in orbits of the appropriate inclinations and periods. If the system had stopped there, it would have differed from the later system of Aristarchus only by the presence of the central fire. But because the Pythagoreans considered the number 10 more perfect than 9, one more body, the dark *antichthon*, or "counter-Earth," was added to the nine already in the system! This counter-Earth moved in a smaller orbit than the Earth and always kept 180° ahead of the Earth in orbital longitude, remaining invisible to us. Thus the antichthon as well as the central fire could never be observed and served only an aesthetic purpose, yet the Pythagorean system was certainly a milestone in the development of planetary systems. We shall not discuss it further, as it was never subjected to quantitative astronomical tests, or, as the Greeks would say, it was never developed to fully "save the celestial phenomena."

Hicetas of Syracuse (c. 350 B.C.)

A follower of Philolaus, Hicetas did not believe in the central fire because it could not be seen in the west, even beyond the Pillars of Hercules (Gibraltar). Yet he held that the apparent diurnal movement of the sky was the result of a rotation or a "fixed revolution" of the Earth about a not too distant axis.

Ecphantus of Syracuse (c. 325 B.C.)

Ecphantus was possibly a student of Hicetas and did not believe in a central fire because Alexander the Great had not seen it, although he had been far enough east to do so. He was probably the first to state unequivocally that the Earth rotates in a 24-hour period about its own axis of figure, not about a central fire or some other parallel axis relatively close by. This for the first time made a rotating Earth central in a world picture.

A third important group of philosophers were representatives of the Eleatic school. These savants, while great as philosophers, held scientific views which to us seem perfectly scandalous even for their own times.

Xenophanes of Colophon (c. 530 B.C.)

A precursor of the Eleatic school, Xenophanes considered the Earth to have no limits, to be flat, and to be rooted in the infinite. This probably meant that it extended indefinitely in all horizontal directions and that the air above was also unlimited. The Sun, stars, and comets were fiery clouds formed by moist exhalations ignited by the friction of their motion. This motion was rectilinear, the circular forms of their daily paths being only an illusion caused by their great distance. New stars were formed every evening and then extinguished at dawn, while the Sun likewise was formed anew every day from small fiery particles which gathered together. The Moon was a compressed cloud that shone by its own light and became extinguished every month. There were many Suns and Moons serving the various climatic zones of the Earth. When the Sun occasionally arrived in an uninhabited region, and thus was in an empty place, it became eclipsed!

Heraclitus of Ephesus (c. 500 B.C.)

The Earth was flat and surrounded by the world ocean Okeanos. On rising from the Earth, moist exhalations turned into fire, which was caught in a hollow basin, a hemispherical bowl, with its cavity turned toward the Earth. The exhalation was ignited as the basin for the Sun rose from Okeanos in the east, later to be extinguished when it set in the west. In this way the Sun was continually renewed and was always young. It shone the more brightly the higher and purer the air in which it moved. The Moon, which was also a bowl with fire, shone more dimly than the Sun because it moved in thicker air. The stars were of the same nature but fainter because they moved at a greater distance. Eclipses, and probably also the phases of the Moon, were caused by the turning of these basins, whereby their other, nonluminous sides became directed toward us. After setting, these boatlike basins sailed behind the world mountain along Okeanos, to be ready in good time for a fresh ascent.

Empedocles of Agrigentum (c. 450 B.C.)

Empedocles is known for his idea of evolution (superficially similar to that of Darwin) whereby man had evolved from centaurs rather than from apes. He was the first to set forth the famous four elements: fire, air, water, and earth. The world was made from these and from compounds of these, held together by love and strife (i.e., by attraction and repulsion), which alternately predominated and thus divided the history of the world into periods of different character. The world was presently in one of the expanding phases, in which repulsion had the upper hand, and had been formed from condensed air. It was finite, solid, and with no void anywhere. In the beginning the Earth was softer and wetter than now, and the heavenly crystalline sphere enclosed it much more tightly. The rotation of the sphere therefore could "get a hold" on Earth and squeezed the water out of it, thus producing oceans and solid ground. Even now there was an air pressure against the crystalline sphere, and the stars were condensations of rarefied air, ether, or invisible fire, into lumps of visible, cold fire (like St. Elmo's fire), held against the sphere by the air pressure. The planets were condensations of visible fire lower down, and moved more freely through the air at their respective layers.

The Earth was flat. Originally the pole of the sky had been directly overhead at the zenith, but now, owing to the rapid motion of the Sun, the air supporting the Earth had yielded around the southern edge. The result was that the Earth tipped so as to elevate the north and depress the south. This was why the pole was halfway between the zenith and the horizon. The Moon was air rolled together and mixed with fire. Empedocles's view that the Moon was air mixed with fire was refuted by the Stoics, who argued correctly that we would not see distinct lunar phases or lunar eclipses if the Moon were thus luminous, for then the dark part should be faintly luminous. They added further that neither could the Moon be all celestial fire, for if it were, it could not produce solar eclipses but would act as a transparent lens rather than as a dark screen. They concluded that the Moon was a totally dark body, receiving all its light from the Sun. (The earthshine phenomenon was not yet discussed, although one may suppose that it had been seen.) The Moon was a flat disk and illuminated by the Sun. Day and night were explained as follows: There were two separated hemispheres rather close to the Earth, one of fire (perhaps not all hot fire) for the day side and one of air (with only a little fire) for the night side. As the celestial sphere rotated, the two halves were in turn above and below the Earth.

The Sun was an image of the fiery daytime hemisphere formed by the daytime crystalline sphere acting as a concave mirror, and therefore the Sun moved as the sphere was displaced. The Sun itself was not of a fiery nature but was a reflection by the concave daytime sphere of the fire surrounding the Earth. Hence Empedocles could make such statements as "the Sun's course is the limit of the world" and "the Sun is at the greatest distance one can see." It is probable that he did not consider the Sun to be hot, but instead believed that the heat of the day came from the sky itself, or rather from the fiery hemisphere just outside the atmosphere. The fiery hemisphere annually occupied first more than and then less than half the celestial sphere. The increments and decrements had to be asymmetrically applied, as they accounted for the seasons by making the Sun (which was a reflection of the hemisphere) move north and south during a year, thus producing the Sun's annual motion in declination. He did not mention the Sun's annual eastward motion among the stars. Eclipses of the Sun were correctly explained as being caused by the Moon passing in front of the Sun. The distance to the Moon was 1/3 the radius of the crystalline sphere. The planets were kept in their orbits beyond the Moon by a sort of centrifugal force. He said that planetary motion was preserved by a spinning of the planets that was "quicker" than their tendency to fall.

Leukippus of Abdera (c. 450 B.C.)

This famous physicist, known as the founder of atomism, had peculiarly primitive cosmological ideas. The Earth was compared to a tympanum: flat on the upper surface, with an elevated rim, and on the bottom rounded to fit into the celestial sphere. There was an air cushion to support the Earth from below, as it rode in the sphere. The inclination of the heavens to the horizon was due to two causes. First, the air around the north side was frozen and very thick, while the air around the south side was heated by the Sun and much thinner. Second, the great mass of vegetation growing on the south side made it the heavier side, and the pole of the heavens was therefore sunk from the zenith halfway toward the northern horizon! The Moon was the nearest body, the Sun farthest, and between them were the planets. The planets did not have any orbital motion of their own, but merely revolved from east to west somewhat slower than the fixed stars. The Sun and Moon were large, solid masses, smaller than the Earth. The markings on the face of the Moon were caused by the shadows of mountains and valleys. The celestial sphere was a spherical membrane on which were placed the stars, consisting of bunched atoms set on fire. Both the Earth and the Sun had originally been dark earths. In falling toward the center, they had collided. We had retained the loose dirt and slag, and the Sun was now a ball of hot, purified iron. The Moon was also on fire but had some dirt left.

Democritus of Abdera (c. 400 B.C.)

This student of Leukippus is even more famous than his teacher, for he worked out the atomic theory in much greater detail. While Leukippus had compared the Earth to a tympanum, Democritus thought it more like a discus, highest at the circumference, lowest in the middle. While Leukippus had the order of the celestial bodies as Moon, planets, and Sun, Democritus had them as Moon, Morning Star, Sun, planets, fixed stars. He had a remarkably correct conception of the Milky Way, the light of which he explained as caused by a great multitude of faint stars.

Metrodorus of Chios (c. 400 B.C.)

Metrodorus was a student of Democritus. He thought that the fixed stars and planets were nearer to us than the Moon and the Sun. Both fixed stars and planets were illuminated by the Sun. The Earth was a deposit of condensed water. The Sun, being a deposit of something lighter, rose high into the air and took fire. When air became condensed, it produced clouds and later water, which put out the Sun. By degrees the Sun in turn dried up and transformed the water into stars. Day and night, as well as eclipses, were produced as the Sun lit up or was extinguished.

Plato of Samos (427–347 B.C.)

Plato, although acknowledged as one of the greatest philosophers of antiquity, may almost be ignored as far as scientific astronomy is concerned. He probably knew that the Sun was always found on the ecliptic, a circle of fixed inclination to the celestial equator, and he may also have realized its slightly variable eastward annual motion along the ecliptic. Beyond these facts, however, his knowledge was very limited. He knew that the planets travel mainly from west to east, but he never mentioned their velocities, orbital inclinations, stations, or retrogressions. Thus he erroneously assumed that all seven "planets" always move eastward in the plane of the ecliptic. The seven planets were (in order) Moon, Sun, Venus, Mercury, Mars, Jupiter, Saturn. (This was the order adopted also by Anaxagoras, the Pythagoreans, Eudoxus, and Aristotle.) He knew that Mercury and Venus are never seen at any great angular distance from the Sun.

Plato's descriptions of his cosmogony in *Phaedo*, *The Republic*, and especially *Timaeus* constitute an orgy of anthropomorphic, metaphysical, ethical, and aesthetic judgments that are scientifically irrelevant and should have been so even in his time. He seems never to be satisfied unless he can make a plausible guess at what the "Great Artificer" of the world has done, and why. Everything in cosmology was either good or not, and for a purpose. The cosmos was set rotating to be "an image of time." It was a spherical shape because a sphere can spin on an axis while always occupying the same space. (But this is true of any figure of revolution, however irregular its meridional section.) Furthermore, a sphere had the advantage of being bounded by only one surface. (This meant that the spherical surface, having a constant curvature, was considered superior to an ellipsoid, say.) The celestial equator was the image of "sameness," and the ecliptic the image of "diversity" or "change," etc., etc. – judgment after judgment of an entirely conjectural and often ridiculous nature.

The world was closed and finite, a cosmos consisting of the celestial sphere with its attached fixed stars, and everything else interior (i.e., the seven spheres of the planets, the sublunar region, and the Earth). The cosmos as a whole had a "soul" and an intelligence, and so did every star, planet, the Earth, and each distinct unit in nature. Most of the souls were "sleeping," but yet they obediently followed the natural movements and tendencies that were granted them by the Great Artificer, or "Prime Mover." Following the Pythagoreans, Plato asserted that the soul of the cosmos had a natural tendency to divide itself into regions corresponding to the two simplest geometrical series. These regions were of course spheres, in fact crystalline spheres carrying the seven planets. The radii of these spheres were chosen by interleaving the numbers belonging to the series 1, 2, 4, 8, ... , with those from the series 1, 3, 9, 27, ... Therefore the distances of the planets were 1, 2, 3, 4, 8, 9, and 27 for the Moon, Sun, Venus, Mercury, Mars, Jupiter, and Saturn. The fixed stars were spherical, rotating balls of cold, celestial fire, each possessing a living, obedient stellar soul that made it forever follow its 24-hour circular course about the center of the cosmos, i.e., the center of the stationary Earth. Each planet also was celestial, i.e., made of cold fire and free from gravity, mass, and other sublunary imperfections. Each had a complex soul that made it obediently, naturally, and effortlessly follow a complicated motion of both sameness and diversity, dictated by the Great Artificer and communicated to the planets through a mechanism containing three principal wheels. The three living intelligences for Past, Present, and Future occasionally touched these wheels with their hands and thus produced the irregular planetary deviations from the uniform celestial motion of the fixed stars!

Aristotle of Stagira (384–322 B.C.)

Aristotle, more than any other philosopher, collected and systematized all the knowledge of his time. The very extent of his knowledge gives his writings the character of summaries, but it would be inaccurate to say that he was always superficial. In his own special field, biology, and especially in his politics, he reveals himself as a forceful thinker, whose writings are useful even today. Although he was a systematizer of logical reasoning, nevertheless his views of astronomy and cosmogony were only slightly less fanciful than those of Plato. However, while Plato's arguments were for the most part only assertions of what he considered "good," "beautiful," "purposeful," or "fitting," Aristotle often supplied correct or slightly better arguments in the form of direct conclusions from crude observations.

Aristotle was well acquainted with the Pythagorean notion of the Earth's rotation on its axis of figure as an explanation of the diurnal movement of the sky. However, he rejected this view because of his conception of the nature of motion. He had two kinds of motion: celestial motion, which applied to the massless celestial bodies, and terrestrial motion, which applied to the ponderous sublunar bodies composed of the four elements. For celestial bodies the natural tendency was to move in circles; for sublunar materials it was to move in straight lines; for heavy bodies it was toward the center of the Universe; and for light bodies it was away from the center, toward the limit of the world. Since bodies on the Earth fell straight down, and fire ascended vertically, the Earth could not be rotating. If it were, the individual particles would have a natural circular motion, which, he said, they did not. He also decided against any orbital motion of the Earth, such as the Pythagorean motion about a celestial fire. His argument here was perfectly sound and ran as follows: If the Earth traveled in a great orbit as Aristarchus believed, we should be brought into different regions of the stars at different times of the year, and this would greatly change the appearances of the constellations. Not being able to perceive any such changes, Aristotle concluded that the Earth did not move. (His fallacy lay in his failure to appreciate the enormous distances of the stars.) Thus he decided against any motion of the Earth, whether of rotation or translation.

Aristotle gave three arguments to prove that the shape of the Earth was spherical: (1) the falling of heavy bodies toward the center of the Universe would have produced a sphere of matter, for heavier particles would have pushed aside lighter ones until the packing was uniform; (2) different stars were observed to be overhead at different latitudes of the Earth; and (3) the edge of the Earth's shadow on the Moon's face at lunar eclipses was always circular in shape, regardless of the orientation of Earth and Moon at the time of eclipse. He also stated that the Moon was the closest celestial body, since it occulted the Sun, stars, and planets. His planetary system was that of Eudoxus and Callippus modified by the addition of 22 crystalline spheres, as discussed in Chapter II.

Aristotle's view concerning the Universe was similar to that of Plato. The Universe was finite, for it would be impossible for a star at an infinite distance to move around the Earth in the finite time of one day. The Universe was spherical because (1) it must be bounded by one simple, "perfect," closed surface, namely a sphere; and (2) the only body which in revolving continually occupies the same space is a sphere (the same argument as Plato).

Besides the above metaphysical arguments, Aristotle also entertained some views no less queer than Plato's. He thought that the celestial motions were uniform because not to be so was "unnatural." Also, the direction of rotation of the celestial vault was in accord with the principle that motion from the right was "more honorable."[1]

Aristotle took the stars and planets to be spherical balls of celestial ether at rest in their respective crystalline spheres. They were spherical in form, as one could see directly in the case of the Moon. The stars twinkled, but not the planets, because light rays became unsteady for distances greater than that of Saturn. (The ancients thought that sight was due to a beam of light sent out by the eye to "feel" the object, a sort of radar effect!) Finally, there was no music of the spheres (thus refuting a Pythagorean doctrine) because if there had been, the spheres (being so great) would send out a noise far exceeding that of thunder in intensity, and we would certainly know about it!

---

[1]Because Aristotle gave primacy to this principle even in the face of the apparent "rightward" diurnal motion (westwardly motion when looking south at the bulk of the sky), he came to the conclusion that in fact Greece must be situated within the *bottom* hemisphere of the Earth! (*On the Heavens*, bk. II, ch. 2.)

Heracleides of Pontus (c. 350 B.C.)

Heracleides, a contemporary of Plato and Aristotle, is often associated with a partly heliocentric planetary system. He is said to have let Mercury and Venus move in circular orbits concentric with the Sun, with Mercury nearest the Sun. The Moon, Sun (carrying these two planets), Mars, Jupiter, and Saturn all moved in orbits concentric with the Earth. This arrangement was often referred to as the Egyptian system, and it has been supposed that Heracleides learned of it during his stay in Egypt. But there is absolutely no independent evidence that the Egyptian priests ever entertained the notion of this system.

Heracleides may have been better versed in astronomy than the great speculative philosophers. Writers several centuries after his time stated that he knew the difference between the sidereal and the solar day, and he may have been familiar with the Sun's variable velocity along the ecliptic in the course of a year. He was even said to have considered the Sun to be moving on a small epicycle centered on a deferent concentric with the Earth, both circles being in the plane of the ecliptic. This arrangement would nicely take care of the annual differences between true and mean celestial longitude of the Sun (about 1.9° near 5 April).

As to the diurnal rotation of the heavens, Heracleides, like a good Pythagorean, let the Earth move in a small circular orbit in the plane of the celestial equator with a fixed rotation with respect to the center of the orbit. The diameter of this orbit was but a small fraction of that of the Earth itself (which he took to be spherical); the Earth therefore had at most a slight daily "wobble," only slightly different from a pure rotation.

Aristarchus of Samos (c. 310–230 B.C.)

Archimedes (287–212 B.C.), a younger contemporary of Aristarchus, states: "Aristarchus supposes that the fixed stars and the Sun are immovable" and that "the Earth is carried around the Sun in a circle in the plane of the ecliptic, but the sphere of the fixed stars, centered on the Sun, is infinitely great as compared to the Earth's orbit." Thus "the Sun is in the center of the Universe." Although nothing is said about the order of the orbits of the planets, it was almost bound to be in accord with the modern view, or otherwise the elongations of the inner planets, as well as the stations and sizes of the retrograde arcs of the outer ones, would have involved inconsistencies too glaring for Aristarchus not to have noticed them.

We do not know to what extent he worked out his system, for his works dealing with this subject are entirely lost, and we have only very brief accounts of his ideas. For example, Plutarch wrote a dialogue in which one of his characters subscribed to Aristarchus's views and was accused of "turning the world upside down." He replied that he was quite content to leave it that way, provided he was not accused of impiety. Also, Cleanthes (3rd century B.C.) advocated that Aristarchus be charged with impiety (and banished) for moving the hearth of the world, because the central fire would be moving annually about the Sun in an oblique orbit, while the Earth would spin about the fire in 24 hours. Aristarchus was the man who in order "to save the phenomena" supposed that "the heavens stand still, and the Earth moves in an oblique circle at the same time it turns round its axis." It may surprise us that Aristarchus's heliocentric system never claimed many adherents among his contemporaries, but was instead soon replaced by either of two geometrical systems: the epicycles of Apollonius or the eccentrics of Hipparchus.

The only surviving work of Aristarchus is his treatise *On the Sizes and Distances of the Sun and Moon.* In this work, starting from certain postulates, Aristarchus provided the first correct geometrical solution of this problem. The data then available were poor, but within a few centuries better measurements of the angular sizes of the Moon and of the Earth's shadow (seen on the Moon during a lunar eclipse) led to excellent values for the size and distance of the Moon. Although Hipparchus and Ptolemy later introduced some innovations, the geometrical approach to this problem remained essentially the one that had been pioneered by Aristarchus. In the case of the Sun, however, Aristarchus very seriously underestimated its size and distance. This situation was not much improved until the seventeenth century.

CHAPTER II

Eudoxus (408–355 B.C.)

In all the planetary systems prior to Kepler we find accepted the Greek notion that all celestial motions, no matter how apparently complicated, must be explained by a system of one or more uniform circular motions, in circles with different radii and periods, about one or more centers. In the earliest attempts, the sphere of the fixed stars, during its daily revolution, dragged all the other spheres along with it.

Eudoxus of Knidus in Asia Minor (408–355 B.C.), a highly skilled mathematician, was the first to give what in his time was a scientific explanation of the motions of the celestial bodies. Although nothing of his own writings has been preserved, his work is known from a short reference to it by Aristotle and from a lengthy commentary on it by Simplicius (6th century A.D.).

In Eudoxus's system of homocentric spheres each celestial body had its own nest of transparent spherical shells, one within the other, all concentric with a fixed Earth, which was itself taken to be spherical in shape. The shells were rotating uniformly at different speeds about variously inclined axes. The ends of the rotation axis of any one sphere were on the surface of the next sphere outside it, with the celestial body positioned on the equator of the innermost sphere. The outermost sphere for any nest of spheres could be the sphere of the fixed stars, but any sphere containing the nest concentrically would do, provided it rotated uniformly westward about the axis of the fixed-star sphere with a period of 24 hours. By varying the angular orientations of one axis to another and the speeds of rotation of the spheres, many complicated motions could be reproduced. To the Greeks this system possessed an aesthetic advantage in containing nothing but "perfect bodies" (spheres) and in forming a symmetrical arrangement.

Eudoxus himself made no attempt to connect the movements of the various groups of spheres with each other. It seems probable that he only regarded his system as providing a plausible physical account of the cosmos and as a field of play for proving interesting geometrical theorems. The thicknesses and distances of the spheres, their method of suspension, and their order of arrangement were immaterial and never needed even to enter the discussion. For example, the spheres might be taken as infinitely thin and of equal (even infinite) radius. There was, as yet, no attempt at numerical accuracy.

In his youth Eudoxus is said to have attended some of Plato's lectures and to have spent time with the Egyptian priests at Heliopolis. He later returned to Athens with some of his students and started a school of his own. According to Aristotle he was perhaps better known for his eudemonistic philosophy than for his mathematical ability, although he was considered one of the two greatest living mathematicians. It is therefore very possible that his astronomical effort was influenced by Plato, who, not himself a mathematician, had suggested to his students that the irregular motions of the celestial bodies should be capable of explanation by some system of uniform circular motions of different periods and amplitudes.

It is fair to assume that Eudoxus, being primarily a philosopher and a mathematician, had only a moderately detailed knowledge of the main phenomena of astronomy. At any rate he does not seem to have known about the nonuniform motion in celestial longitude of either the Sun or the Moon. In the case of the Moon, for instance, the daily increments in celestial longitude vary from 11° to 15°, a difference easily noticed by the attentive naked-eye observer. In the case of the Sun, although the effect is smaller and cannot be easily observed without the use of instruments, it is definitely implied in the unequal lengths of the seasons. He did know that the Moon's orbit makes an angle of about 5° with the ecliptic. But he erroneously assumed that the Sun's path makes a similar, smaller angle, perhaps $\frac{1}{2}^\circ$ to 1°, with the ecliptic! For the Sun and Moon he cooked up theories, that for the Sun explaining both too much and too little, and that for the Moon explaining too little. In spite of these shortcomings we will outline his theories, since they were the first serious attempts to explain the retrograde motions of the celestial bodies in a rational manner, that is, free from mythological concepts.

Eudoxus's Lunar Theory

Eudoxus was correct in ascribing an inclination of about 5° to the lunar orbit and a regression of its nodes on the ecliptic with a period of about 19 years. In his system both these facts, as well as the rising and setting, are taken care of by placing the Moon on the equator of the third, or innermost, of his nest of three lunar spheres.

Thus in Figure II.1 the motion of the Moon M about the Earth T was produced by the combined, simultaneous motion of three concentric spheres:

- The first and outermost sphere L_1 rotated westward about a north-south axis NS (the axis of the celestial poles) in a period of 24 sidereal hours, thus producing the diurnal motion of the system.
- The middle sphere L_2 turned very slowly from east to west around ZZ', the axis of the ecliptic, inclined $23\frac{2}{3}^\circ$ to NS, in a period of 18.6 years (modern value), thus producing the regression of the nodes of the Moon's orbit.
- The third sphere L_3 turned eastward about OO', the perpendicular to the Moon's instantaneous orbit, in a period of 1 sidereal month. OO' is constantly inclined to ZZ', the axis of the zodiac, at an angle OTZ equal to 5.1° (modern value). Therefore $\angle ETM = 5.1^\circ$ represents the highest celestial latitude reached by the Moon, when the ascending node is at the vernal equinox. The Moon itself is placed on the equator MM' of this third sphere.

Eudoxus's Solar Theory

Eudoxus erroneously assumed that the Sun followed an orbit inclined about $\frac{1}{2}^\circ$ to the ecliptic, with an advance of the nodes of 360° in 2922 years. Thus his solar theory was exactly like his lunar theory if we make the following five changes. In Figure II.1, let the Sun replace the Moon; change $\angle ETM$ to $\frac{1}{2}^\circ$; reverse the direction of motion about OO';

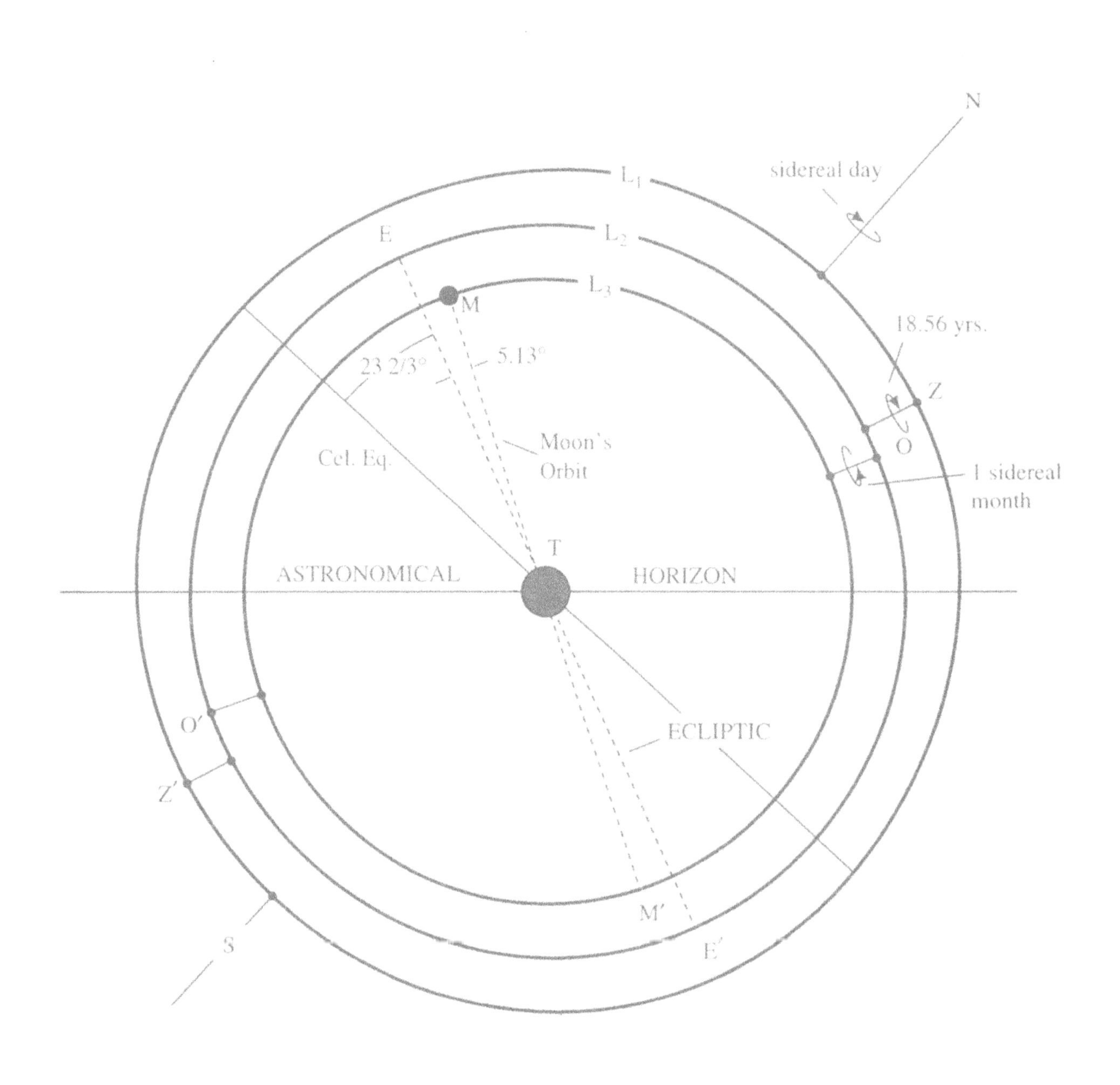

Figure II.1 — Eudoxus's concentric spheres of the Moon (seen from the outside and from the east).

change the period about OO' from 18.6 to 2922 years; and change the period about OO' to 1 sidereal year.

Thus Eudoxus's system of the Sun's motion, besides failing to provide a mechanism for the unequal lengths of the seasons, also erred in requiring the numerical values of the declinations of the solstices to vary periodically by 1° in about 3000 years. This produced a displacement of the equinoxes, which was not the precession (discovered later by Hipparchus) but was similar to the (nonexistent) "trepidation" introduced by Arab astronomers during the early Middle Ages (although the period and amplitude for the trepidation were quite different). His system of the Moon's motion, on the other hand, was deficient in not providing for any variable motion in lunar celestial longitude or any variation in inclination (discovered much later by Tycho).

Eudoxus's Planetary Theory

As an example of Eudoxus's planetary theory, let us describe his mechanism for explaining the stations and retrograde motions of Jupiter.

In Figure II.2 consider a nest of four closely fitting concentric spheres K', K, K_1, and K_2. The planet P is always on the equator of the innermost sphere K_2. The combined action of the two innermost spheres (K_1 and K_2) is to produce a bow-like path which is dragged along the ecliptic by sphere K. The successive positions of P, projected on the sky, then reproduce phenomena similar to the stations and retrogressions of the planet in various parts of the zodiac. In this model K' rotates westward in a period of 24 sidereal hours, on an axis NS coinciding with the axis of the sky's diurnal motion. This produces the diurnal motion of all the other spheres. The second sphere K is pivoted on two opposite points Z and Z' (on the surface of K), the axis ZZ' being perpendicular to the plane of the ecliptic. K rotates eastward on this axis with a period equal to the sidereal period of the planet (11.86 years for Jupiter). Its motion leads to periodic, bow-like excursions of moderate size from a mean position. By the action of the two innermost spheres these excursions always occur near the equator of K (the ecliptic). Successive excursions always occur about one sign (30°) further east in the zodiac, i.e., about 1 year and 1 month later. Eudoxus called such an excursion, shown in Figure II.3, a *hippopede* (a horse fetter or hobble). For Jupiter, this 398-day curve, when combined with the eastward movement of 11.86 years' period due to the action of K, results in a path projected on the celestial sphere as depicted in Figure II.4.

The rigorous parametric equations for the hippopede on a spherical surface were first derived by Giovanni V. Schiaparelli (1835–1910). Building on his work, Herz (1887) gave a close convenient approximation to the formulas by projecting the spherical coordinates on a plane tangent to the celestial sphere. For example, using this method Herz found $\lambda = 230°$, $\beta = +31.5'$ for the mean celestial longitude and latitude of Jupiter as observed from Alexandria at midnight on 5 August A.D. 133.[1] He found overall that the hippopede method was very inaccurate in celestial latitude and totally unreliable in longitude. For instance, retrograde arcs were 20° too long for Jupiter and 16° too short for Saturn. For Mars Eudoxus assumed a rotation of the spheres K_1 and K_2 with one-third the synodic period. While this explained the main retrogression, it also created two nonexisting, smaller

[1] A quick check made by the present author using the five-place Ahnert (1971) tables gives $\lambda = 230.1°$, $\beta = 0.0°$.

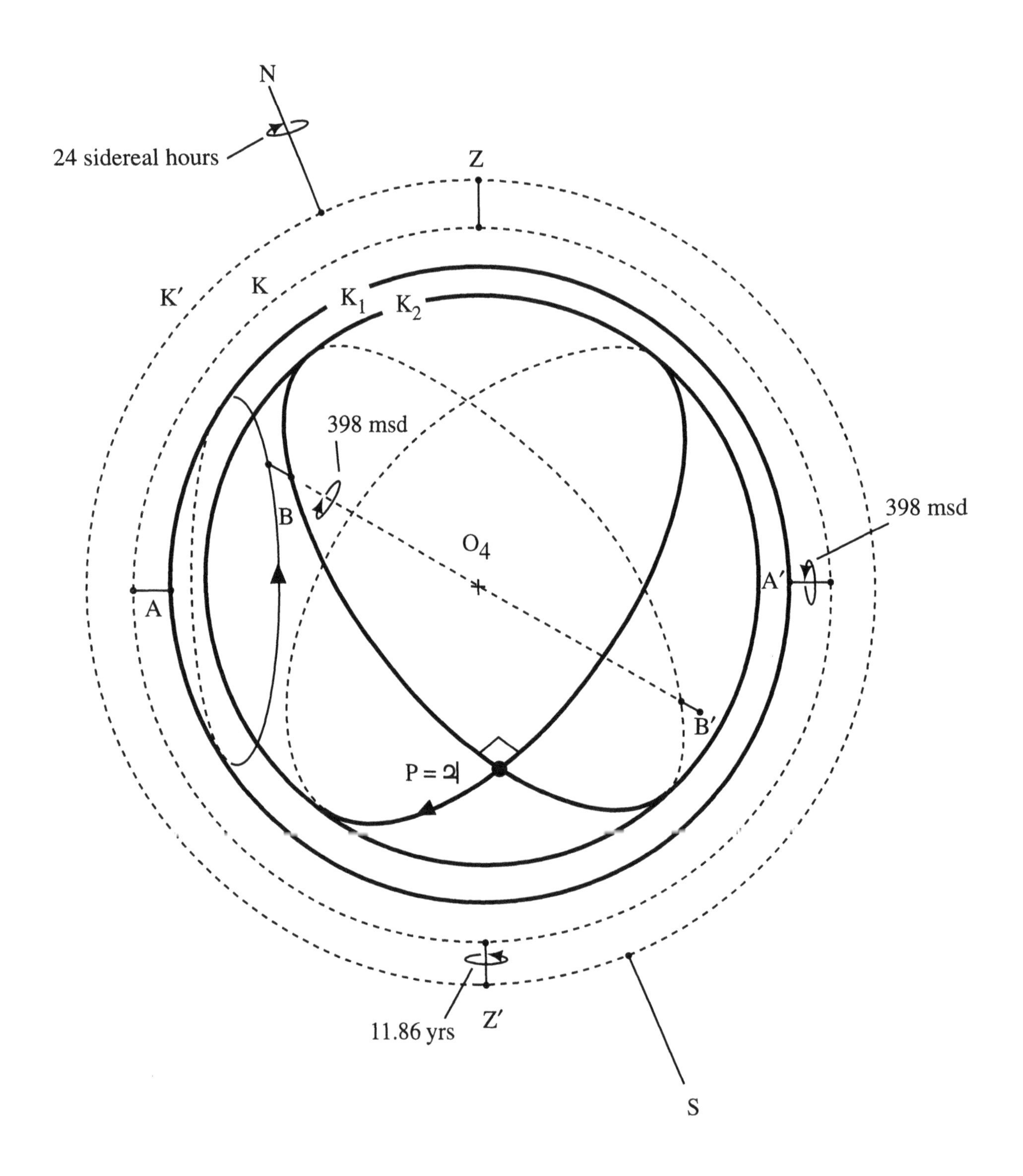

Figure II.2 — Eudoxus's system of concentric spheres for Jupiter (seen from the outside and from the west).

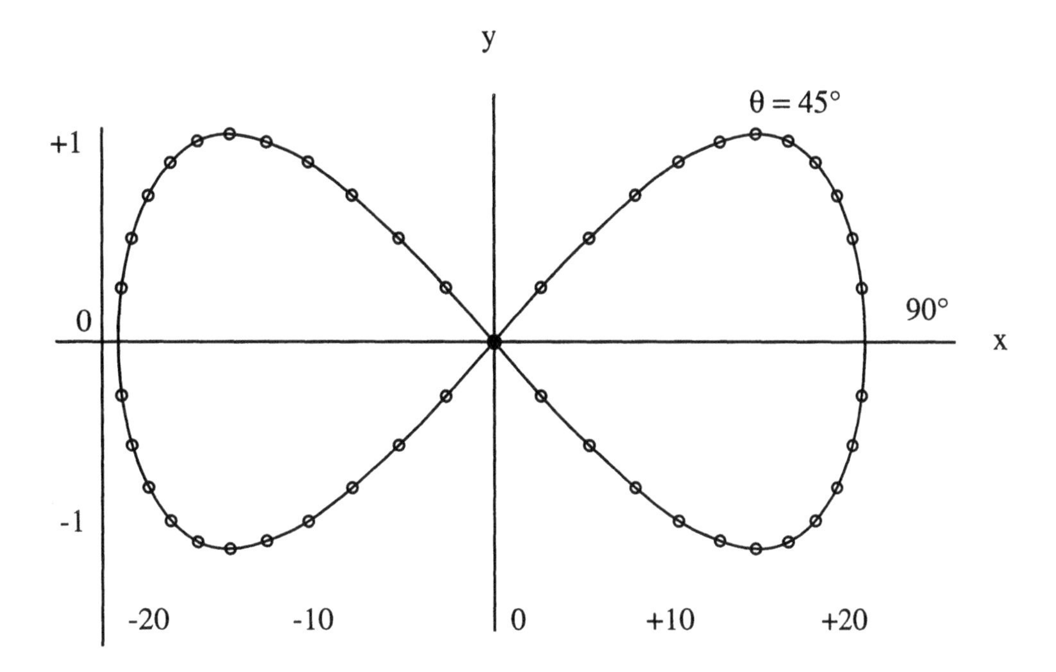

Figure II.3 — Eudoxus's hippopede for Jupiter, shown on a scale of 10:1 for y and x. Motion resulting from movement of K_1 and K_2 only.

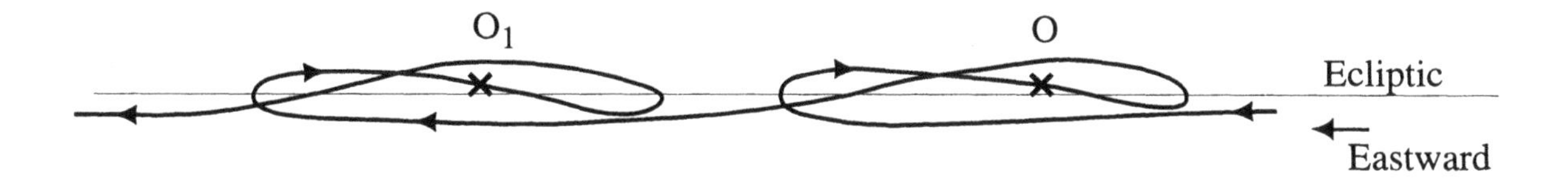

Figure II.4 — A possible apparent motion of Jupiter resulting from motion in the hippopede. O, O_1 are positions of Jupiter at opposition approximately 1 year and 1 month apart ($\angle OO_1 = 30.35°$).

retrogressions at longitudes of 130° and 40°. Again this was a theory that explained both "too little" and "too much." Finally, for Mercury and Venus Herz found a moderately good representation for the greatest elongations using $i = 24°$ and 46°, yet no satisfactory values for the retrograde arcs were found, even by the rough standards of antiquity.

The Systems of Spheres

Eudoxus used three spheres each for the Sun and the Moon, four each for the five planets, and one for the fixed stars: a total of 27 spheres.

According to Simplicius, Callippus (370–300 B.C.), one of Eudoxus's students, introduced two new spheres for the Sun to account for its nonuniform motion along the ecliptic, which, it is said, Callippus discovered. This enabled him to represent the motion of the Sun with an accuracy almost as great as that of Hipparchus, who used an eccentric circle. Callippus also employed two more spheres for the Moon to represent its larger nonuniform motion. In addition he improved the very defective representations of the motions of Mars and Venus by using one more sphere for each planet. Thus the Callippic system contained in all 33 spheres.

Eudoxus and Callippus, being essentially mathematicians, probably regarded their systems of spheres as mere aids to the description and prediction of celestial motions. Aristotle, however, being more physically inclined, regarded the cosmos as an "animal" in which there could be no empty spaces. This required the existence between the planetary spheres of material, which also permitted motion to be transmitted from one sphere to the next by a sort of friction. In the "cosmic animal" he therefore introduced "reacting spheres" between the crystalline spheres, starting from the "Prime Mover," the sphere of the fixed stars, and ending with the innermost lunar sphere. (The Moon did not drive anything inside its fifth sphere.) For a number n of crystalline spheres in his system, there would always be $n - 1$ reacting spheres. Thus, Aristotle's cosmos required the following number of spheres:

Planet	Crystalline Spheres	Reacting Spheres
Saturn	4	3
Jupiter	4	3
Mars	5	4
Venus	5	4
Mercury	5	4
Sun	5	4
Moon	5	
Sum	33	22
Total number of spheres		55

CHAPTER III

Hipparchus (fl. 146–126 B.C.)

Hipparchus (fl. 146–126 B.C.) was one of the most original and creative astronomers of the Greek tradition, and indeed, he is worthy of comparison with the greatest astronomers of all time. Hipparchus's work fundamentally altered the character of Greek astronomy and put astronomy on the path that was to lead ultimately to the beautifully successful synthesis by Ptolemy three centuries later. Hipparchus's most decisive innovations were the following:

1. He invented or greatly advanced trigonometry, which radically improved the methods available for numerical computation of the sides and angles of geometrical figures, whether in a plane or on a sphere.
2. He played a large role in making Babylonian observations available to Greek astronomy and showed how to use comparisons between old and new observations to reveal astronomical changes too slow to be detected within a single lifetime.
3. Although Greek astronomers had begun to pay serious attention to observation shortly before Hipparchus's time, Hipparchus fundamentally changed the role of observation in Greek astronomy, both by making extensive series of observations himself and by insisting on the importance of numerical accuracy and precision.
4. He was the first to systematically employ eccentrics and epicycles for representing the motion of the Sun and Moon and to show how to deduce the numerical values of the elements from suitably selected observations.

We have very little information about Hipparchus's life. He was probably born at Nicaea in Bithynia and probably spent most of his adult life at Rhodes. The observations of Hipparchus cited by Ptolemy in the *Almagest* range from 162 to 127 B.C. The only surviving work of Hipparchus is his *Commentary on the Phenomena of Aratus and Eudoxus.* Although containing many data involving positions of the fixed stars, it was almost certainly written before Hipparchus made his great discovery of the precession of the equinoxes. Our knowledge of Hipparchus's other astronomical works – their titles and their partial contents – is based almost entirely on Ptolemy's use of them in the *Almagest.* For Hipparchus's work in geography our best source is Strabo.

Hipparchus's greatest contributions to astronomy resulted from a merging of the geometrical-philosophical approach of earlier Greeks such as Eudoxus and Apollonius of Perge with the numerical-observational approach pioneered by the Babylonians. Hipparchus turned Greek astronomy into a fundamentally different enterprise.

Hipparchus's Main Astronomical Contributions

1. He developed trigonometry into a science useful for computations.
2. He observed the positions of more than one thousand fixed stars.
3. He introduced the use of stellar magnitudes.
4. He discovered the precession of the equinoxes and estimated its rate.
5. He greatly improved on Aristarchus's estimate of the Moon's distance.
6. He determined the length of the tropical year (to within 6 minutes' accuracy).
7. He described the Sun's variable annual motion and determined the unequal lengths of the seasons.
8. He discovered the second inequality in the Moon's motion (the evection) and determined its maximum amount to be $1\frac{1}{4}^{\circ}$. (Modern value is $1^{\circ}16'$.)
9. He predicted the time of eclipses to within 1 hour.
10. He predicted the terrestrial latitude wherein a particular eclipse would be visible.
11. He represented the Sun's motion in the ecliptic by an eccentric.
12. He measured the inclination of the lunar orbit to be $5^{\circ}8'$ and found it to be variable.
13. He treated the regression of the nodes of the Moon's orbit on the ecliptic, with a period of about 19 years.
14. He treated the advance of the apsides of the lunar orbit, with a period of about 9 years.
15. He observed and described a new star (a nova of about 134 B.C.).

Hipparchus's Solar Theory

As noted in Chapter I, the rate of the Sun's apparent eastward motion in the ecliptic is smoothly variable in a period of one anomalistic year. This variability of the Sun's progress along the ecliptic is said to have been known to some astronomers before the time of Eudoxus, although he himself did not mention it. But his successor Callippus is said to have been aware of it and to have also discovered the variable motion of the Moon in celestial longitude. Nevertheless, among the Greeks it was Hipparchus who first definitely recognized, observed, and determined these inequalities of motion in some detail. In the case of the Sun, this was accomplished by his discovery of the unequal lengths of the seasons.

The Apparent Nonuniform Motion of the Sun in the Ecliptic

Figure III.1 illustrates schematically the present-day situation with respect to the Sun's annual apparent motion in the ecliptic:

E = the Earth,

EA = a reference direction, that of the Sun's apogee as observed in A.D. 1960,

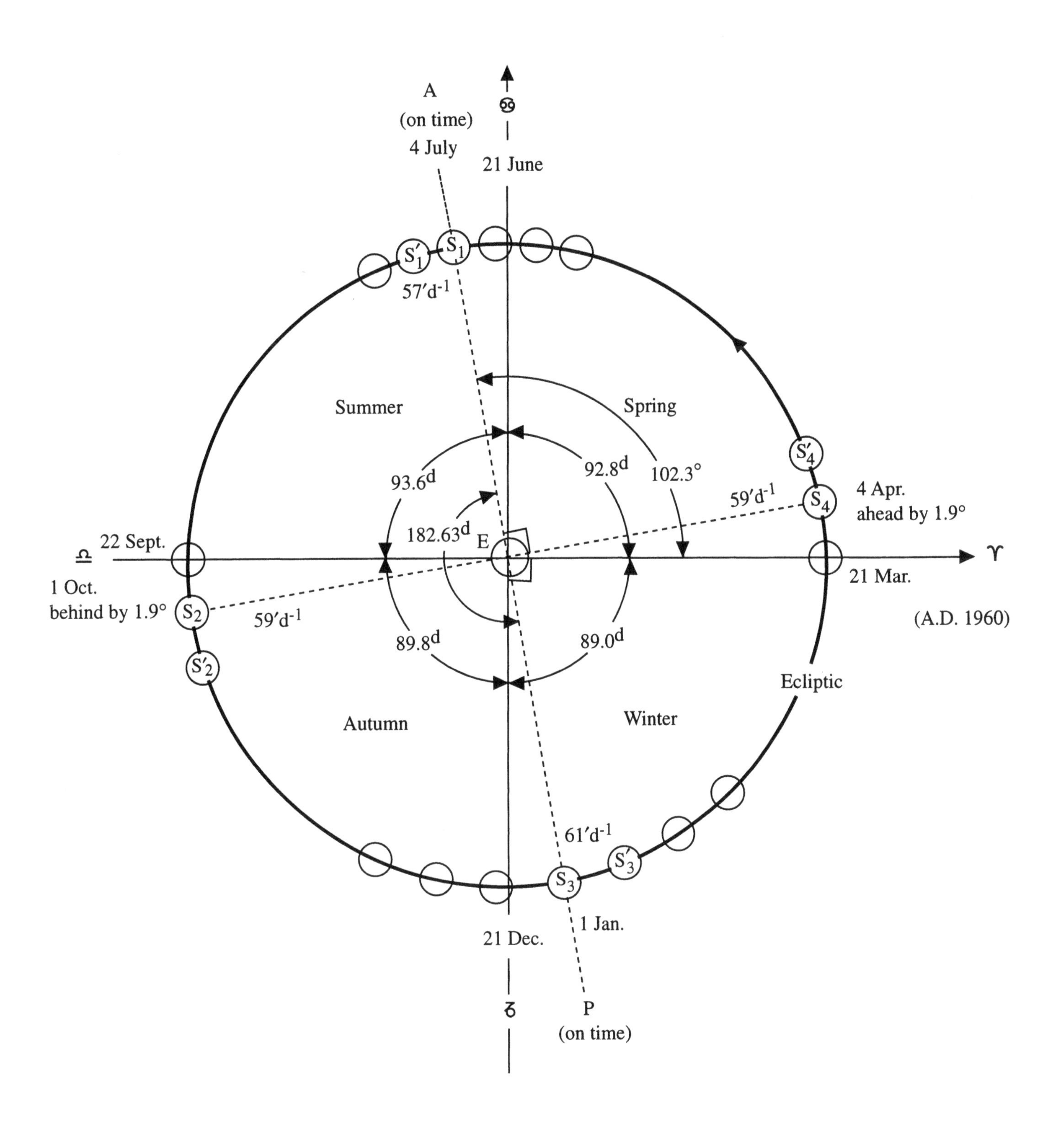

Figure III.1 — The apparent motion of the Sun.

EP = direction of the Sun's perigee (1960),

AP = line of apsides,

ES_1, ES_3 = directions of the Sun at apogee and perigee, as observed in a given year,

ES_2, ES_4 = directions of the Sun at right angles to the line of apsides,

E-♈ = direction of the Sun at the vernal equinox,

E-♎ = direction of the Sun at the autumnal equinox,

E-♑ = direction of the Sun at the winter solstice,

E-♋ = direction of the Sun at the summer solstice,

S'_n = direction of the Sun one day after its passage at S_n ($n = 1, 2, 3, 4$).

Elliptic motion, we now know, requires symmetrical daily increments of true anomaly about the line of apsides, S_1S_3, and average motion (59′/day) at points approximately 90° from the line of apsides, near S_2 and S_4. Today, daily increments of solar celestial longitude vary from *(a)* about 61′/day on 1 or 2 January (depending upon leap year), when the Sun is at S_3, to *(b)* 57′/day on 4 or 5 July, when the Sun is at S_1. Average values of 59′/day occur at about 4 April and 1 October ($S_2S'_2$ and $S_4S'_4$) when the Sun is near the perpendicular to the line of apsides AP. Near these positions the equation of the center (which is true anomaly minus mean anomaly) reaches a maximum of +1.92° on 4 April and a minimum of −1.92° on 1 October. Present-day observations give the following lengths of the seasons:

Winter = Time between Sun's passing ♈. and Sun's passing ♑,
= 89.0 mean solar days,

Spring = Time between Sun's passing ♋ and Sun's passing ♈,
= 92.8 mean solar days,

Summer = Time between Sun's passing ♎ and Sun's passing ♋,
= 93.6 mean solar days,

Autumn = Time between Sun's passing ♑ and Sun's passing ♎,
= 89.8 mean solar days.

In principle one should be able to find from a table of daily observed apparent longitudes the direction of the line of apsides, AP, by arranging the smoothly varying daily increments of longitude over one year symmetrically about AP, so that the maximum and minimum values fall at P and A, respectively. But a more reliable and elegant method followed by Hipparchus required only three observed quantities: the times and positions of the equinoxes and of a single solstice. The solstices could be determined from solar declinations observed by a gnomon at noon some days before and after the dates of their occurrence. But Hipparchus realized that such a determination did not admit of much accuracy, and consequently he also used quadrants to observe the solstices and equatorial rings to observe the equinoxes. The equatorial ring was placed accurately in the plane of the celestial equator. At the moment of equinox (assuming that it occurred during daylight hours) the shadow of the upper part of the ring fell exactly on the lower part of the ring.

Hipparchus's Method of Finding the Line of Apsides and the Eccentricity of the Sun's Orbit (Considered as an Eccentric Circle)

Hipparchus's method of finding the Sun's orbit consisted in observationally determining the exact lengths of the seasons by fixing the dates and moments when the Sun passed the directions E-♈, E-♎, and E-♋ (see Fig. III.1) in its yearly progress along the ecliptic. He thus acquired the following data:

length of spring $= 94\frac{1}{2}$ days = passage through ♋ minus passage through ♈,

combined length of summer $= 92\frac{1}{2}$ days = passage through ♎ minus passage through ♋,

combined length of autumn and winter $= 178\frac{1}{4}$ days.

Given these data, his method may be outlined with reference to Figure III.2, in which

$E =$ the Earth,

$O =$ the center of the Sun's path, considered as an eccentric circle QNP of radius $r = 1$,

$A_O = \angle N'EO =$ the celestial longitude of the line of apsides EO belonging to the eccentric QNP (to be found),

$\rho =$ the eccentricity of the Sun's (circular) path QNP (to be found),

N', P', M', $Q' =$ the Sun's positions at the equinoxes and solstices,

$PQ =$ the diameter of the Sun's path parallel to the line of solstices $P'Q'$,

$NM =$ the diameter of the Sun's path parallel to the line of equinoxes $N'M'$,

$R =$ the intersection of PQ and $N'M'$,

$\alpha = \angle QON'$,

$\beta = \angle N'OP$,

$a = \angle N'OP'$,

$b = \angle P'OM'$,

$c = \angle M'ON'$,

$d = \angle POP' = \angle EP'O$.

From the observed data (ca. 150 B.C.):

$$a = 2\pi\frac{94.5}{365.25}, \qquad b = 2\pi\frac{92.5}{365.25}, \qquad c = 2\pi\frac{178.25}{365.25}.$$

By geometry:

$$\alpha = \tfrac{1}{2}c,$$

$$\beta = \pi - \alpha = \pi - \tfrac{1}{2}c = \tfrac{1}{2}(a+b),$$

$$d = a - \beta = a - \tfrac{1}{2}(a+b) = \tfrac{1}{2}(a-b).$$

By trigonometry:

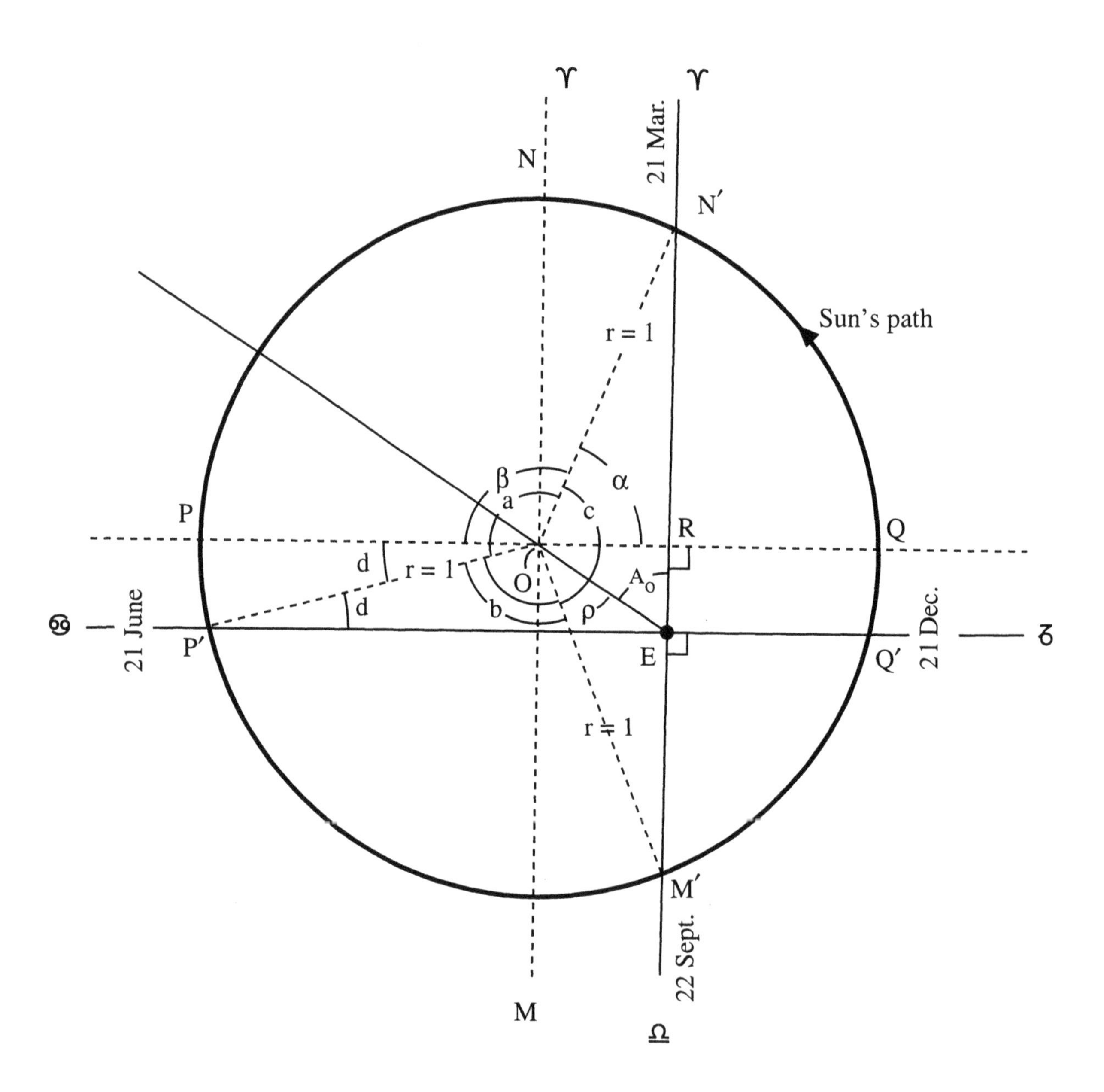

Figure III.2 — Hipparchus's method of finding the line of apsides and eccentricity of the Sun's circular orbit.

$$\rho \sin A_O = r \cos \alpha = 1 \cos \alpha = \cos \alpha,$$

$$\rho \cos A_O = r \sin d = 1 \sin d = \sin d,$$

$$\tan A_O = \cos \alpha \csc d,$$

from which A_O and $\rho = \sqrt{\cos^2 \alpha + \sin^2 d}$, the eccentricity of the Sun's circular eccentric. Substituting numerical values, Hipparchus found $A_O = 65\frac{1}{2}^\circ$ and $\rho = \frac{1}{24}$, both results fairly in agreement with modern calculations for 150 B.C. of 66.1° and 1/28.4. In 150 B.C. the true eccentricity of the Earth's orbit, 1/28.4, yields a maximum value of the equation of center $(v - M)_{max} = 2.0°$, whereas Hipparchus's value of the eccentricity, 1/24, corresponds to 2.4°.[1] This slight error in Hipparchus's eccentricity made a maximum error in the Sun's longitude through the year of only 22′.

It should be noted that in spite of the Earth's orbit being elliptical rather than circular eccentric, Hipparchus's solar theory can be fitted to the real movement of the Sun to an accuracy of about 1′ in celestial longitude. Not until Kepler's exacting and precise work on "the bisection of the eccentricity" for the Earth's motion about the Sun did it become necessary to distinguish between an eccentric circle and an ellipse (with half the eccentricity).

Hipparchus's Method of Predicting the Sun's Place at Any Instant

An immediate use of the elements A_O (longitude of apogee) and ρ (eccentricity) of the Sun's eccentric circle is to predict the Sun's geocentric celestial longitude L for any time. In Figure III.3, let E be the Earth, O the center of the Sun's eccentric, C the Sun's place at any time T, L its corresponding longitude, and M its mean anomaly. Let A be the apogee of the orbit (at longitude A_O) and T_O the moment the Sun is at A. Let v be the Sun's true anomaly and x its equation of center, both at T. Then the problem may be stated as follows: given T, T_O, $\mu = 0.99°/\text{day}$, A_O, and ρ, to find L.

By definition: $M = \mu(T - T_O)$, from which M. From ΔOEC:

$$\frac{\sin x}{\sin v} = \frac{\rho}{1} \qquad \text{or} \qquad \sin x = \rho \sin v.$$

Also, $v = M - x$. Substituting:

$$\sin x = \rho \sin (M - x),$$

$$\sin x = \rho (\sin M \cos x - \cos M \sin x),$$

$$\sin x + \rho \sin x \cos M = \rho \sin M \cos x,$$

$$(1 + \rho \cos M) \sin x = \rho \sin M \cos x,$$

$$\tan x = \frac{\rho \sin M}{1 + \rho \cos M},$$

from which x. Finally, $v = M - x$ and, from the figure, $L = A_O + v$.

[1] The A.D. 1960 value of the eccentricity (of a circle adopted to represent the Sun's path) is $e = 1/29.9$. This produces a present maximum value of the equation of center of 1.92°.

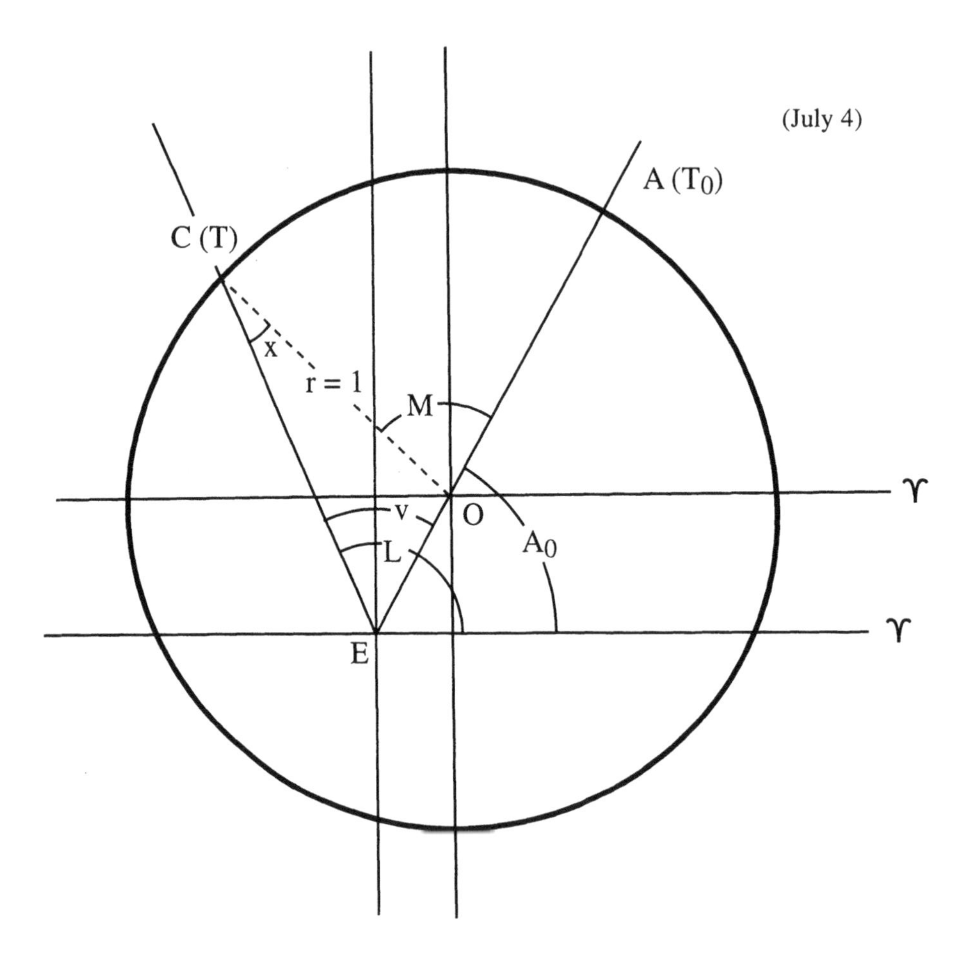

Figure III.3 — Hipparchus's method of predicting the Sun's geocentric longitude at any time.

Equivalence of Epicyclic and Eccentric Motion

Hipparchus noted that the annual motion of the Sun along the ecliptic could be represented not only by the above method but, equally well, by a direct uniform motion of a point on a circle in the plane of the ecliptic, but eccentric to the Earth. This device, discussed in detail by Apollonius, was a uniform retrograde motion of a point on a circular epicycle whose center was endowed with a uniform direct motion on a circular deferent concentric with the Earth – all circles being in the plane of the ecliptic, and all motions having a period of one anomalistic year. In Figure III.4 let

T = Earth,

A, P = apogee and perigee of the Sun's path,

S = a general position of the Sun in its orbit K,

K = a circular eccentric (center C, radius R, motion uniform and direct),

D = a circular deferent (center T, radius R, motion uniform and direct),

E = a circular epicycle (center S', radius e, motion uniform and retrograde),

e = eccentricity of $K = TC = SS'$,

E_1 = position of E at moment of average rate of increase of geocentric longitude S_1 of Sun,

E_2 = position of E at moment of average rate of increase of eccentric longitude S_2 of Sun,

D', S', S'_1, S'_2 = centers of the epicycle when the Sun is at A, S, S_1, S_2, respectively,

$\lambda_0 = \angle \Upsilon TA$ = geocentric longitude of Sun's apogee,

L = intersection with E of TS' (extended),

$\angle ACS = \angle D'TS'$ = mean anomaly of Sun when at $S = M$,

$\angle ATS$ = true anomaly of Sun when at $S = v$,

$\angle \Upsilon CS$ = the Sun's eccentric celestial longitude,

$\lambda' = \angle \Upsilon TS' = \angle \Upsilon S'L$ = geocentric celestial longitude of epicyclic center,

$\lambda = \angle \Upsilon TS$ = the Sun's geocentric celestial longitude.

Let the motion start with the Sun at the apogee A and let the period be the same on all the circles. Inspection shows that in any general position S of the Sun, $\angle ACS = \angle ATS' = \angle CTS' = \angle SS'L = \angle(\lambda' - \lambda_0) = M$ = Sun's mean anomaly. It follows that $SS'TC$ is always a parallelogram, and it is therefore immaterial whether the motion of S be described by a single uniform direct motion on the eccentric K or by a combination of retrograde motion on the epicycle SLE with uniform direct motion on the concentric deferent D. (Hipparchus chose the former representation as being the simpler of the two.) The above discussion illustrates the geometrical result that a uniform direct circular motion on an eccentric circle can be equally well described by a uniform retrograde motion on a circular epicycle that slides with uniform direct motion on a concentric circular deferent, the periods of all motions being equal.

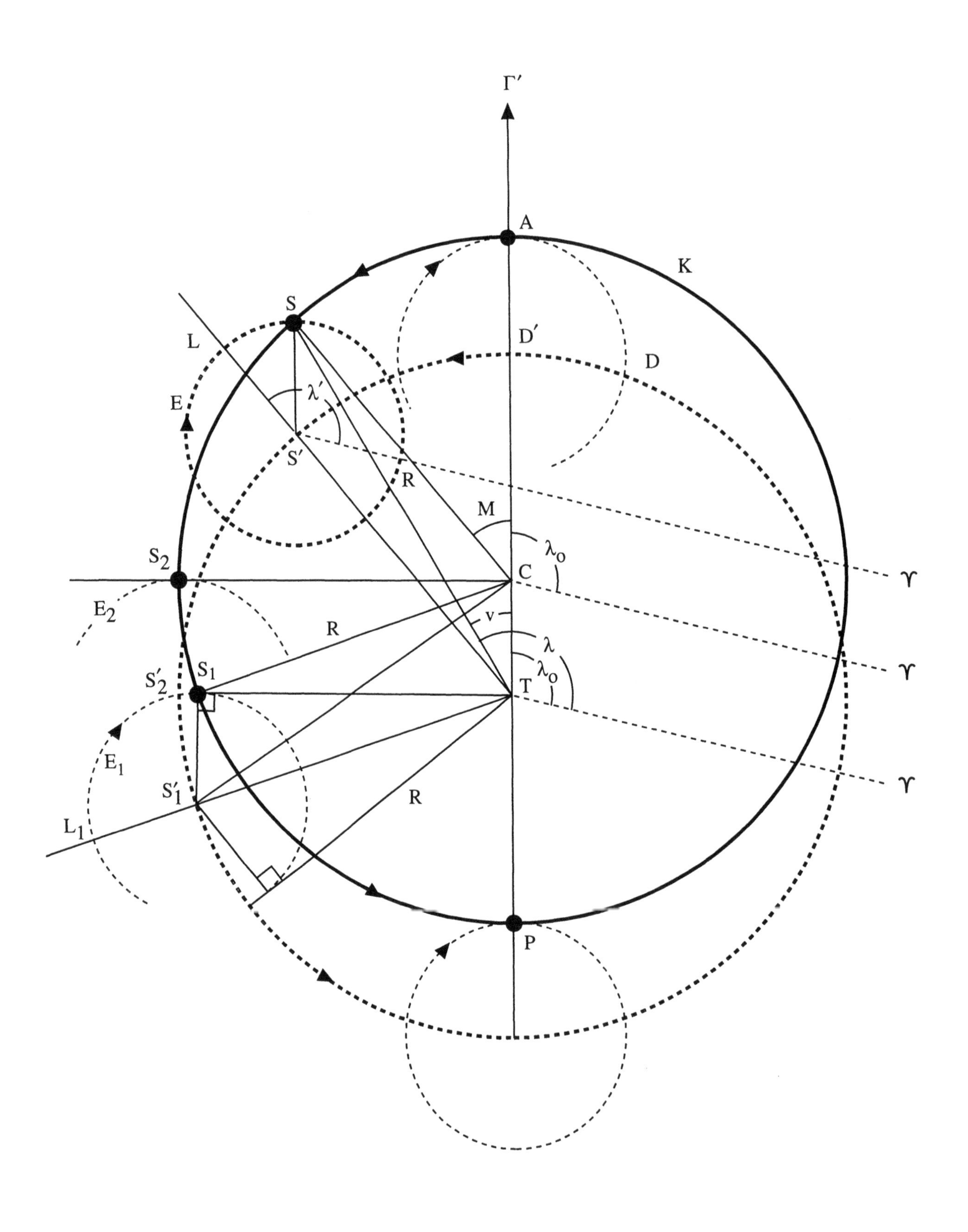

Figure III.4 — Equivalence of epicyclic and eccentric motion.

In this connection it should be noted that a uniform advance or regression of the line of apsides can be achieved by choosing different periods on the epicycle and on the deferent. This valuable property of epicyclic motion may have partly influenced Ptolemy in his choice of an epicyclic scheme for the Moon's motion, so that he could elegantly effect the considerable rate of advance of the Moon's line of apsides, well known to the ancients. For the Moon, as we shall see in Chapter IV, Ptolemy used an epicyclic system as described above, except that the epicycle was moving on a deferent whose center was itself moving, together with certain other motions.

The average drift in apparent celestial longitude (as seen from T in Fig. III.4) occurs at S_1, exactly when the planet is moving instantaneously on the tangent S_1T to the epicycle on the following side of the epicyclic center. At this moment the planet is exactly 90° from the line of apsides ($\angle ATS_1 = 90°$). Within the observational accuracy possible to the ancients, uniform motion in an eccentric circle gave a very good simulation of the Sun's motion in celestial longitude (to an accuracy slightly better than $1'$), and thus Ptolemy kept intact Hipparchus's eccentric orbit of the Sun. As we shall see, however, in the case of the planets Ptolemy found there was a major advantage of the epicyclic representation over the simple eccentric one.

Hipparchus's Lunar Theory

Hipparchus assumed for the Moon a simple epicycle model, similar to the epicycle version of the solar theory. Thus, the epicycle's center moves eastward at a uniform angular speed around a deferent circle centered on the Earth. The epicycle's center makes one circuit of the deferent in a sidereal period of the Moon, or about $27\frac{1}{3}$ days; i.e., it has a uniform eastward motion of 13.2°/day. The Moon itself uniformly revolves on the circumference of this epicycle, in a retrograde direction with a slower angular motion than that on the deferent, the result being a uniform advance of the line of apsides of 360° in 8.86 years, i.e., 40.6°/year, or 3.3°/sidereal month. Thus the period on the epicycle is $27\frac{1}{3} \times \frac{360}{(360-3.3)}$ $= 27.51$ days. This model was undoubtedly inspired by the work of the mathematician Apollonius, but its detailed astronomical application to the motion of the Moon – finding the necessary periods and constants – was definitely the work of Hipparchus.

Naturally, there exists an equivalent eccentric-circle model, which Hipparchus also analyzed. We shall find the eccentric model simpler to use in making a more detailed examination of Hipparchus's lunar theory. Thus, assume the lunar orbit to be a circle $ABCD$ (center O) eccentric to the Earth E by e, its plane being inclined to the plane of the ecliptic by 5°, intersecting it in E_1E_2 passing through the Earth's center E (see Fig. III.5). The orbit plane rotates in the retrograde direction around the axis of the ecliptic through E (perpendicular to the plane of the page) so that the nodes E_1 and E_2 of the orbital plane on the ecliptic plane retrograde one complete revolution in $18\frac{2}{3}$ years. On this circle the Moon moves uniformly eastward with the sidereal period of the Moon, or about $27\frac{1}{3}$ mean solar days. The center O of the eccentric, and therefore the whole system, shifts uniformly eastward with a period of 8.86 years on a circle centered at E, in order to account for the advance of the line of apsides of the Moon's orbit. The rate of progression of this gyrational shift was thus 40.6°/yr, or 3.3°/month.

These two versions of the theory should be equivalent. Somewhat surprisingly, Ptolemy tells us that Hipparchus found an eccentricity e in the eccentric-circle version that differed

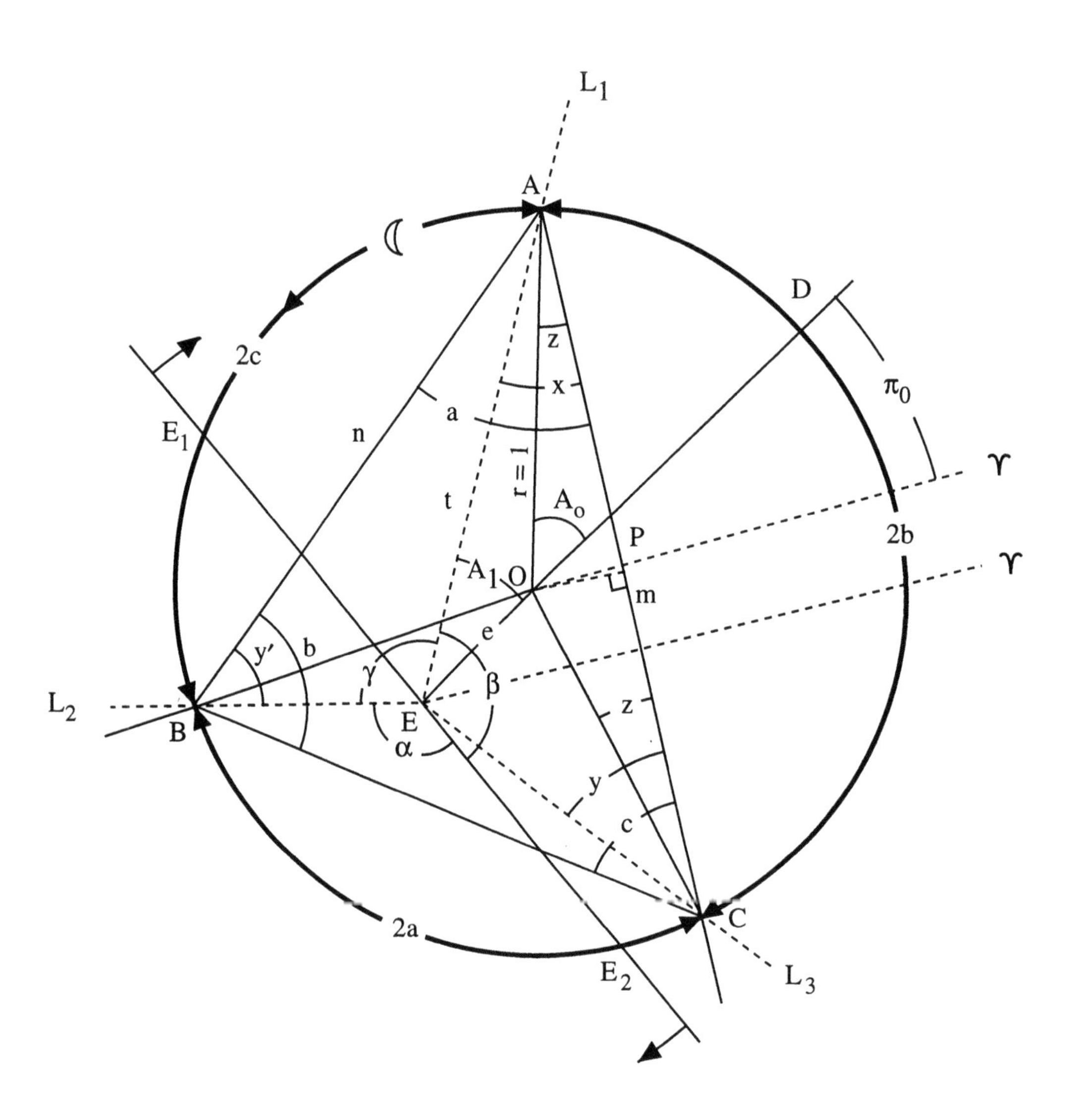

Figure III.5 — Hipparchus's method for finding the constants of the Moon's orbit.

from the radius ρ of the epicycle in the epicycle-plus-concentric version. By Apollonius's equivalence theorem, these two should be exactly the same. However, Ptolemy resolves the dilemma by showing that Hipparchus worked from different triplets of eclipses in the two calculations and also made some slips when calculating the time intervals between eclipses. In any case, Ptolemy says that Hipparchus found, in the eccentric-circle model, an eccentricity

$e = 6\frac{15}{60}$, where the radius r is 60, or

$e = 0.104$, if we put the radius $r = 1$.

But, in the epicycle model, Hipparchus found for the radius of the epicycle

$\rho = 4\frac{46}{60}$, where the deferent's radius is 60, or

$\rho = 0.0794$, if we put $r = 1$.

Ptolemy remarks that Hipparchus's first result gives a maximum lunar equation of about $5°49'$, while the second gives $4°34'$. (These are slightly inaccurate and should be $5°59'$ and $4°33'$ respectively.) In his own version of Hipparchus's lunar theory Ptolemy adopted an epicycle radius

$\rho = 5\frac{1}{4}$, where the radius of the deferent is 60, or

$\rho = 0.0875$, if we put the radius $r = 1$.

This produces a maximum lunar equation of about $5°$.

Hipparchus founded his theory of the Moon's motion mainly on Babylonian and Alexandrian lunar eclipse observations. His orbit fit the motion very well at the syzygies and so was useful in eclipse prediction. He also examined whether the Moon at other points of its orbit conformed to his calculations, especially at the quarters. He found that sometimes the observed place agreed with his theory, but at other times it did not; however, he had to leave to his great successor and admirer, Ptolemy, the detailed investigation of this effect.

Hipparchus's Method of Finding the Line of Apsides and Eccentricity of the Moon's Orbit

Hipparchus determined the eccentricity and other constants of the lunar orbit by a method which may be considered an elaboration of the one he had used for the Sun's annual (eccentric) orbit. A possible method is outlined below. In our representation we use simple equations involving unknowns for the sake of simplicity, well knowing that such analytical expressions were not available to Hipparchus. Also, for the sake of simplicity, we shall use the eccentric-circle version of the theory. In Figure III.5 let

E = the Earth,

A, B, C = three positions of the Moon, at corresponding times T_1, T_2, T_3, lying on the eccentric circle with center O and radius $r = 1$,

Υ = the vernal equinox,

L_1, L_2, L_3 = the observed (from eclipses, etc.) true geocentric celestial longitudes of A, B, C, respectively, corrected for a uniform advance of the lunar apsides,

D = the apogee of the lunar orbit considered as the eccentric $DABC$,

π_0 = the geocentric celestial longitude of the lunar apogee at some arbitrary epoch T_0, to which all constants and observations are referred,

$a = \angle BAC$,

$b = \angle ABC$,

$c = \angle ACB$,

P = the midpoint of AC,

A_0 = the mean anomaly of A,

A_1 = the true anomaly of A,

$\alpha = \angle BEC = L_3 - L_2$,

$\beta = \angle AEC = L_1 - L_3$,

$\gamma = \angle AEB = L_2 - L_1$,

n = chord AB,

m = chord AC,

$e = EO$ = the eccentricity of the Moon's circle,

t = distance EA,

$x = \angle EAC$,

$y = \angle ECA$,

$y' = \angle EBA$,

$z = \angle OAC = \angle OCA$,

μ = the mean daily motion of the Moon with respect to the fixed direction OD.

The problem may now be stated: given L_1, L_2, L_3, T_1, T_2, T_3, and μ, find e, π_0, A_0. By definition:

$$2a = \mu(T_3 - T_2), \qquad 2b = \mu(T_1 - T_3), \qquad 2c = \mu(T_2 - T_1), \qquad \text{and}$$

$$\alpha = L_3 - L_2, \qquad \beta = L_1 - L_3, \qquad \gamma = L_2 - L_1.$$

From ΔAEC:

$$t = m\,\frac{\sin y}{\sin \beta}. \tag{1}$$

From ΔAEB:

$$t = n\,\frac{\sin y'}{\sin \gamma}. \tag{2}$$

From ΔABC:

$$\frac{m}{n} = \frac{\sin b}{\sin c}. \tag{2'}$$

By division:

$$\frac{(1)}{(2)} = \frac{m \sin y \sin \gamma}{n \sin \beta \sin y'} = 1.$$

Using equation (2′):

$$\frac{\sin y \sin b}{\sin \beta} = \frac{\sin y' \sin c}{\sin \gamma}. \tag{3}$$

The sum of the angles in ΔAEB and $\Delta AEC = 360° = y + y' + a + \beta + \gamma$. (4)

Let $d = 360° - (a + \beta + \gamma)$, from which d. From equation (4):

$$d = y + y', \tag{5}$$

from which $y + y'$. By trigonometry eliminate y' from equations (3) and (5), obtaining a solution for y:

$$\tan y = \frac{\sin d \, \frac{\sin c \sin \beta}{\sin b \sin \gamma}}{1 + \cos d \, \frac{\sin c \sin \beta}{\sin b \sin \gamma}} \tag{6}$$

From ΔAEC: $x = 180° - (\beta + y)$.

From ΔAOC: $z = 90° - b$.

From ΔOAE: $\angle OAE = x - z$.

From ΔAEC: $t = m \frac{\sin y}{\sin \beta}$.

From ΔAPO: $r = \frac{m}{2 \sin b}$.

From the last two expressions:

$$t = 2r \frac{\sin b \sin y}{\sin \beta}.$$

From ΔAOE:

$$e \sin A_1 = r \sin (x - z) \qquad \text{and} \qquad e \cos A_1 = t - r \cos (x - z),$$

from which, simultaneously, the eccentricity e and A_1. Finally,

$$A_0 = A_1 + (x - z) \qquad \text{and} \qquad \pi_0 = L_1 - A_0,$$

the longitude of apogee.

The same method can be used for finding an eccentric orbit for any planet. In that case the values of the geocentric longitudes L_1, L_2, and L_3 are usually obtained from observations of three oppositions.

Hipparchus's Method of Predicting the Moon's Place at Any Time

Hipparchus had at his disposal very accurate values of the lengths of the various months (sidereal, anomalistic, nodical) and thus of the mean daily motions of the Moon with respect to the fixed stars, with respect to the mean line of apsides, and with respect to the mean nodes of the lunar orbit. He derived the following constants:

$$\mu = \text{mean daily motion (sidereal)} = 47434.976'',$$

$$\mu_1 = \text{mean daily motion (anomalistic)} = 47033.942'',$$

$$\mu_2 = \text{mean daily motion (nodical)} = 47625.661''.$$

Disregarding at first the Moon's celestial latitude, we may represent the lunar orbit by an eccentric circle in the plane of the ecliptic. For simplicity we will first consider the orbit fixed in position. To predict the Moon's place L at the moment T we start from a known position, e.g., a given longitude L_0 at T_0. In Figure III.6, let

E = the Earth,

O = center of the circle representing the lunar orbit (eccentricity $= EO = e$, radius $r = 1$),

A = lunar apogee,

$E\Upsilon = O\Upsilon$ = direction of the vernal equinox,

X = position of the Moon at eccentric longitude L at time T (geocentric longitude $\lambda_{☾}$),

P = the foot of a perpendicular EP to XO (extended),

M = the Moon's mean anomaly at T,

v = the Moon's true (geocentric) anomaly at T,

$\lambda_{☾}$ = the Moon's geocentric celestial longitude at T,

L = the Moon's eccentric longitude at T,

μ = the Moon's mean daily sidereal motion in celestial longitude,

Ψ = the Moon's prosthaphaeresis at T,

d = the Moon's geocentric distance at T,

C, ρ = the center and radius of the Moon's equivalent epicycle at T.

The problem may now be stated: Given T, μ, e, L_0, T_0, find $\lambda_{☾}$. From the figure, and by definition of mean daily motion:

$$M = \angle L_0OX = L_0 + \mu(T - T_0).$$

From ΔEOX:

$$d \sin \Psi = e \sin M,$$

$$d \cos \Psi = r + e \cos M,$$

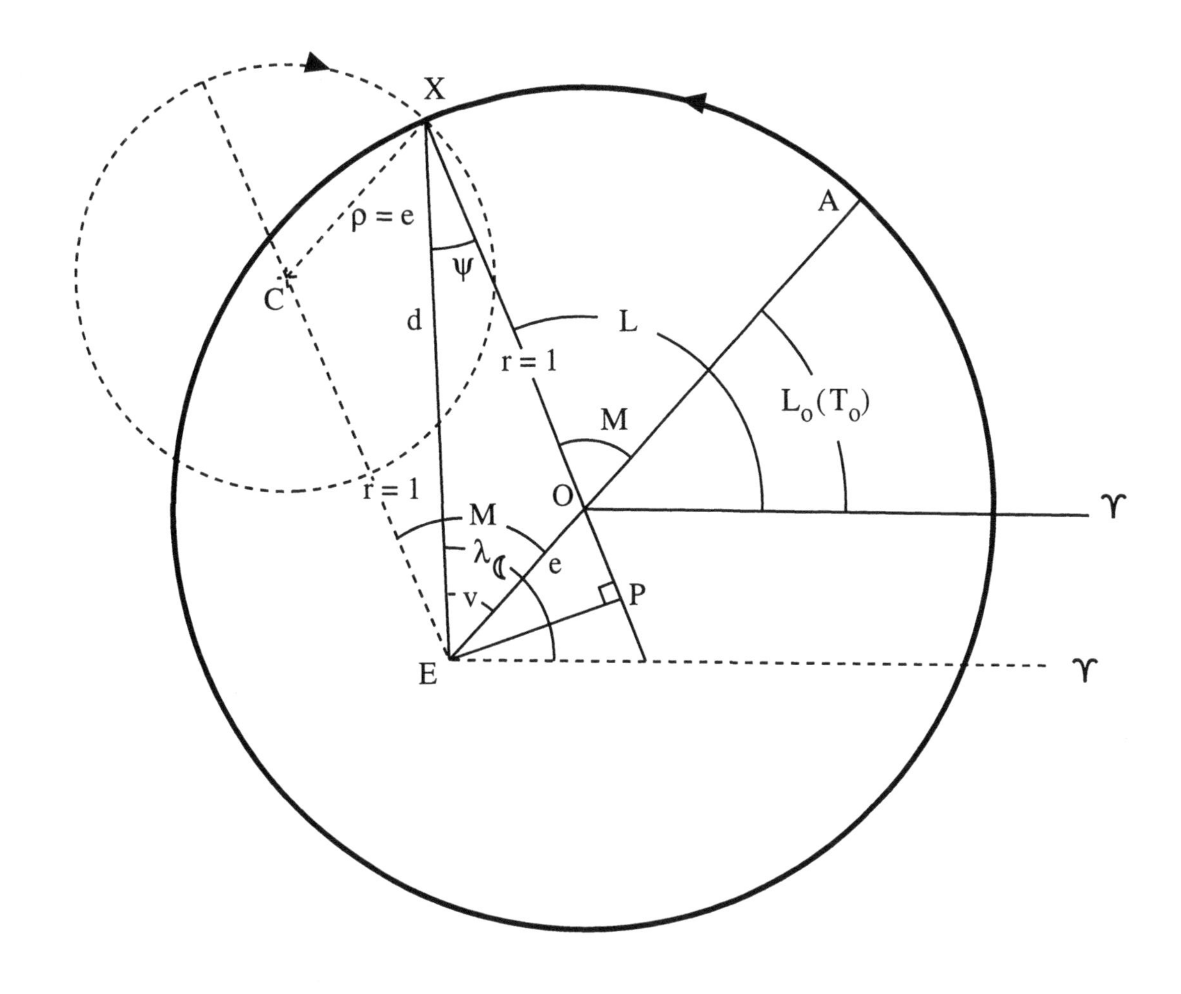

Figure III.6 — Hipparchus's method of predicting the Moon's celestial longitude.

from which d and Ψ. Also,

$$v = M - \Psi,$$

from which v. By definition:

$$\lambda_☾ = L_0 + v,$$

from which $\lambda_☾$.

A more direct formula for Ψ, to be used as a check, or if only the prosthaphaeresis itself is required, may be derived from ΔEPX as follows:

$$\tan \Psi = \frac{\text{EP}}{\text{XP}} = \frac{e \sin M}{r + e \cos M}\ .$$

In the case of the Moon, however, it will not suffice to use, as we did for the Sun, the simple hypothesis that the motion is in a fixed eccentric circle. This is because the sidereal motion of the Moon, $\mu(T - T_0)$, takes place in $(T - T_0)$ days, whereas the change in anomaly is $\mu_1(T - T_0)$. Thus the advance of the perigee in this interval is $(\mu - \mu_1)\ (T - T_0)$, where $\mu - \mu_1 = 401.034''$. Recognizing this drift of the apogee of the eccentric circle used above to approximate the lunar orbit, Hipparchus assigned the origin of mean anomaly to a moving point L' rather than L_0, where $L' = L_0 + (\mu - \mu_1)(T - T_0)$. In other words he used $M = L' + \mu_1(T - T_0)$ to find the mean anomaly. With this change of procedure, the above discussion becomes rigorous, except for the neglect of the orbit's 5° inclination to the plane of the ecliptic.

To place the Moon in celestial latitude Hipparchus proceeded in principle as follows. In Figure III.7, let (in addition to the quantities defined in Fig. III.6):

$♈Q$ = the celestial equator (here considered fixed),

$♈L$ = the ecliptic (here considered fixed),

β = the celestial latitude of the Moon,

$☊_0$ = the position of the ascending node at T_0,

$☊$ = the position of the ascending node at T,

i = the inclination of the lunar orbit to the plane of the ecliptic,

μ_2 = the Moon's mean daily motion with respect to the node of its orbit.

Given the quantities i, μ_2, L_0, $☊_0$, $\lambda_☾$ (either from general astronomical knowledge or from the type of computation outlined above), find β.

Now, from the definition of μ_2:

$$☊_0☊ = (\mu_2 - \mu)\ (T - T_0).$$

To find $☊L$:

$$☊L = ♈L - ♈☊,$$

where $♈L$ is $\mu_2(T - T_0)$, and $♈☊ = ♈☊_0 - ☊☊_0$, both of which are known. In right spherical $\Delta☊L☾$, by Napier's rules:

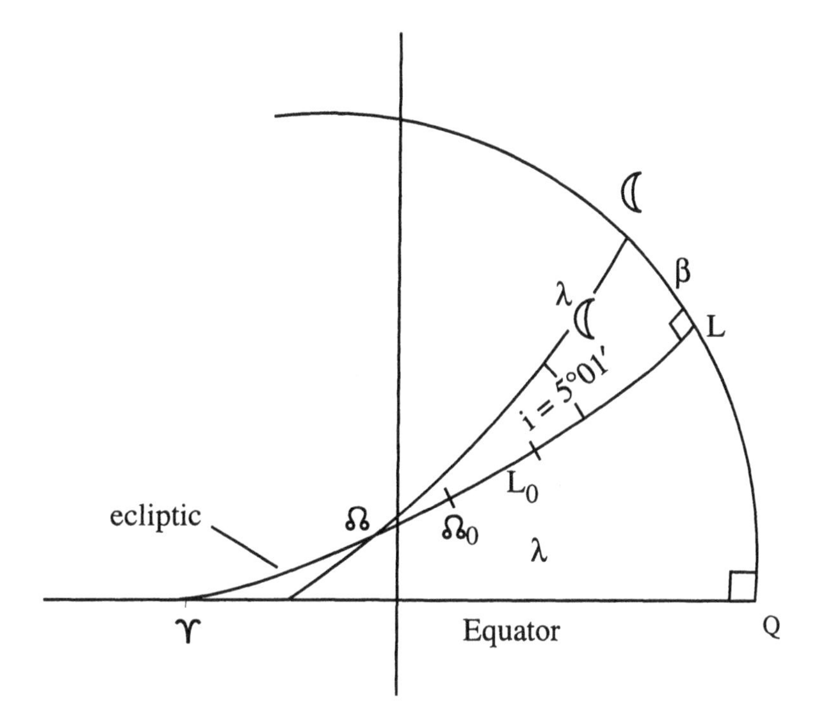

Figure III.7 — Hipparchus's method of predicting the Moon's celestial latitude.

$$\sin\beta = \cos(90^\circ - \lambda_{☾})\ \cos(90^\circ - i),$$

$$\sin\beta = \sin\lambda_{☾}\sin i,$$

$$\beta = \sin^{-1}(\sin\lambda_{☾}\sin i),$$

from which β, the celestial latitude.

Hipparchus's (Abortive) Theory of Planetary Motion

Hipparchus himself left no detailed theory of the motions of the planets. Except in the case of the Sun, for which he found a good circular, only slightly eccentric orbit, he generally favored the theory of epicycles over that of movable eccentrics, saying that it seemed more credible that the whole system of the heavenly bodies was arranged symmetrically with regard to the Earth, the center of the world. According to Ptolemy, one reason why Hipparchus did not make much headway with planetary theory was that he had very few accurate observations from his predecessors. Like Tycho Brahe, he realized that unless one had observational material with which to check the planetary positions all the way around the sky, one could make little theoretical progress. To supply such material would take at least 12 years in the case of Jupiter, and 30 years for Saturn.

Earlier observers were mainly interested in producing planetary observations at first or last visibility, or at the beginnings and ends of retrograde motion. But as Ptolemy later pointed out, these kinds of observations are fraught with uncertainty – first and last visibilities because they are horizon phenomena, and the stations because they extend over many days. Nevertheless, by Hipparchus's time it was clear that the retrograde arcs were of different lengths when the oppositions occurred in different constellations, a consequence (as we now know) of the elliptical shapes of the physical orbits. But these differences proved difficult to elucidate in terms of an eccentric-plus-epicycle model. Hipparchus therefore concentrated on producing reliable observations and on analyzing the older (mainly Babylonian) observational material that had come into his possession. These and other observations enabled Ptolemy, about 300 years later, to produce a very satisfactory planetary theory.

Hipparchus's Method of Finding the Stationary Points and Arcs of Retrogression of a Planet (a Method Originally Due to Apollonius)

In Figure III.8, let a planet Pl move uniformly and directly with angular speed μ' on an epicycle of radius r. The epicycle's center C slides uniformly and directly with angular speed μ on a deferent of radius R and center O, the Earth. Consider a small time interval t during which the planet moves from the stationary point G to the neighboring point G_1. Draw line OF, which intersects the epicycle at G; draw also the perpendicular from C to P, and line OG_1, which intersects the perpendicular to CG at G'. Let the interval t be taken so small that G_1 and G' may be considered coincident.

Now rigorously, $\angle GCG_1 = \mu' t$ and $\angle GOG_1 = \mu t$, since the planet appears stationary. Let G'' be the intersection of OG' with the perpendicular to OG at G. Then, from the figure, the condition for a stationary point at G occurs when GG'' equals the projection of GG' onto the line GG''.

By circular measure, $GG'' = OG(\mu t)$, and $GG_1 = GC(\mu' t) = r(\mu' t) \approx GG'$. Let α be the angle between OF and GG'; then $\angle G'GG'' = \angle FGC = 90^\circ - \alpha$. Carrying out the

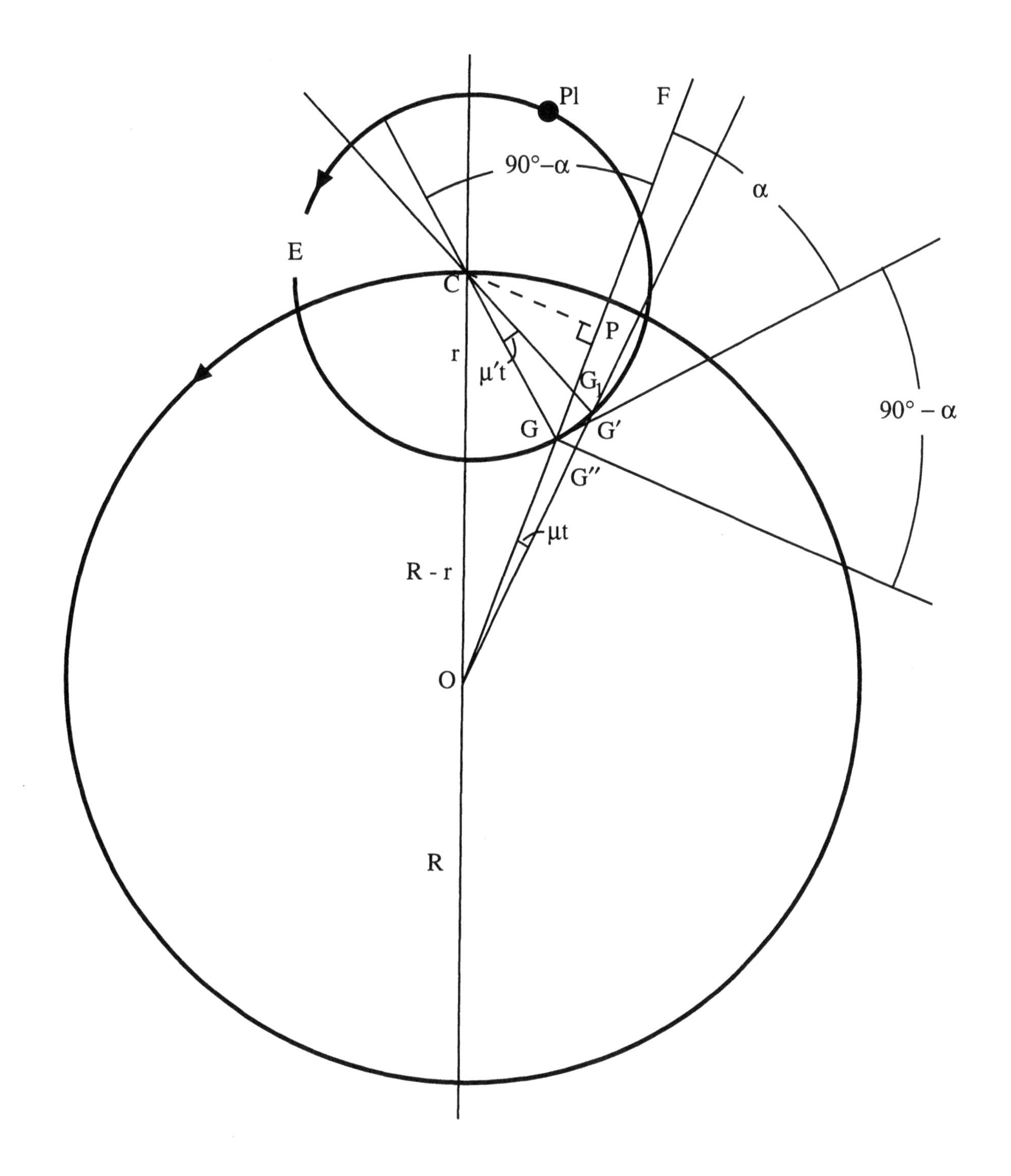

Figure III.8 — Hipparchus's method of finding the stationary points and arcs of retrogression of a planet.

projection above and disregarding the sign:

$$GG'' = OG(\mu t) = r(\mu' t) \cos(90^\circ - \alpha).$$

Letting $r = 1$,

$$OG(\mu t) = (\mu' t) \cos(90^\circ - \alpha) \qquad \text{or} \qquad OG(\mu t) = (\mu t)(GF/2),$$

as seen from ΔGPC. Finally,

$$\frac{GF/2}{OG} = \frac{\mu}{\mu'}.$$

This is called "Apollonius's condition" for the occurrence of a stationary point of a planet orbit.

If one knows the orbit, r, R, μ, and μ' are all given constants. From a plot of the planet's position in the orbit, GF and OG can be found by measurement at any time. A table of the ratios of $GF/2$ to OG for any direction from OC between zero and $\tan^{-1}(r/R)$, with argument t, can easily be constructed. From such a table the angle COG and the moment when $\frac{GF/2}{OG} = \frac{\mu}{\mu'}$ can be found. Thus one finds $\angle MCG$, which divided by μ' gives the interval T from conjunction (or opposition) to the stationary point. Twice T is the total interval of retrogression, and twice μT is the arc of retrogression. Further, if the time T_0 and celestial longitude λ_0 of the conjunction or opposition of the planet are given, the time of occurrence $T_0 \pm T$ and the celestial longitude $\lambda_0 \pm \mu T$ of the stationary points can be found. (For more detail, see Ptolemy's work on the retrogression of the planets.)

Hipparchus's Eclipse Method of Finding the Actual Distances and Diameters of the Sun and Moon

According to Ptolemy, Hipparchus made an estimate of the distances and sizes of the Sun, the Moon, and their orbits by an ingenious method probably originated by Aristarchus of Samos (fl. 280–264 B.C.). From estimates of the times of the Moon's new, first quarter, and full phases, Aristarchus had derived that at first quarter the angle Sun-Earth-Moon was on the average 87° (see $\angle SEM$ in Fig. III.9(a)). Hence he estimated the Moon-Earth distance to be about 1/19 of the Earth-Sun distance. Since the apparent diameters of the Sun and Moon are nearly the same, Hipparchus concluded that the Sun was about 19 times as distant as the Moon.

In Figure III.9(b), let

S = the Sun's center,

E = the Earth's center,

M = the Moon's center,

T = the intersection of SEM (extended) with a common tangent $S_1E_1M_1$ (extended) to the apex of the Earth's umbra,

σ = the angular semidiameter of the Moon and of the Sun,

$R_☾ = MM_2$ = the linear semidiameter of the Moon,

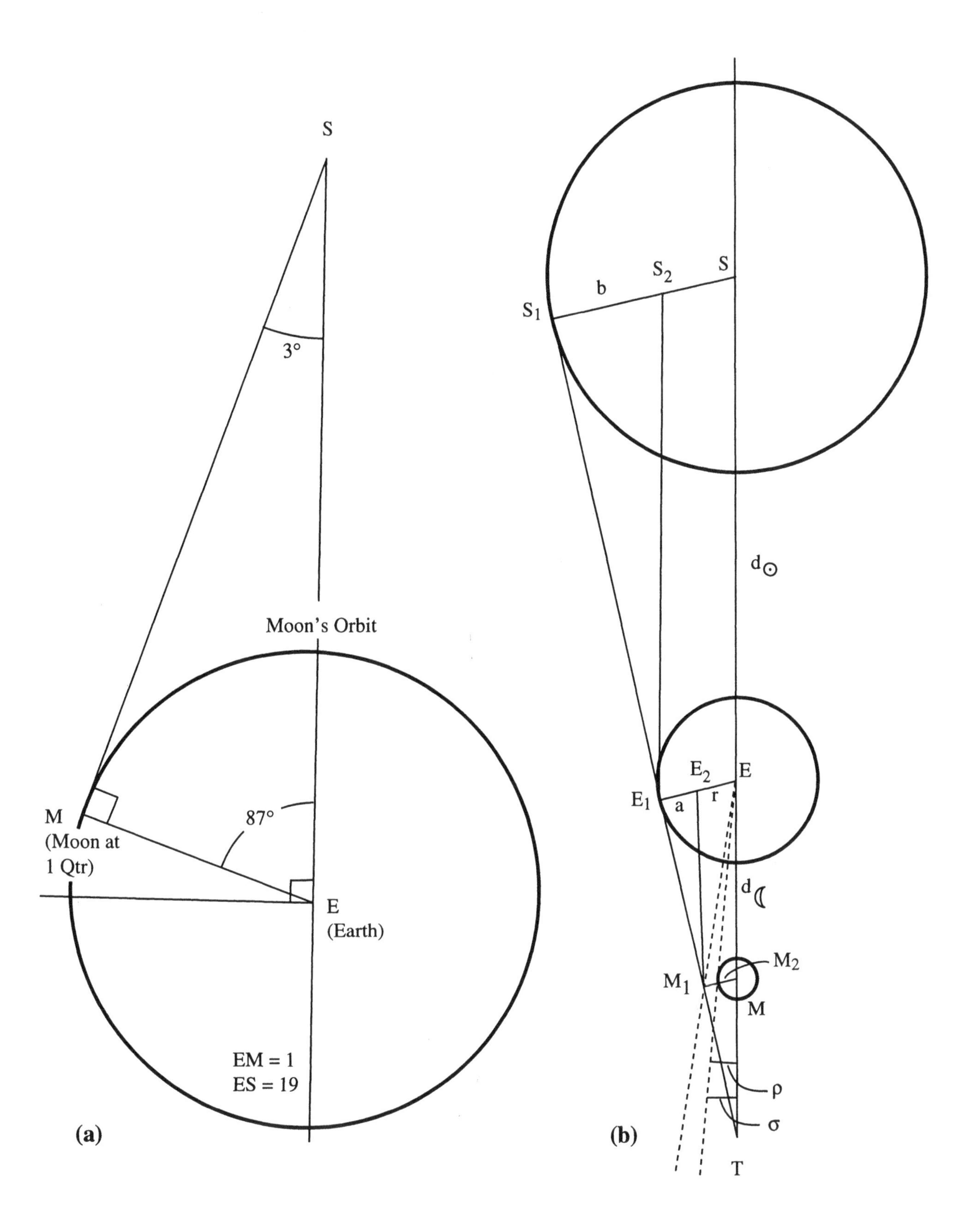

Figure III.9 — Hipparchus's eclipse method of finding the distances and diameters of (a) the Moon and (b) the Sun.

$\rho =$ the angular semidiameter of the total lunar eclipse region,

$r = MM_1 = EE_2 =$ the linear semidiameter of the total lunar eclipse region,

$R_\oplus \ = \ EE_1 =$ the linear radius of the Earth,

$R_\odot \ = \ SS_1 =$ the linear radius of the Sun,

$a \ = \ E_1E_2 \ = \ R_\oplus - r,$

$b \ = \ S_1S_2 \ = \ R_\odot - R_\oplus,$

$d_\odot \ = \ ES \ = \ E_1S_2 =$ the linear distance of the Sun from the Earth,

$d_☾ = \ EM \ = \ M_1E_2 =$ the linear distance of the Moon from the Earth.

In order to find $d_☾$, let $n \ = \ b/a$. Then from similar triangles $E_1S_1S_2$ and $M_1E_1E_2$,

$$b/a = d_\odot/d_☾ \equiv n. \tag{7}$$

From equation (7): $\sigma d_\odot \ = \ n\sigma d_☾$, or (if we express σ in radian measure)

$$R_\odot \ = \ nR_☾\,. \tag{7a}$$

By definition: $b \ = \ R_\odot - R_\oplus \ = \ nR_☾ - R_\oplus$, and $a \ = \ R_\oplus - r$. Now equation (7) becomes

$(nR_☾ - R_\oplus)/(R_\oplus - r) = n/1$ or

$nR_☾ - R_\oplus = nR_\oplus - nr$ or

$n(R_☾ + r) = nR_\oplus + R_\oplus$ or finally,

$$R_☾ + r = R_\oplus(1 + \tfrac{1}{n})\,. \tag{8}$$

Observations of the Sun, Moon, and total lunar eclipses gave the values $\rho = 40'$ and $\sigma = 15'$. Thus

$$\frac{R_☾}{r} = \frac{15}{40} = \frac{3}{8} \quad \text{or} \quad r \ = \ \frac{8}{3}\,R_☾\,. \tag{9}$$

Using equation (9) in equation (8), we have

$$R_☾ + \frac{8}{3}R_☾ = R_\oplus(1 + \frac{1}{n}) \quad \text{or} \quad \mathrm{R}_☾ = \frac{3}{11}(1 + \frac{1}{\mathrm{n}})\mathrm{R}_\oplus\,. \tag{9a}$$

By definition: $\sigma d_☾ \ = R_☾$, or since $\sigma = 15'$,

$$d_☾ = \frac{57.3}{1/4}\,R_☾ = 229\ R_☾.$$

Using this in equation (9a),

$$\frac{d_☾}{229} \ = \ \frac{3}{11}(1 + \frac{1}{n})R_\oplus \quad \text{or} \quad d_☾ = 62.5\ (1 + \frac{1}{n})\ R_\oplus\,. \tag{10}$$

Using Aristarchus's value of n, Hipparchus found from equation (10):

$d_☾ = 62.5 \times \dfrac{20}{19}\ R_\oplus = 65.7\ R_\oplus\,.$ (The modern value is 60.4 $\mathrm{R}_\oplus$.)

We note in passing that if n is large, the result is only slightly influenced by an appreciable error in n.

From equation (9a) Hipparchus found

$$R_{☾} = \frac{3}{11}(1 + \frac{1}{n})R_{\oplus} = \frac{3}{11} \times \frac{20}{19}R_{\oplus} = 0.30\ R_{\oplus}\,. \quad \text{(The modern value is 0.27 R}_{\oplus}.)$$

Also, from equation (7a):

$$R_{\odot} - nR_{☾} = 19 \times 0.297\ R_{\oplus} = 5.64\ R_{\oplus}\,. \quad \text{(The modern value is about 109 R}_{\oplus}.)$$

Finally, from equation (7):

$$d_{\odot} = nd_{☾} = 19 \times 65.5\ R_{\oplus} = 1240\ R_{\oplus}\,. \quad \text{(The modern value is 23,500 R}_{\oplus}.)$$

Knowing $R_{\oplus}$ from the work of Eratosthenes of Cyrene (c. 280–195 B.C.), who found its value to be about 6000 km,[2] all the dimensions of the Sun-Moon-Earth system became known in absolute terms. Hipparchus's value for the Sun's distance was used extensively during the following centuries – in fact, until refinement of measurements permitted a direct estimate of the solar parallax. For instance, as late as 1610 Kepler used the value $d_{\odot} = 1200\ R_{\oplus}$.

Hipparchus's Discovery of the Precession of the Equinoxes

After his observation of the appearance of a new star (nova), Hipparchus, working on the island of Rhodes, commenced extensive observations of the fixed stars. Comparing his own observations with those of the Alexandrian astronomers Timocharis and Aristyllus, made at Alexandria about 150 years earlier, Hipparchus realized that the celestial longitudes had all increased by about the same amount, while the celestial latitudes remained the same. This was most readily apparent for stars near the ecliptic such as Spica, which showed an increase in longitude of 2° in about 150 years, or an average increase of 48″/year.

At first, Hipparchus could not be sure whether this motion was parallel to the ecliptic or to the equator. Moreover, it was not clear whether it was a motion common to all the stars or only a peculiar motion of the stars in the zodiac band. To help future generations of astronomers decide the latter question, he recorded a large number of star *alignments*, i.e., three stars observed to be on a straight line. Each one of Hipparchus's alignments connected a zodiac star with stars outside the zodiac. According to Ptolemy, Hipparchus strongly suspected that the precession motion was parallel to the ecliptic and common to all stars. By Ptolemy's time, three centuries later, the accumulated position shifts had become large enough to show that Hipparchus had indeed been right. Although there is some evidence that Hipparchus had originally proposed a precession rate fairly close to the value now accepted (50″/year), apparently he later was content to state a lower limit of 1° per century (> 36″/year). This slower rate, which was adopted by Ptolemy and later became canonical, turned out to have unfortunate consequences. In particular, Arab astronomers

[2]Eratosthenes found a value of 5000 *stadia* for the $7\frac{1}{4}°$ arc of terrestrial circumference between Alexandria and Syene. The conversion of the ancient measurements into modern measure is, however, quite uncertain. The chief problem is that several different lengths of the *stade* were in common use and we have no way of knowing which one Eratosthenes used. A good guess is that 5000 *stadia* were equivalent to ~800 km, in which case his estimate of the Earth's circumference was about 40,000 km (fortuitously agreeing very closely with the modern value), from which $R_{\oplus} = 6400$ km.

working during the ninth-century Islamic renaissance measured a faster precession rate and concluded that the rate was variable. They then devised a complex mechanism to explain such a supposed variable rate.

Hipparchus's Discussion of Errors

An important reason for Hipparchus's greatness was his clear realization of the limits of accuracy of his instruments and his discussions of possible errors in his derived results caused by unavoidable observational uncertainties. Thus in discussing the accuracy of his value of the length of the tropical year, he pointed out that the time of a solstice might be erroneous to the extent of $\frac{3}{4}$ day, while that of an equinox may be expected to be within $\frac{1}{4}$ day of the correct value. This would indicate a possible error of $1\frac{1}{2}$ days over a period of 150 years, or about 15 minutes in the length of a year. Actually, his estimate of the length of the year was only about 6 minutes too large.

CHAPTER IV

Ptolemy (fl. 125–150)

The greatest theoretical astronomer of antiquity was undoubtedly Claudius Ptolemaeus of Alexandria, the author of a monumental work, the *Almagest*, which contained a summary of ancient scientific astronomy as well as his own astronomical ideas. Ptolemy refined spherical trigonometry and improved the computation of chords. His book brought the ancient planetary theory into its final form. The *Almagest* represented the epitome of geometrical astronomy for about 1400 years. Regiomontanus, Copernicus, Tycho, and Kepler studied it in great detail, with respect and admiration.

Little is known about Ptolemy's life or personality. He lived and worked at Alexandria in the Roman province of Egypt during the second century A.D. Besides the *Almagest*, Ptolemy wrote the oldest extant treatment of mathematical geography, as well as works on philosophy, music theory, optics, and astrology. In the Middle Ages he was often confused with the kings of the Ptolemaic dynasty and was sometimes mistakenly pictured wearing a royal crown.

Ptolemy's Main Contributions to Astronomy

Perhaps the best way to introduce Ptolemy's work is to outline the contents of the *Almagest*, but we shall here first separately call attention to his two greatest contributions, the *evection* and the *equant*, either of which would have ensured the immortality of his name.

The Evection. The evection is the greatest of the approximately monthly perturbations of the Moon's orbit about the Earth. It is caused by the Sun and alternately advances and retards the Moon in its orbit by about $1\frac{1}{4}^\circ$. Although Hipparchus had discovered the effect of the evection on the timing of lunar and solar eclipses, and correctly estimated its greatest amount, it remained for Ptolemy to study it in detail, especially between the syzygies, to call attention to the form of its variation with lunar phase angle, and to determine its amount almost exactly. (He found its amplitude to be $1^\circ 19' 30''$, whereas the modern value is $1^\circ 16'$.)

The Equant. To simulate the nonuniform motion of the planets better than could be done by Hipparchus's eccentric circle theory, Ptolemy introduced a *punctum equans* in each planetary orbit, a displaced center about which the uniform circular motion took place (Fig. IV.1). The equant point E was situated on the line of apsides, toward the apogee Ap. The distance of E from the geometrical center C of the deferent was equal to the distance of C from the Earth T. The motion of the planet's epicycle around the deferent took place at uniform angular speed, not as viewed from C or T, but as viewed from E. Physically this means that the epicycle as seen from T travels more slowly near apogee Ap and more rapidly near perigee Pe. Ptolemy's introduction of nonuniformity of motion represented the

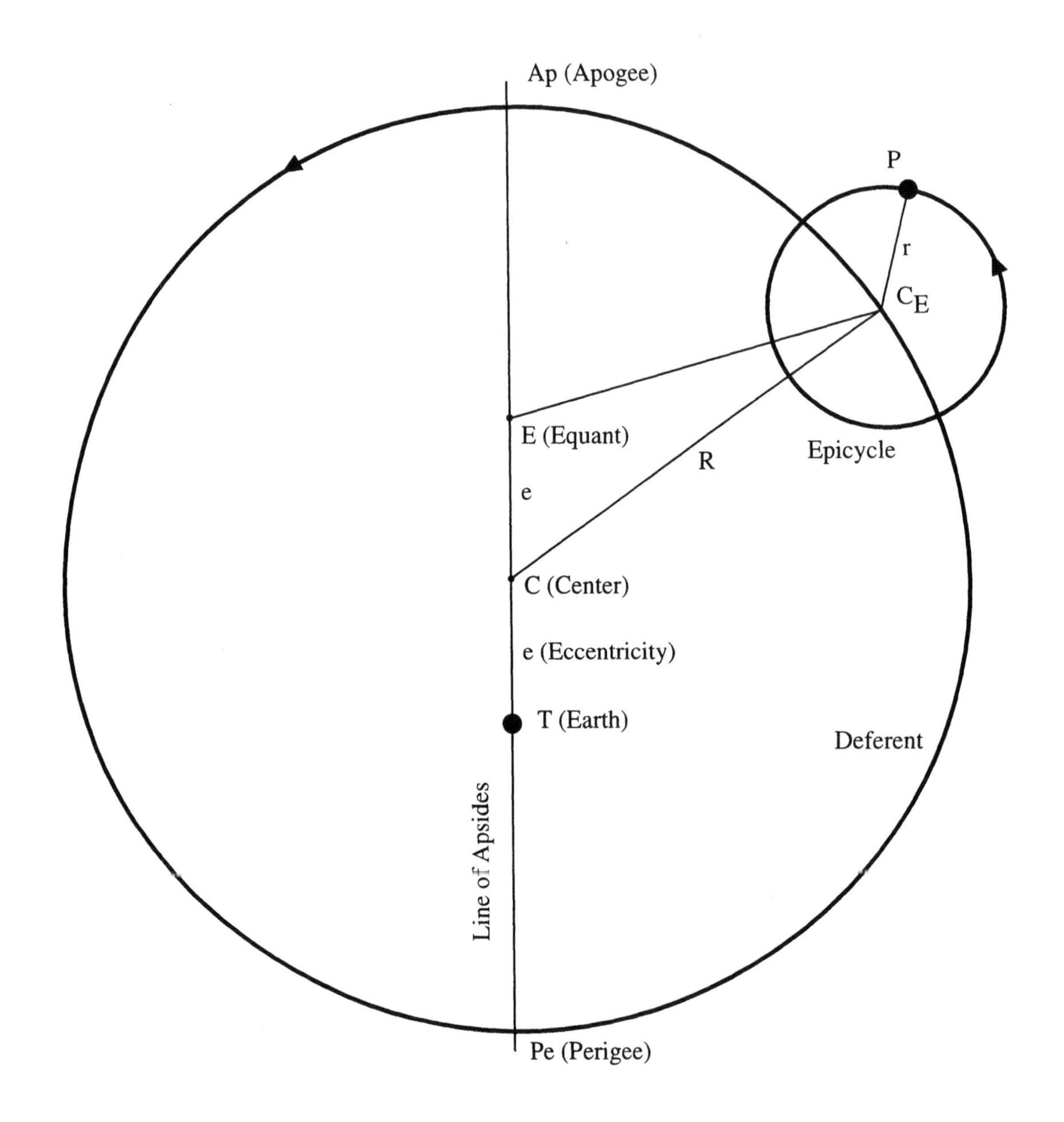

Figure IV.1 — Ptolemy's general planet orbit showing position of the planet P, the Earth T, the center C, and the equant point E.

deepest insight into the nature of planetary motion before Kepler.

The general arrangement of Ptolemy's geocentric system for the Moon and planets, leaving out some details such as the positions of the equants and the latitude mechanisms (which will be treated later), is shown in Figure IV.2.

CONTENTS OF THE *ALMAGEST*

Books 1 and 2 describe the cardinal circles and points used on the celestial sphere and discuss the trigonometry necessary for accurate work in astronomy and geography. An excellent table of *chords* for every half degree is given. This table can be used as a table of natural sines, since by definition: $\text{chord}\,\alpha = 2\sin\frac{1}{2}\alpha$. Its great accuracy (about equivalent to that of a five-place logarithmic table) arises partly from the checks Ptolemy applied to it, using the "Ptolemaic Theorem" of plane geometry. This theorem (sometimes called the theorem of Menelaus, after Ptolemy's predecessor in trigonometry) asserts the following. If a quadrilateral is inscribed in a circle, the product of the diagonals equals the sum of the products of the opposite sides. In trigonometric computations, Ptolemy has constant recourse to this theorem.

Book 2 also contains Ptolemy's method of determining the geographical latitude of a place by observing semidiurnal arcs of the Sun. The method is briefly as follows.

From the "sunset formula": $\cos t = -\tan\phi\tan\delta$, where

t = the hour angle of the setting or rising body,

δ = the declination of the body, and

ϕ = the astronomical latitude.

It follows that, for sunset on the longest day of the year,

$$\tan\phi = -\cot\delta_\odot\cos t_\odot = -\cot 23^\circ51'20''\cos t_\odot = 2.26\sin(t_\odot - 90^\circ)$$
$$\approx \tfrac{9}{4}[\tfrac{1}{2}\,\text{chord}\,(2t_\odot - 180^\circ)] = \tfrac{9}{8}\,\text{chord}\,(2t_\odot - 180^\circ),$$

from which ϕ may be computed. Ptolemy constructs a table for $t_\odot$ at various latitudes on 21 June, when according to his measurements $\delta_\odot = 23^\circ51'20''$. The times of Alexandria may have been measured by observing meridian passages of bright stars shortly after sunset or before sunrise, and running a sand clock or a water clock for the short interval between the moments of these observations and that of sunrise or sunset. (For the correction of clocks Hipparchus had supplied a list of stars which culminated at 1-hour intervals throughout the night.)

Book 3 contains Ptolemy's eccentric-circle theory of the annual motion of the Sun. Ptolemy's solar theory was basically the same as Hipparchus's and thus did not involve much original work. However, in his first practical application of the solar theory, Ptolemy showed great originality in his detailed explanation of the difference between mean and apparent solar time. Ptolemy's is the first known treatment of the effect today called the "equation of time."

Book 4 contains Ptolemy's work on the Moon's orbit, which was by far the greatest contribution to that subject until Tycho's painstaking modifications. Ptolemy's theory for the Moon's motion gave an excellent account of the evection. It could usually predict the Moon's position with an accuracy of about 10′, a small quantity in the astronomy of his

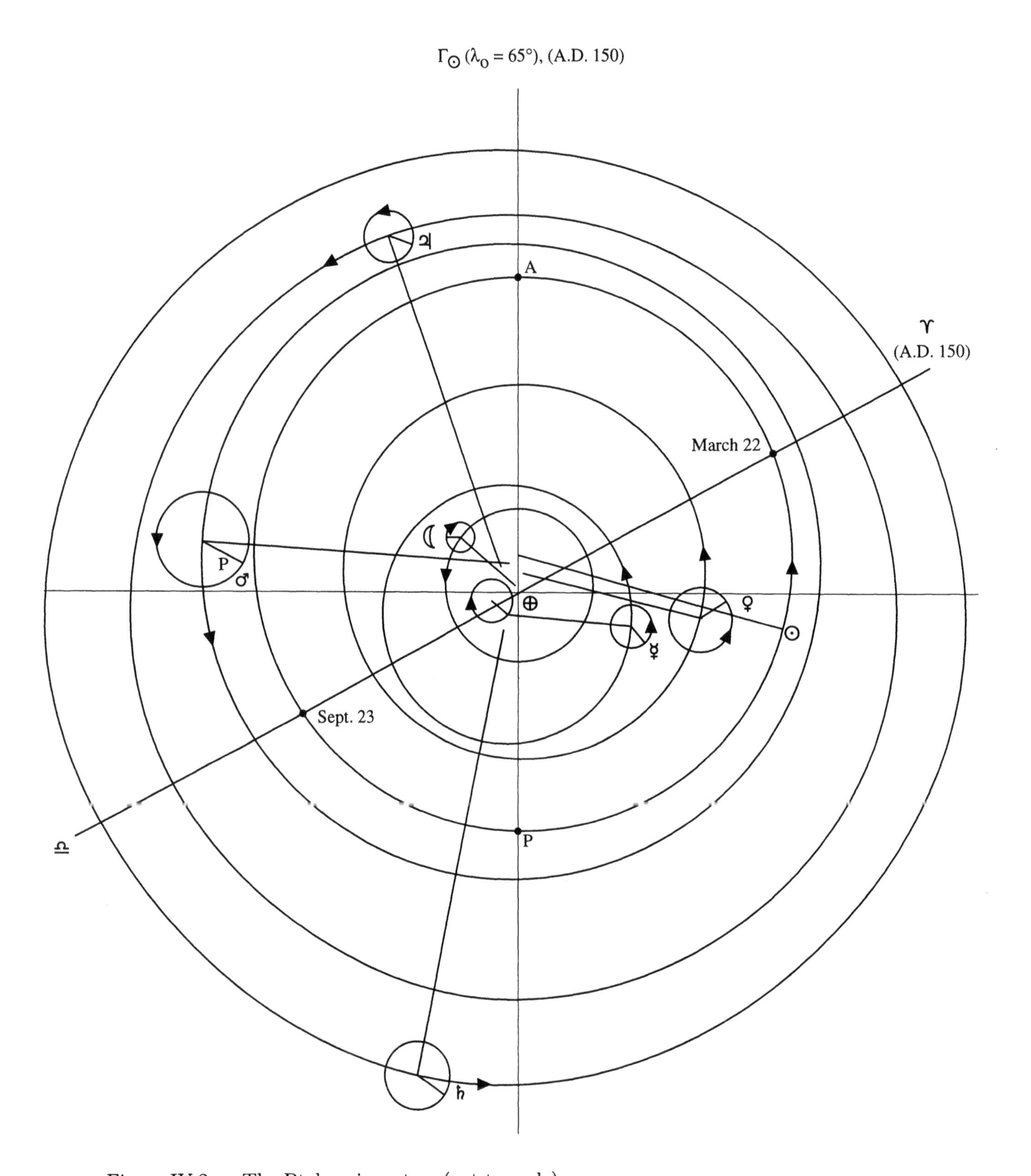

Figure IV.2 — The Ptolemaic system (not to scale).

time. As we shall see, this was accomplished at the expense of introducing the completely artificial supplementary effect called the prosneusis. The system required a variation of the apparent lunar diameter by a factor of nearly 2. This was a serious drawback to his theory that Ptolemy should easily have noted!

Book 5 of the *Almagest* contains an account of the construction and use of Ptolemy's chief astronomical instruments, which included armillary spheres. He clearly discusses the parallaxes of the Sun and Moon. The lunar parallax is fairly well determined, but Ptolemy could not measure the position of the Sun with sufficient accuracy to tell anything about its parallax or distance. He does obtain a good estimate of the Moon's distance from its measured parallax.

In Book 6, Ptolemy discusses eclipses, but he adds hardly anything to the results of Hipparchus. He explains how the observation and timing of eclipses at two different stations permit their longitude difference to be found. In fact this is Ptolemy's recommended method for determining longitude, although the rarity of timed eclipse observations taken simultaneously in separate locations prevented this method from being used very often by ancient geographers.

Books 7 and 8 contain a star catalogue listing the longitudes, latitudes, and magnitudes of 1022 stars, as well as a discussion of precession. For more than two centuries a dispute has raged among historians of astronomy about the origin of this star catalogue. Some have maintained that the catalogue was really compiled by Hipparchus and that Ptolemy merely updated Hipparchus's longitudes for precession. The issue would, of course, be easy to decide if Hipparchus's star catalogue (assuming he really compiled one) still existed. But all we have of Hipparchus's work on the stars is his *Commentary on the Phenomena of Aratus and Eudoxus.* This work, although it contains many numerical data on star places, is very far from being a systematic and comprehensive star catalogue such as we find in the *Almagest.* Thus, if we want to make the *Almagest* catalogue dependent on Hipparchus, we must postulate the existence of a large Hipparchian catalogue which is now presumably lost. A huge literature pro and con has accumulated around this issue. The evidence is complex and the arguments often indirect. Not all scholars are in agreement. It appears most likely that Ptolemy made some substantial use of Hipparchus's star data, probably recasting it into more convenient form, and that he added data for stars he had observed himself.

The last five books of the *Almagest* (books 9–13) contain Ptolemy's theories of the planetary orbits and derivations of their elements. Although his system is no longer regarded as correct, his prediction mechanisms were quite successful in describing the apparent motion of each body. For each planet, the relative size of the epicycle and the deferent was determined with excellent accuracy. Table IV.1 shows a comparison between the relative dimensions of the actual elliptic orbits, described either by their semimajor axes a or by the ratio of the radius r of the epicycle, responsible for the retrograde motion, to the radius R of the deferent in the Ptolemaic system. Under "Ptolemy," for Mercury and Venus we give Ptolemy's value for r/R; for the outer planets we give Ptolemy's value for R/r. For comparison, the results for the systems of Copernicus and Kepler are also given. Under "Copernicus" we give Copernicus's value for the radius of the planet's heliocentric circular orbit, expressed in units of the radius of the Earth's orbit.

Table IV.1

Planet	Modern	Ptolemy	Copernicus	Kepler
Mercury	0.387	0.371	0.376	0.388
Venus	0.723	0.719	0.719	0.724
Earth	1.000	—	1.000	1.000
Mars	1.524	1.519	1.520	1.524
Jupiter	5.203	5.216	5.219	5.196
Saturn	9.539	9.234	9.174	9.510

Ptolemy's Solar Theory

Ptolemy, being primarily a mathematician, may have thought it too little of a challenge to observationally check the Sun's orbit in detail, since, being relatively simple, it was closely enough represented by Hipparchus's theory. This theory had direct uniform motion on a single eccentric circle. (Equivalently, one may use a direct uniform motion, on a *concentric* circular deferent, of an epicycle with retrograde uniform motion of the same period.) Ptolemy therefore used no equant for the motion of the Sun. This uncritical adoption by Ptolemy of Hipparchus's solar theory after so many years (to Ptolemy, Hipparchus was one of "the ancients"!) was, however, somewhat unfortunate. Thus a position of the Sun taken from Ptolemy's table may be in error by as much as 1°40′, due mainly to the accumulated effects over 300 years of a small error in Hipparchus's length of the tropical year. By retaining Hipparchus's values for the lengths of the seasons, Ptolemy also missed discovering the advance of the line of apsides.

Ptolemy's Work on the Lunar Orbit

As in the case of the Sun, for the Moon Ptolemy started out with Hipparchus's primitive theory. But while he left the solar theory practically unchanged, he greatly improved that of the Moon. He worked out the lunar orbit alternatively *(a)* as an eccentric circle, *(b)* as a concentric circular deferent plus a circular epicycle, and *(c)* as an eccentric circular deferent plus an epicycle. He finally preferred to work with *(b)*, first finding from it the main features of the model, then using a modification of *(c)* to explain certain deviations from *(b)* in longitude, observed partly by Hipparchus and partly by himself.

Thus, in outline, Ptolemy's improvements on Hipparchus's lunar theory were:

1. Introducing a uniform motion in a geo-eccentric circular deferent that varies in position to better simulate the lunar motion in all parts of the orbit (note: the resulting motion of the Moon on this shifting deferent was not uniform with respect to a fixed center);
2. More fully elucidating and determining the evection, especially between the syzygies and at the quarters; and
3. Attempting to find more accurate values of the evection by inventing the "prosneusis."

Ptolemy's View of the Regression of the Nodes and the Advance of the Apsides of the Lunar Orbit

As explained in Chapter III, Hipparchus had simply assumed that the Moon moved uniformly in the retrograde direction on a coplanar epicycle, which in turn glided uniformly eastward on an inclined geocentric circular deferent. The pole of this deferent rotated with a period of $18\frac{2}{3}$ years uniformly in the retrograde direction in a small circle of 5° radius, centered on the pole of the ecliptic. This arrangement obviously took care of a constant 5° inclination of the lunar orbit, as well as a constant regression of its nodes in a period of $18\frac{2}{3}$ years. Furthermore, since the period of the Moon on the epicycle was taken to be 27.51 days, while the period of the epicycle on the deferent was $27\frac{1}{3}$ days, this slight difference nicely accounted for the average advance of the line of apsides by about 3.3° per month. Hipparchus had also discovered an evection of ±1.3° at the syzygies, as well as a longitude inequality at the quarters (which, however, he did not study in detail). Ptolemy determined the precise amplitude of the evection and attempted to derive its law of variation in all positions of the Moon.

Thus Ptolemy's lunar theory (see Fig. IV.3) utilizes one circular deferent AB, with direct motion eccentric to the Earth T. However, the deferent circle shifts in position, its center moving retrograde around circle CC_1N. The Moon itself moves on a circular epicycle with retrograde motion counted from a variable origin on the epicycle, while the center of the epicycle moves uniformly eastward on the (shifting) deferent. The whole system lies in one plane inclined 5° to that of the ecliptic. The line of nodes of this orbital plane retrogrades in the ecliptic by 1.6° per month, with the line of apsides advancing by 3.3° per month.

Before taking up Ptolemy's detailed work on the lunar orbit, it is instructive to show how Ptolemy determined the evection to get a rough first approximation to the orbit.

Ptolemy's Preliminary Derivation of the Elements of the Lunar Orbit

Granted that the simple model of a direct concentric deferent and a coplanar retrograde epicycle of the same period will approximately represent the Moon's orbit, only three observations are required to solve the problem of finding the relative sizes and orientations of these circles. (The method was due to Apollonius, and well known by Hipparchus.) Ptolemy first derived the constants of this model from three eclipse observations (giving precise geocentric celestial longitudes). From the description of the more complex model it appears that the complexities reduce to zero at the syzygies; thus eclipse observations are desirable for this reason, even more than for their reputed greater accuracy. In the following we shall treat the general problem encountered in ancient Greek astronomy: how to find a geocentric circular deferent and a coplanar circular epicycle to fit a set of three observations. A good example is Ptolemy's derivation of the lunar orbit.

In Figure IV.4, at time t_1', let one of the observed positions of the eclipsed Moon be A', whose true geocentric longitude is λ_1, with corresponding mean geocentric longitude λ_1 (mean). The longitudes are all measured counterclockwise from some arbitrarily chosen direction, denoted λ_0. Let α_1 be the epicyclic anomaly, measured clockwise from the apogee L_1' of the epicycle. Now, if at time t the Moon was at the apogee L of the epicycle, at longitude $\lambda_0 = 0$, then λ_1 (mean) $= \mu(t_1' - t)$, where μ = mean daily angular motion = $2\pi/P$, where P is the sidereal period of the Moon. Thus, at the moment t_1' of the eclipse,

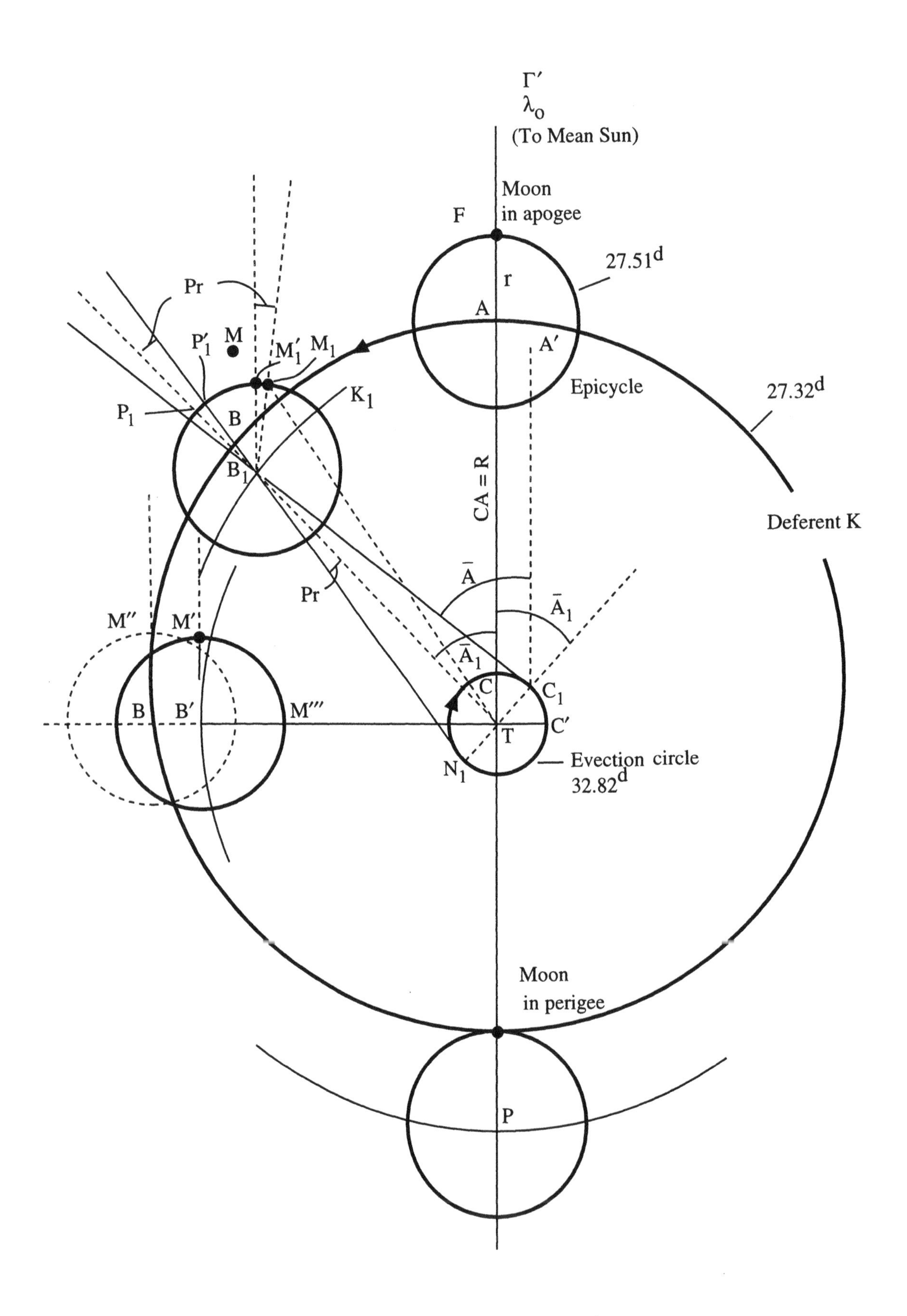

Figure IV.3 — Ptolemy's orbit of the Moon (not to scale).

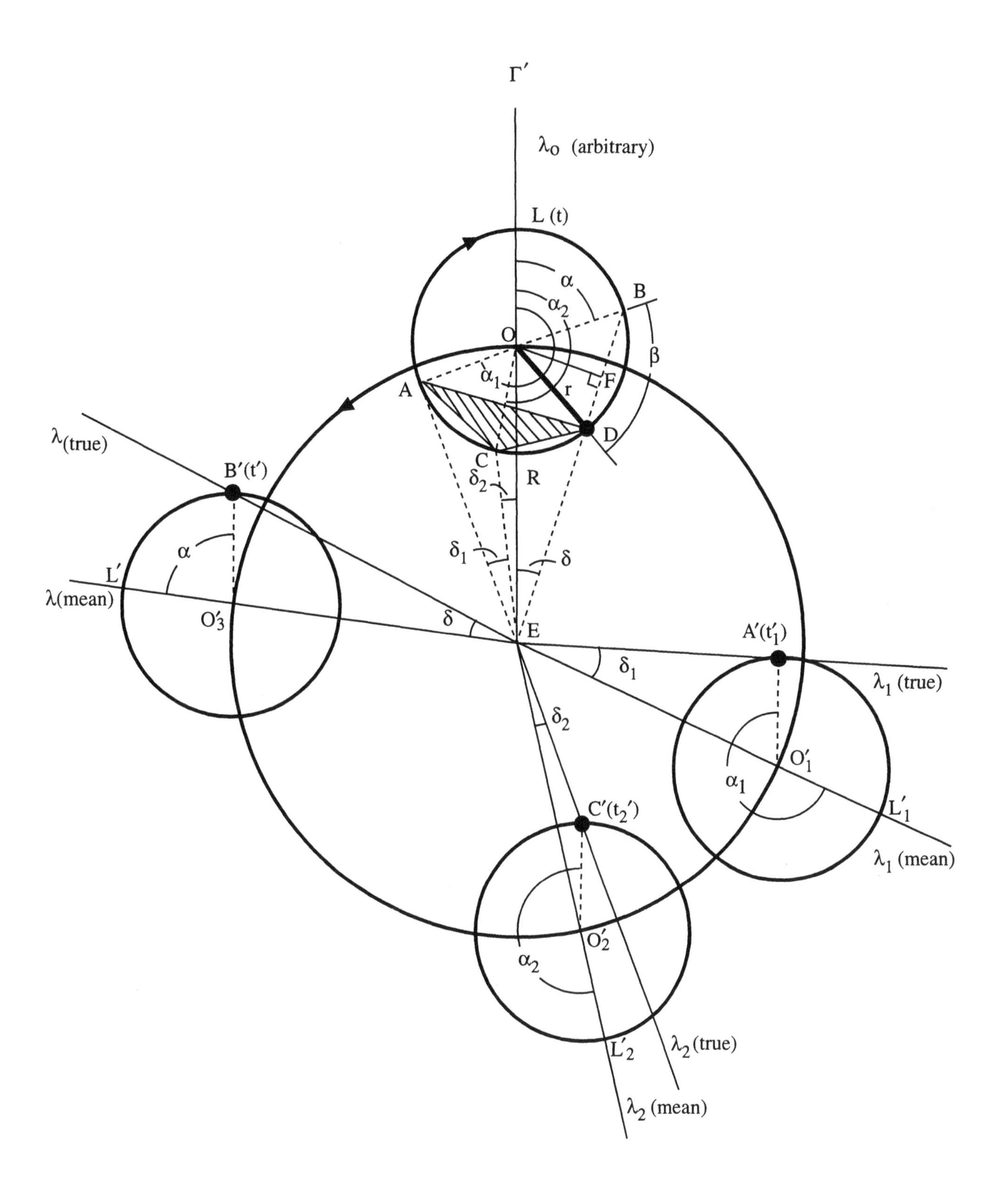

Figure IV.4 — Ptolemy's preliminary orbit of the Moon (from three eclipses with Sun at Γ′).

when the Moon is at A', the true longitude λ_1 (true) is found by observation and the mean longitude λ_1 (mean) is found by calculation. At this moment the prosthaphaeresis is $\delta_1 = \lambda_1$ (true) $-\lambda_1$ (mean). Finally, since by hypothesis the direct motion on the deferent and the retrograde motion on the epicycle take place with the same period, we must have $\alpha_1 = \lambda_1$ (mean).

In the same way let B' and C' be the positions of the Moon at two other eclipses, at times t' and t'_2, with geocentric longitudes λ and λ_2, with corresponding mean longitudes λ (mean) and λ_2 (mean). At these moments the apogee of the epicycle is at L' or L'_2, and the epicyclic anomaly is α or α_2. The calculated mean longitudes are λ (mean) $= \mu(t' - t)$ and λ_2 (mean) $= \mu(t'_2 - t)$. The prosthaphaereses are $\delta = \lambda$ (true) $- \lambda$ (mean) and $\delta_2 = \lambda_2$ (true) $- \lambda_2$ (mean). And by hypothesis $\alpha = \lambda$ (mean) and $\alpha_2 = \lambda_2$ (mean). Thus, α, α_1, α_2 and δ, δ_1, δ_2 are given by a combination of observation and computation. The problem can now be stated: given these six angles, find the ratio $\kappa = r/R$ (the radius of the epicycle, expressed as a fraction of the deferent's radius). We proceed as follows.

Imagine the three epicycles superposed by moving them backward on the deferent through the angles λ_1 (mean), λ_2 (mean), λ (mean), respectively (or, equivalently, through α_1, α_2, α), but leaving the points on the epicycles as they were when observed. Then on a standard epicycle, centered anywhere, as at O, say, the three thus reduced epicycles are made to coincide, and the points A, C, B will retain the epicyclic angular spacing of the observed points. Also draw EB, intersecting the standard epicycle at D. Draw AB, AC, AE, AO, OC, OB, and OD. Drop $OF \perp BD$. Let $\angle BOD = \beta$ and $\angle OEF = \delta$.

By inspection, from ΔOFE:

$$\delta + 90° = \alpha + \tfrac{1}{2}\beta \qquad \text{or} \qquad \beta = 2(\delta - \alpha + 90°), \tag{1}$$

from which β. From ΔOED:

$$\frac{r}{R} = \frac{\sin\delta}{\sin ODE} = \frac{\sin\delta}{\sin ODB} = \frac{\sin\delta}{\cos FOD} \qquad \text{or} \qquad \kappa = \frac{r}{R} = \frac{\sin\delta}{\cos\frac{1}{2}\beta}, \tag{2}$$

from which R. Using lunar eclipse observations (of geocentric celestial lunar longitudes) extending back to 720 B.C., Ptolemy found at syzygy the value

$$\kappa = \frac{r}{R} = 0.087\,(= \tan^{-1} 5°)\,.$$

This geometrical method of finding the epicycle radius from three observations of longitude works equally well for a case in which the periods on the epicycle and the deferent are unequal.

Thus the maximum prosthaphaeresis at syzygy is 5°. Similarly, according to Hipparchus (and Ptolemy) the maximum prosthaphaeresis at quadrature is, by observation, 7.6°. Ptolemy accordingly constructed and used tables equivalent to the following expressions for the prosthaphaeresis:

$$P_s = 5° \sin \overline{A}_1 \tag{3}$$

near syzygy, and

$$P_q = 7.6° \sin \overline{A}_1 \tag{4}$$

near quadrature, where $\overline{A}_1$ is the Moon's epicyclic mean anomaly measured from the epicycle's apogee. He found that both of these formulas worked well at the syzygies from which they were developed but failed miserably elsewhere.

The Motions in Ptolemy's Lunar Orbit

Figure IV.3 shows the general idea of the Ptolemaic lunar theory, though not to scale. The motions in the diagram are as follows. Let us start with the Moon in double apogee (i.e., the epicycle in apogee A, on the deferent centered at C, and the Moon in apogee on this epicycle) and in conjunction with the mean Sun. (Thus it is approximately new moon.) The eastward motion of A (the starting position of the epicyclic center) on the deferent (center C, radius R) is smooth, though nonuniform since it moves uniformly about the Earth T, not about its center C. This lack of uniformity means that $\angle ATB$, not $\angle ACB$, increases uniformly. Its sidereal period is 27.32 mean solar days. (Due to the shifting of the deferent, A never arrives at B, nor the Moon at M. Rather, the Moon is at M_1, M' at the moments when the center of the epicycle is at B_1, B'.) The motion of C, the center of the deferent for the epicyclic position A, is a retrograde circular uniform motion of radius e centered on the Earth T. The radius of the deferent always remains equal to R, but the deferent itself is constantly shifting in position. Thus $AC = B_1C_1 = B'C'$ $(= R)$, etc. The uniform retrograde motion of C about the Earth in its "evection circle" CC_1N (center T, radius e) is such that at any time $\angle CTC_1 = \angle CTB_1 = \angle \overline{A}_1$. The motion remains constantly uniform with respect to the direction of the mean Sun, which moves uniformly eastward at 0.99°/day. Hence, the period of C in the evection circle is 32.821 mean solar days (Dreyer 1905, p. 194). As a result of the numerical equality of the retrograde angular velocity of C and the direct angular velocity of A with respect to the mean Sun, we have at all times $\angle B_1TC_1 = 2\angle ATB_1$, etc., for all subscripts. The epicycle is thus closest to the Earth at the quarters, as at B', and farthest from it (distance $= R + e$) at the syzygies, A and P. This device fairly well accounts for the main features of the second inequality, called the evection.

The Effects of Evection

To visualize the maximum and minimum values of the prosthaphaeresis, we consider the principal modern terms in the Moon's celestial geocentric longitude, which are as follows:

L = mean longitude of mean Moon (uniformly increasing),

v = angular distance of mean Moon from mean lunar apogee (the "anomaly effect"),

ϕ = angular distance of mean Moon from Sun (the "phase effect"),

ℓ' = angular distance of mean Sun from apogee (the "annual effect"), and

λ = longitude of actual Moon.

Then

$$\begin{array}{cccccccc} \lambda = & \underset{\text{I}}{L} & - & \underset{\text{II}}{6^\circ 17' \sin v} & + & \underset{\text{III}}{13' \sin 2v} & - & \underset{\text{IV}}{1^\circ 16' \sin(2\phi - v)} \\ & & + & \underset{\text{V}}{40' \sin 2\phi} & + & \underset{\text{VI}}{11' \sin \ell'} & - & \underset{\text{VII}}{2' \sin \phi} \quad + \ldots \end{array} \qquad (5)$$

Here

I represents uniform circular motion.

II and III are called "the elliptical terms."

I + II + III represent simple elliptical (Keplerian) motion.

II, or II + III, is called "the first inequality" or "equation of center."

IV represents the "evection."

V represents the "variation."

VI represents the "annual equation."

VII represents the "parallactic inequality."

Equation (5), which contains an approximation to elliptic motion as well as the four largest perturbations, gives the lunar longitude to an accuracy comparable with the keenest naked-eye observations.

Now, since at the four cardinal points here considered (the syzygies and quarters), *(a)* III and V are zero, *(b)* VI is always varying slowly, and *(c)* the coefficient of VII is small, the prosthaphaeresis P (i.e., the correction whose application to a mean position results in a true position) at these points is given approximately by

$$P = -6.3^\circ \sin v - 1.3^\circ \sin(2\phi - v), \tag{6}$$

measured from apogee. Here the first term, often referred to simply as the first inequality, may be said to represent the approximate deviation, at the syzygies and the quarters, of elliptic orbital motion from uniform circular motion about the Earth. The second term, or evection (called the second inequality by Ptolemy), is the only appreciable modification of elliptic motion at the four cardinal points considered – hence the rationale for equations (3) and (4).

The general effect of the evection term on the celestial geocentric lunar longitudes at the four cardinal points here considered is shown in Figure IV.5 for one lunation. Neglecting relatively long period changes such as the annual motion of the Sun, the regression of the lunar nodes, and the advance of the lines of the lunar and terrestrial apsides, the value of the evection term, $\pm 1.3^\circ \sin(2\phi - v)$, is noted at each of the syzygies and quadratures. Hence one sees its effects in advancing or retarding the Moon with respect to its undisturbed elliptic motion in longitude, given by I, II, and III of equation (5). At the syzygies the results were obtained by Hipparchus and Ptolemy by timing the moments of central eclipses, or the expected moments of conjunctions of the Moon with the Sun or the anti-Sun. At the quarters both men determined, by using an astrolable and a clepsydra (water clock) or other time-measuring device, the deviation from 90° of the difference in longitude of the Moon and the Sun at the moment of expected quadrature of the epicycle.

In Figure IV.5 the maximum effect of the evection term IV in equation (5) in advancing or retarding the Moon in longitude is illustrated for two orientations of the line of apsides of the lunar orbit with respect to the direction of the Sun. In Figure IV.5(a) the line of apsides coincides with the Earth-Sun line. A study of Figure IV.5(a) reveals that, in this configuration, the new and full moons are on time, while first quarter occurs about $2\frac{1}{2}$ hours later (and last quarter similarly earlier) than would be the case in the absence of any evection term. Similarly, when, as in Figure IV.5(b), the major axis of the lunar orbit is at

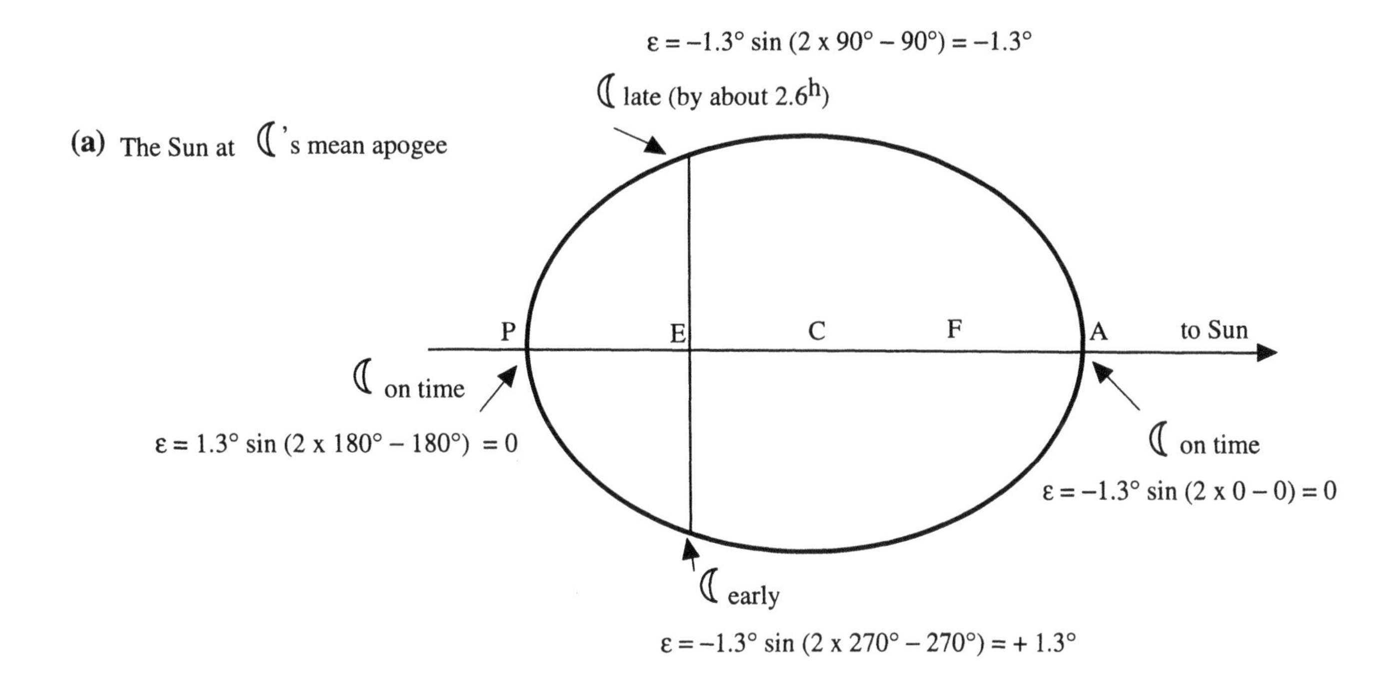

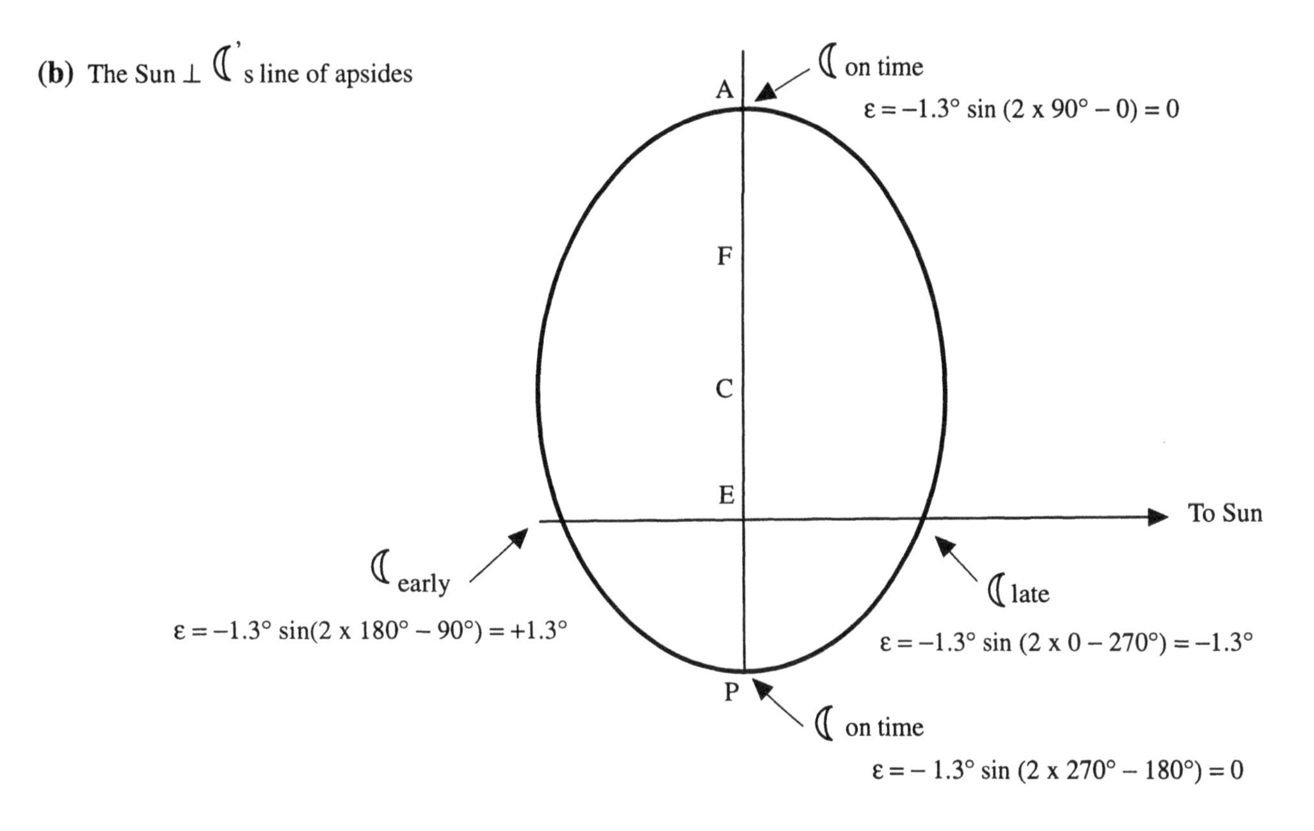

Figure IV.5 — Approximate quantitative effect of evection in perturbing the Moon's motion in celestial longitude. (a) The Sun at the Moon's mean apogee. (b) The Sun perpendicular to the Moon's line of apsides.

right angles to the Earth-Sun line, the new moon is late and the full moon early by this approximate amount, while the quarters are on time. The timing of eclipses with known orientation of the lunar orbit gave discrepancies which led Hipparchus to the discovery of evection, as noted in connection with equations (3) and (4).

To further illustrate the effect of evection on the Moon's position, let all terms in equation (5) beyond term III be neglected, so that in Figure IV.5(a) the Moon's orbit is represented as a nearly undisturbed ellipse of eccentricity about 1/18. (The eccentricity of the osculating ellipse best fitting the Moon's actual orbit varies from a maximum of 1/15 to a minimum of 1/22. The excess or defect of the Moon's longitude from that attained in an ellipse of average eccentricity $= 1/18$ caused by this variation of eccentricity is small compared to the effect in longitude due to the evection.) The orbit shown is thus supposed to be represented by the first three terms of equation (5).

In Figure IV.6(a) let E be the focus occupied by the Earth of this ellipse of eccentricity 1/18 (not shown to scale), C its center, F its second focus, P its perigee, A its apogee, $\ell E\ell'$ and $\ell_1 F\ell_1'$ its *latera recta.* Also let Em, Em' be the directions of the mean Moon at the critical moments of quarters and syzygies of the apparent Moon A (considered in Figs. IV.5(a) and 5(b), respectively). Let Ea, Er, Ea', Er' be the directions of the apparent Moon when, according to term IV in equation (5), the angles aEe, rEe, $a'Ee'$, $r'Ee'$ reach their maximum and minimum values (of $\pm 1.3°$). Then, considering the properties of elliptic motion the order of the points ℓ, r, e, a, m and m', r', e', $a' = \ell'$ is shown in Figure IV.6(a). In each group all the points except the ℓ's will coincide at A and P.

Thus Em turns uniformly about E, and Ee moves with the pure elliptic motion of e in the orbit. Using Ward's Principle (see section on Ptolemy's introduction of the equant) and noting that the eccentricity is small, Ee lags behind Em from A to about ℓ_1, the intersection of the orbit with the latus rectum through the empty focus, and gains on uniform angular revolution from ℓ_1 to P. The motion of e, then, is a standard of relatively undisturbed elliptic motion (the "variation" is zero at all critical points considered) against which one can mentally compare motion that includes evection.

Thus in Figure IV.6(a), starting from A, the Moon falls behind e until the evection term reaches its maximum negative value ($-1.3°$) at ℓ. Then it gains on e, enough to coincide with e and m at P. The motions in the orbit are symmetrical about AP. Hence, the relative disposition of the primed letters and the circumstances of the related motions can be immediately visualized. The result is that the greater value of the maximum prosthaphaeresis, 7.6°, occurs only with the Moon at the quarters, when the Sun is in the line of apsides of the lunar orbit; and the smaller value of the maximum prosthaphaeresis, 5°, occurs only at the syzygies of the Moon, when the Sun is at right angles to the line of apsides of the Moon's orbit. These observational facts were recognized by both Hipparchus and Ptolemy, and illustrate why Ptolemy used equations (3) and (4).

Ptolemy's Explanation of the Evection

The preceding discussion permits one to form a mental picture of the approximate effect of the evection in perturbing the Moon's simple elliptic motion in celestial longitude, or (which is the same to within a few arc minutes at the cardinal points considered) the orbit as described by epicycles equivalent to terms I through IV of equation (5). The evection is explained by Ptolemy as being due to an apparent increase in the radius of the Moon's

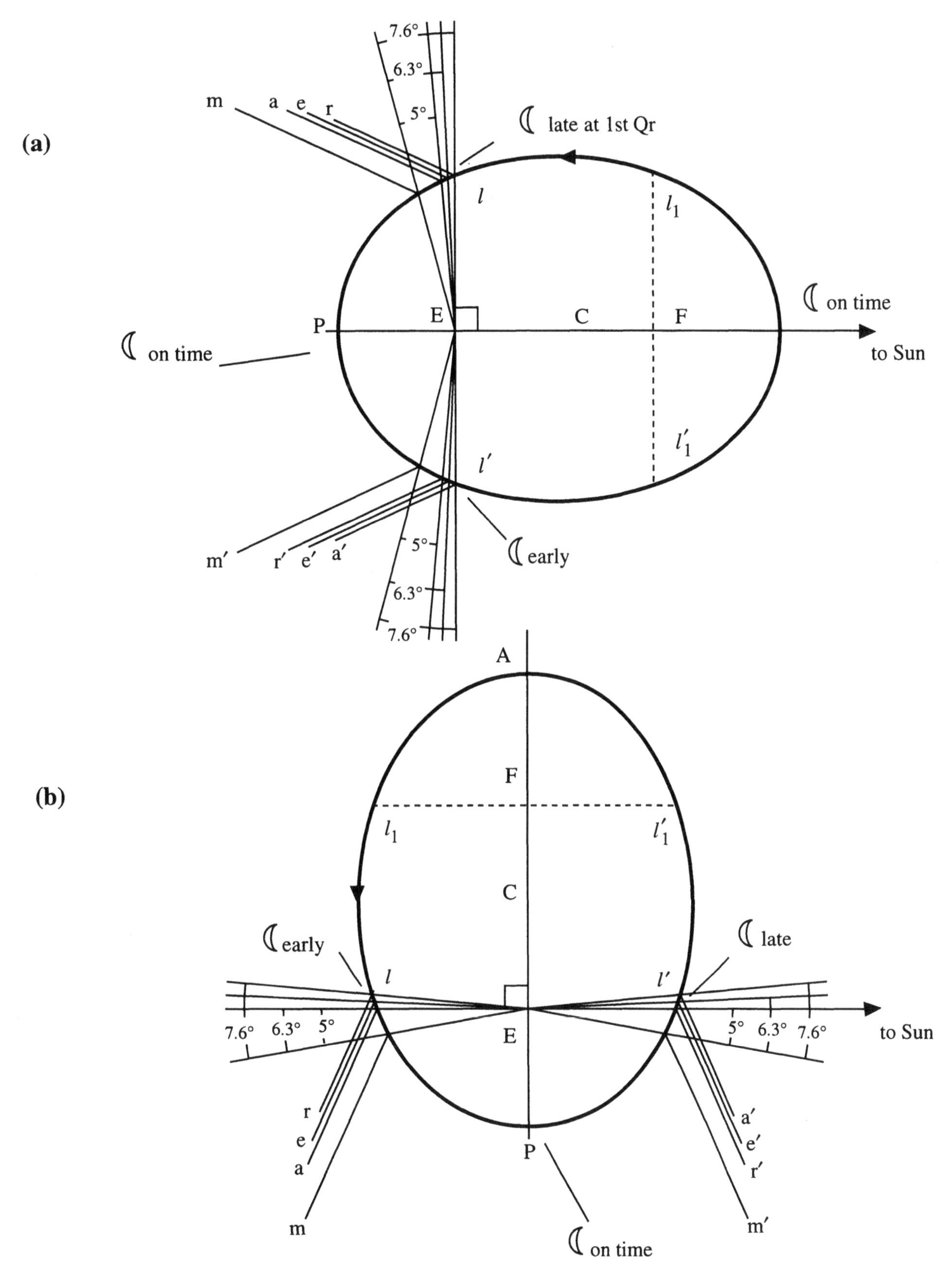

Figure IV.6 — The Moon's maximum prosthaphaereses at the cardinal points of the Moon's orbit. (a) Sun on lunar line of apsides and Moon at a quarter. (b) Sun perpendicular to lunar line of apsides and Moon at syzygy.

epicycle due to its approach to the Earth caused, as it were, by the gyration of the whole plane of the deferent about T in the orbit CC_1N (see Fig. IV.3), as was shown in the section discussing the motions in Ptolemy's lunar orbit. Since his measurements resulted in relative dimensions for the system of $R = 1$, $r = 0.105$, and $TC = 0.21$, the reduction of the distance of the epicycle center from the Earth's center at the quarters would be in the ratio $TA/TB' = 1.21/0.79 = 1.53$. During a whole lunation, the change of distance of the Moon from the Earth's center ($TF/TM''' = 1.315/0.685 = 1.92$) would cause a change of the Moon's apparent diameter of nearly a factor of 2, a consequence that Ptolemy does not mention. However, he certainly could not have been unaware of this.

Ptolemy's Determination of the Evection at the Quarters

Procedure: In Figure IV.7 Ptolemy observed the Moon's angular distance from the Sun and found the largest deviations $(\lambda_☾ - L_☾)$ (max) $= a$ or b when the motion of the Moon was straight toward the Earth along the tangent to the epicycle, so that the motion in geocentric lunar celestial longitude was then equal to that of the center of the epicycle, and therefore was a mean value. Let

$\lambda_☾$ = the true celestial longitude of the Moon,

$L_☾$ = the mean celestial longitude of the Moon = celestial longitude of the Moon's epicycle,

$\overline{A}_1$ = the epicyclic anomaly of the Moon,

$R = 1$ = radius of the Moon's deferent,

r = radius of the Moon's epicycle,

ρ = eccentricity of the Moon's deferent.

Then, from Figure IV.7,

$$\frac{r}{R-\rho} = \frac{r}{1-e} = \sin a\,,$$

$$\frac{r}{R+\rho} - \frac{r}{1+e} - \sin b\,, \qquad \text{and}$$

$$\frac{1+\rho}{1-\rho} = \frac{\sin 7\frac{2}{3}^\circ}{\sin 5^\circ} = \frac{0.1334}{0.0872} = 1.530\,,$$

from which $\rho = 0.209$, and $r = (1-\rho)\sin a = 0.791 \times 0.1334 = 0.106$.

Hence the maximum evection is $\pm 1\frac{1}{3}^\circ$ near the quarters and syzygies, and the formulas for the prosthaphaereses P_q and P_s close to these points are approximately

$$P_q \approx 7.6^\circ \sin \overline{A}_1 \qquad \text{and} \qquad P_s \approx 5^\circ \sin \overline{A}_1\,.$$

The Prosneusis

The combination of uniform circular motions so far described was not considered by Ptolemy to be sufficient to predict the Moon's position between the syzygies and the quarters. As we have seen (Fig. IV.3) the motion of the Moon M on the circular epicycle of radius

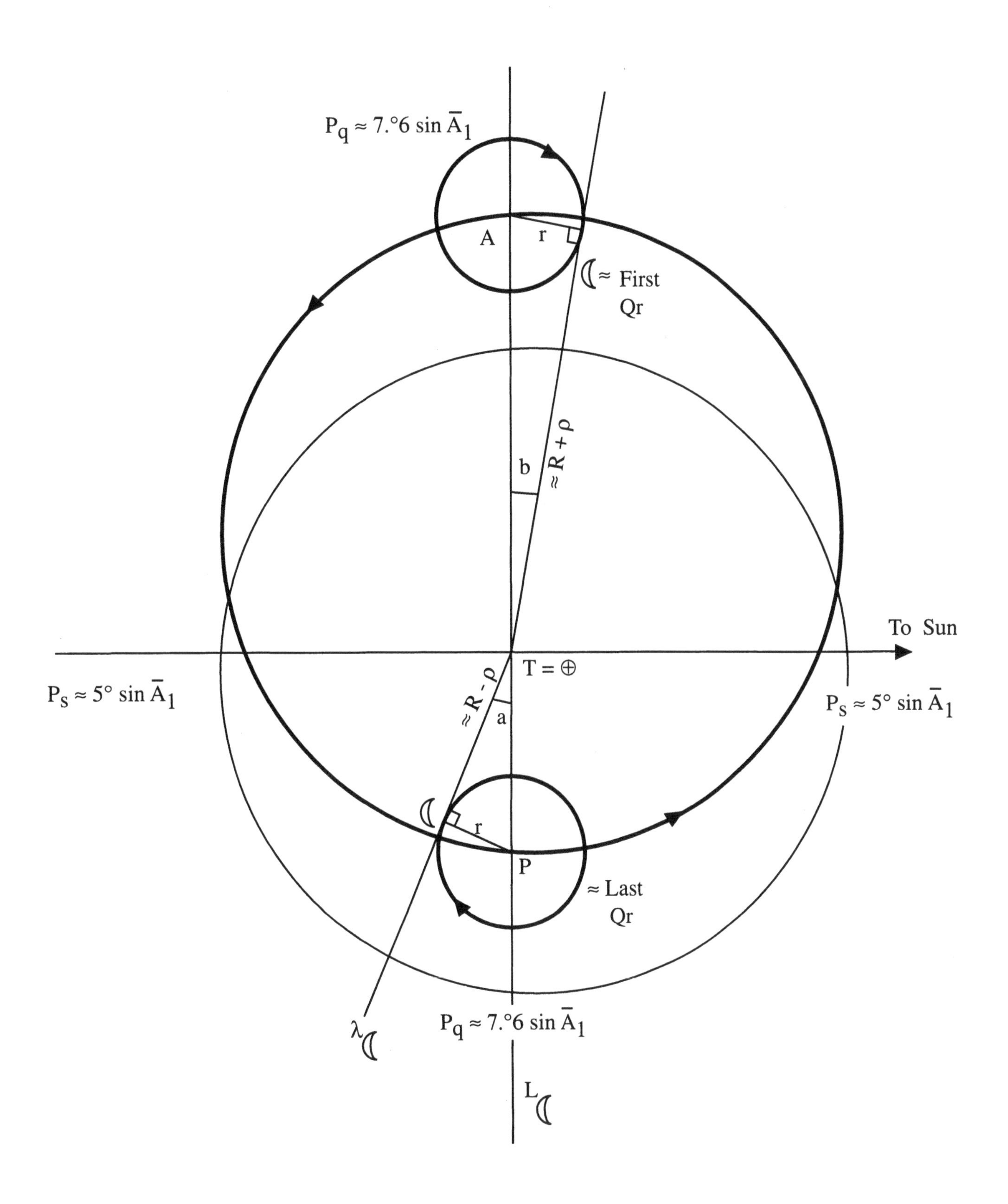

Figure IV.7 — Ptolemy's determination of the Moon's second inequality (the evection) at the quarters.

r in a period of 27.51 days is retrograde and uniform about the moving center of the epicycle, which takes up positions successively at A, B_1, B'. Since this motion is uniform, $\angle P_1B_1M_1 = \frac{360°}{27.51^d}T_d$, where T_d denotes the time in mean solar days elapsed since the moment of conjunction with the mean Sun. Thus $\angle P_1B_1M_1$ lags behind $\angle B_1TA$, which equals $\frac{360°}{27.32^d}T_d$. (The difference between 27.32^d in the last expression and 27.51^d in the earlier one produces the advance in the line of apsides of the lunar orbit.) However, Ptolemy holds that for the purpose of computing the evection it is not sufficient to figure the position of M_1 from the point P_1. The epicyclic anomaly must be reckoned from P_1', the intersection with the instantaneous epicycle of a straight line from N_1 to the epicyclic center B_1 (extended), where N_1 is the point on the evection circle diametrically opposite C_1, the instantaneous center of the deferent K_1. Thus B_1M_1 is drawn in the figure slightly *converging* toward TA, whereas the slight inequality of the periods involved would require it to be slightly *divergent* from TA.

In other words, the epicyclic anomaly is measured, not from the true instantaneous apogee P_1 (in Fig. IV.3), but from a fictitious apogee P_1' "as seen from N_1." The correction to the epicyclic anomaly due to this change of origin for the epicyclic anomaly, $\angle TB_1N_1 = \angle P_1B_1P_1' = \angle M_1B_1M_1'$, is called the prosneusis Pr and may amount to about $\frac{1}{4}°$ at maximum, near the trines. Hence the definition: "The prosneusis is the angle subtended by TN_1 at the epicyclic center B_1," etc., for all subscripts. A less exact but useful definition (derived from Dreyer's 1905 description of the theory) is as follows: "The prosneusis is a correction ($\angle P_1B_1P_1'$) applied to the epicyclic mean anomaly ($\angle P_1B_1M'$) before using it to figure the evection." Thus the prosneusis is not to be confused with "equantal motion" (uniform angular motion about a fixed point on TA), because it refers entirely to a correction of the epicyclic (and not the deferential) anomaly. After the epicyclic anomaly of M_1 is thus corrected, all the necessary data are at hand to compute the prosthaphaeresis of the Moon, $\angle B_1TM_1$, and then the true lunar longitude, $\angle ATM_1 + \lambda_0$, where $\lambda_0 =$ the (given) true longitude of A.

Although, as we shall see, the prosneusis is essentially different from "the third inequality" (also called the "variation"), its effect in combination with the equation of center and the Ptolemaic evection (but in the absence of the variation) does somewhat improve the representation of the Moon's orbital motion and permits an accuracy of about $10'$ in the predicted place of the Moon at times not too far removed from any epoch for which the position is correctly given. Since ancient observations were not reliable to any better accuracy, Ptolemy's lunar theory can be considered perfectly adequate for its time.

Ptolemy's Computation of the Prosneusis and an Example of His Prediction of the Moon's True Longitude

To find the true longitude of the Moon in the sky, Ptolemy of course does not use formulas. Rather, he provides a set of five tables, giving various quantities in terms of the elongation of the mean Moon from the mean Sun. To the modern mind this procedure is, to say the least, complicated. A better perspective of the process may perhaps result from a set of simple formulas, involving mathematics no more advanced than plane trigonometry. This process, outlined below, is rigorous if the Sun is in the line of nodes of the lunar orbit, and if the circles are all in the instantaneous plane of the lunar orbit. It is approximately true at all other times if one neglects the effect of the slight orbital inclination (5°) of the lunar orbit to the ecliptic and the changes in orientation of the lunar orbit, i.e., in longitude of the node and longitude of the apogee.

Before proceeding with this geometrical version of Ptolemy's method for computing the Moon's orbital position, we will briefly outline a method for ascertaining the value of the prosneusis itself. In Figure IV.8, let

$\oplus = E =$ the Earth,

$\angle AEC = \angle E_T =$ the instantaneous elongation of the center of the Moon's epicycle from the mean Sun (at the epoch T),

$\overline{A} =$ the epicyclic mean anomaly of the Moon $= \angle LC\overline{☾}$.

Now we have given

$R =$ radius of deferent $= 1$,

$\rho = EN = EO = 0.210$, and

$r = CL =$ epicyclic radius $= 0.105$.

By definition,

$$E_T = (T - T_0)\frac{360°}{27.32^{\mathrm{d}}}, \qquad (7)$$

where $\mathrm{T_0}$ is any given epoch for which $E_T = 0$.

In Figure IV.8, draw NF_1 and OF_2 both $\perp$ CE. Then from right triangle CF_2O,

$$(EC - \rho\cos 2E_T)^2 + (\rho\sin 2E_T)^2 = 1,$$

$$EC = \rho\cos 2E_T + \sqrt{1 - \rho^2\sin^2 2E_T}.$$

From right triangle CF_1N,

$$\tan Pr = \tan MCL = \tan NCE = \sqrt{\frac{\rho\sin 2E_T}{EC + \rho\cos 2E_T}},$$

where Pr is the prosneusis. Then

$$Pr = \tan^{-1}\left[\rho\sin 2E_T/(2\rho\cos 2E_T + \sqrt{1 - \rho^2\sin^2 2E_T}\,)\right]. \qquad (8)$$

To develop a formula for the Moon's instantaneous celestial longitude one may proceed as follows. The prosneusis Pr and the instantaneous distance EC having been found, we next find a formula for the prosthaphaeresis P. This requires a knowledge of the epicyclic anomaly $\angle LC☾$. The given data will include an epicyclic anomaly A_0 at T_0, and hence we find immediately the mean epicyclic anomaly $\overline{A}$ at T from

$$\overline{A} = A_0 + \mu(T - T_0) = A_0 + \frac{360.51°}{27°}(T - T_0)^{\mathrm{d}}.$$

From Figure IV.8, $\angle LC☾ = \angle MC☾ + Pr$ or, measuring counterclockwise,

$$\angle LC☾ = 360° - (\overline{A} - Pr).$$

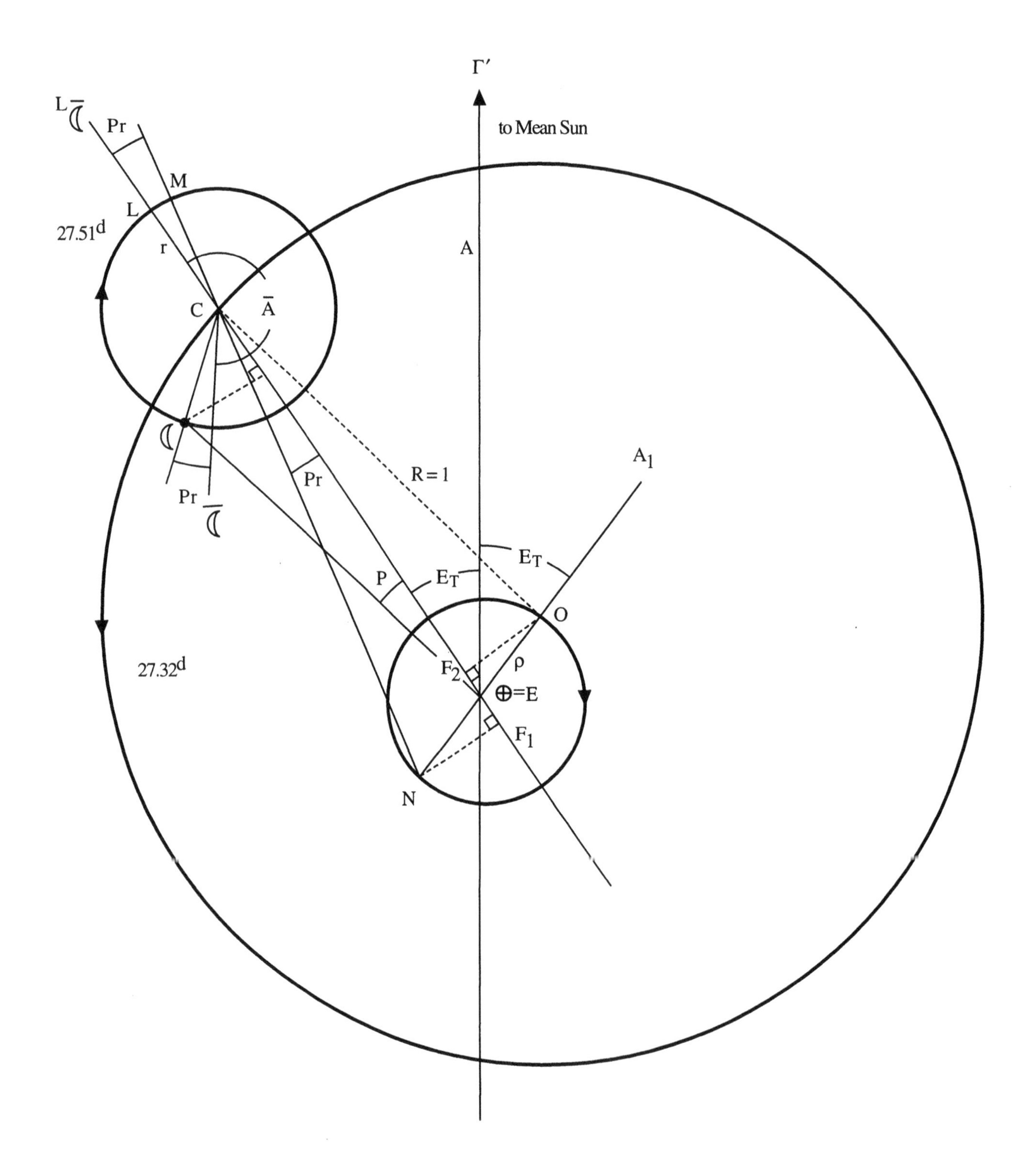

Figure IV.8 — Computation of Ptolemy's prosneusis and prediction of the Moon's true longitude, ∠OE☾.

Draw $☾F_3 \perp EC$; then from triangle $CE☾$,

$$\tan CE☾ = \tan P = \frac{r \sin LC☾}{EC \ + r \cos LC☾}, \tag{9}$$

from which P. Now if the Moon's mean longitude $L_{\overline{☾}}$ at time T is known, then $\lambda_☾ = L_{\overline{☾}} + P$, which solves the problem.

Equation (9) is general and independent of the specification of T_0 as the moment of syzygy. However, if we do choose to let T_0 be the moment of syzygy, then at an apogee E_T will be given, as always, by equation (7), and we have further

$$\lambda_☾ = E_T + \tan^{-1}\left(\frac{r \sin LC☾}{EC + r \cos LC☾}\right),$$

or in still more explicit form,

$$\lambda_☾ = E_T + \tan^{-1}\left[\frac{r \sin(\overline{A} - Pr)}{\rho \cos 2E_T + \sqrt{1 - \rho^2 \sin^2 2E_T} - \cos(\overline{A} - Pr)}\right], \tag{10}$$

from which $\lambda_☾$.

Summary:

T = epoch of interest,

T_0 = epoch for which data are given,

$\lambda_☾$ = true longitude of Moon (from apogee),

E_T = instantaneous elongation of mean Moon from mean Sun (at epoch T),

$\overline{A}$ = epicyclic mean anomaly of Moon,

r = radius of lunar epicycle = 0.105 (for unit radius R of deferent),

ρ = radius of evection circle of Moon = 0.21 ($R = 1$),

Pr = instantaneous value of the prosneusis of the Moon.

Procedure: Compute E_T by equation (7), Pr by equation (8), then $\lambda_☾$ by equation (10).

Approximate Elementary Derivation of the Longitude Correction to the Moon's Position Caused by the Prosneusis

It is difficult for modern readers to visualize the effect of Ptolemy's prosneusis, especially as it does not correspond to the orbital perturbations called the variation and the evection (see eqn. (5)). In this section, we shall try to clarify the prosneusis using a simple construction. As we shall see, the prosneusis differs in its characteristics from the variation, as later discovered by Tycho Brahe (see Chapter VI).

Though not to scale, Figure IV.9 shows Ptolemy's construction of the evection ε, the prosneusis Pr, and the correction to the celestial longitude $\Delta\varepsilon$ caused by Pr. The prosthaphaeresis and the final apparent lunar longitude $\lambda_☾$ at epoch T are also shown.

Figure IV.9 provides a simplified view of the celestial longitude effect of the prosneusis in order to illustrate qualitatively the difference between this and the effect of the variation.

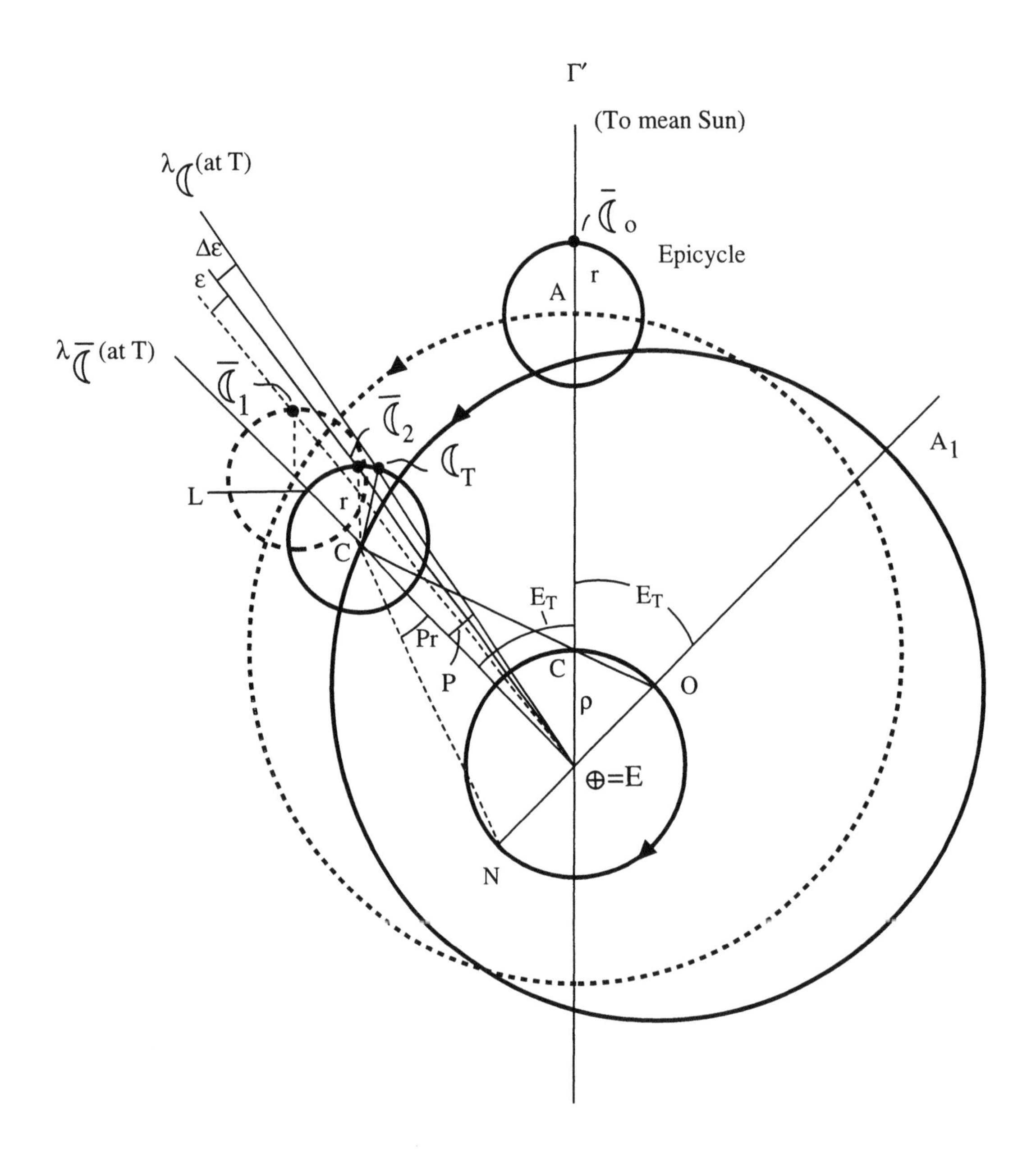

Figure IV.9 — Qualitative construction of Ptolemy's evection, prosneusis, and prosthaphaeresis of the Moon (not to scale; Moon in first octant).

Let it be assumed, for simplicity, that the motion starts with the Moon at new and a "double" apogee (i.e., Moon in apogee on its epicycle and the epicycle in apogee on its deferent). Further, let the motions of the Sun and of the line of apsides of the lunar orbit be neglected for 1 month so that the synodic and anomalistic periods are equal. Thus in Figure IV.9, let

$\oplus = E =$ the Earth,

$EA =$ the line of apsides of the lunar orbit,

$CO = R = 1 =$ the radius of the lunar deferent,

$EO = 0.21 = e =$ the radius of the evection circle,

$r = 0.105 =$ the radius of epicycle,

$L =$ the apparent instantaneous epicyclic apogee with no evection,

$M =$ the mean instantaneous epicyclic apogee with no evection,

$E_T =$ mean geocentric anomaly of the Moon, including the first three terms in equation (5),

$\Delta\varepsilon =$ longitude effect of prosneusis,

$\varepsilon =$ evection at T (negative here),

$\overline{☾}_0 =$ mean Moon at T_0,

$\overline{☾}_1 =$ mean Moon at T,

$\overline{☾}_2 =$ mean Moon at T plus evection,

$☾_T = \overline{☾}_2 + \Delta\varepsilon = \lambda_{\overline{☾}} + P =$ true Moon at T.

$Pr = \angle LBM = \angle NBE \sim 12° =$ the prosneusis at the octants in the mean orbit (estimated by protractor).

Now, by redrawing Figure IV.9 to scale, the longitude effect of the prosneusis $\Delta\varepsilon$ may be constructed at octants 1, 3, 5, and 7 (it is zero at the quarters). Measurements by a protractor give values for the prosneusis $Pr \sim 12°$, and for its longitude effect, $\Delta\varepsilon \sim \frac{3}{4}°$. Also by construction, the algebraic signs of these longitude effects of Pr are determined as given in the third column of Table IV.2. The algebraic signs of the variation are as in the fourth column, which is based upon term V in equation (5). The evection from term IV is entered in the fifth column.

This short table illustrates the entirely erroneous nature of the view that the prosneusis is merely another name for the variation, which seems to have been the main issue in arguments regarding this subject during much of the nineteenth century. Although, as mentioned above, the numerical maxima of the prosneusis occur approximately at the trines rather than at the octants, the algebraic signs are as given in Table IV.2 for all values in the regions between consecutive zeros.

Table IV.2 — Approximate Corrections to the Moon's Celestial Longitude

Phase	Octant	Prosneusis[a]	Variation	Evection
New	0	0	0	0
	1	$-\frac{3}{4}^\circ$	$-40'$	-0.9°
First quarter	2	0	0	-1.3°
	3	$-\frac{3}{4}^\circ$	$+40'$	-0.9°
Full	4	0	0	0
	5	$+\frac{3}{4}^\circ$	$-40'$	$+0.9^\circ$
Third quarter	6	0	0	$+1.3^\circ$
	7	$+\frac{3}{4}^\circ$	$+40'$	$+0.9^\circ$
New	8	0	0	0

[a]From construction of Figure IV.9 redrawn to scale.

Introduction to Ptolemy's Planetary Theory

Although proceeding from completely different premises, a parallelism may nevertheless be drawn between ancient Greek and Renaissance developments regarding the explanation of planetary motions. With admiration we contemplate the introduction of the heliocentric system by Copernicus, the lifelong toil by Tycho resulting in an accurate and prolific accumulation of data, and Kepler's masterly, patient, and ingenious analyses of this material, resulting in the revolutionary introduction of a simple elliptic motion for the several uniform circular motions postulated in earlier systems. In a similar manner Apollonius, Hipparchus, and Ptolemy have been considered an intellectual triumvirate of paramount importance to the development of Greek astronomy. During the earliest scientific age most astronomers thought of the planetary motions in terms of a nest of concentric physical spheres, a model introduced and elaborated by Eudoxus and Callippus (Chapter II). Then Apollonius (the "Copernicus of Antiquity") developed the mathematics of epicycles and eccentrics, and showed that they could better represent the observed celestial motions. He may have had a "system" of the world, but he probably never extensively tested it against observed facts. However, he gave us all the necessary mathematics for describing retrograde motion, stationary points, and the properties of epicyclic and eccentric motion. But being primarily a mathematician (sometimes called the father of the study of conic sections), he was probably content with a Universe "on paper."

Not so Hipparchus (the "Tycho of Antiquity"), who produced numerous and, for his time, exceedingly accurate observations by which to test his postulated eccentric orbital movements of the Sun, Moon, and planets (Chapter III). His mathematics was for the most part based on Apollonius's work, which indeed was easily adequate to his needs. But being himself a great observer he needed certain simplifications in handling his many data. These he developed himself, as did Tycho later in simplifying the computations at Uraniborg.

But Ptolemy (the "Kepler of Antiquity") is deservedly considered the greatest Greek astronomer of ancient times. Using both the old and his own new ingenious methods, from astronomical observations he determined the constants of the celestial motions. In addition, as a consummate mathematician, he developed whatever mathematics he needed for the success of his work. (Incidentally, he is also often considered the father of scientific geography.)

It was Ptolemy who brought Greek planetary theory into its final, successful form. In the case of the Sun, the motion around the ecliptic is very simple: a slightly varying apparent speed and a motion that never becomes retrograde. This "first," or "zodiacal," inequality was well represented by Hipparchus's eccentric circle. In the case of the planets there is indeed a zodiacal inequality, but it is not the most striking aspect of the planets' behavior. The striking reversals of direction associated with the "second," or "synodic," inequality were the most important phenomena to be explained by a geometrical theory of planetary motion. Retrograde motion was explained in a satisfactory, qualitative fashion by Apollonius's model of epicycle motion. The first, or zodiacal, inequality of the planets shows up in a subtle way: the retrogradations of the planets are not equally spaced around the zodiac.

Astronomers between the times of Apollonius and Hipparchus modified the former's model in an attempt to account for the zodiacal inequality. The planet moved uniformly on an epicycle (which accounted for the synodic inequality). But now the epicycle moved uniformly on a deferent circle that was eccentric to the Earth. This epicycle-plus-eccentric model thus succeeded in explaining how the retrogradations could be spaced unevenly about the zodiac. But, as Hipparchus demonstrated, it could not account for the observed, variable lengths of the individual retrograde arcs. Even Hipparchus, who had had great successes with the Sun and Moon, was unable to provide a theory of the planets that worked in detail.

This then was Ptolemy's great achievement. No geometrical planetary theory before his time gave a satisfactory account of the chief manifestations of the planets' zodiacal inequality – the uneven spacing and lengths of the retrograde arcs. As we shall see, this required the introduction of nonuniform motion into the heavens.

Ptolemy's Planetary Theory

Ptolemy started from Apollonius's assumption of a planet moving uniformly eastward in its synodic period on a circular epicycle K_1 of radius r and center C (Fig. IV.10). This epicycle slides with uniform eastward velocity in the sidereal period of the planet on a circular deferent K (radius $R = OC'$, center O) eccentric to the Earth T by the eccentricity e. He further assumes that the radius vector r, from the center of the epicycle to the planet (for Mars, Jupiter, and Saturn), is always parallel to the line from the Earth's center T to the *mean* place of the Sun. The eccentric position of K helps to account for the planet's nonuniform progress around the zodiac. The motion on the epicycle reproduces retrogradation and other synodic phenomena.

For Mercury and Venus the center of the epicycle itself lies on the line from the Earth to the mean Sun. The planet moves uniformly counterclockwise on the epicycle, completinga revolution in its synodic period, while the center of the epicycle slides uniformly eastward on the eccentric (circular) deferent in the zodiacal period of the planet. So far, except for the eccentric position of the deferents, the arrangements were identical to the system of Apollonius. This fairly well represented the stations and the retrograde motions of an inferior planet as seen from the Earth. Nevertheless, the "first inequality" (the main part of the elliptic motion) was very poorly represented.

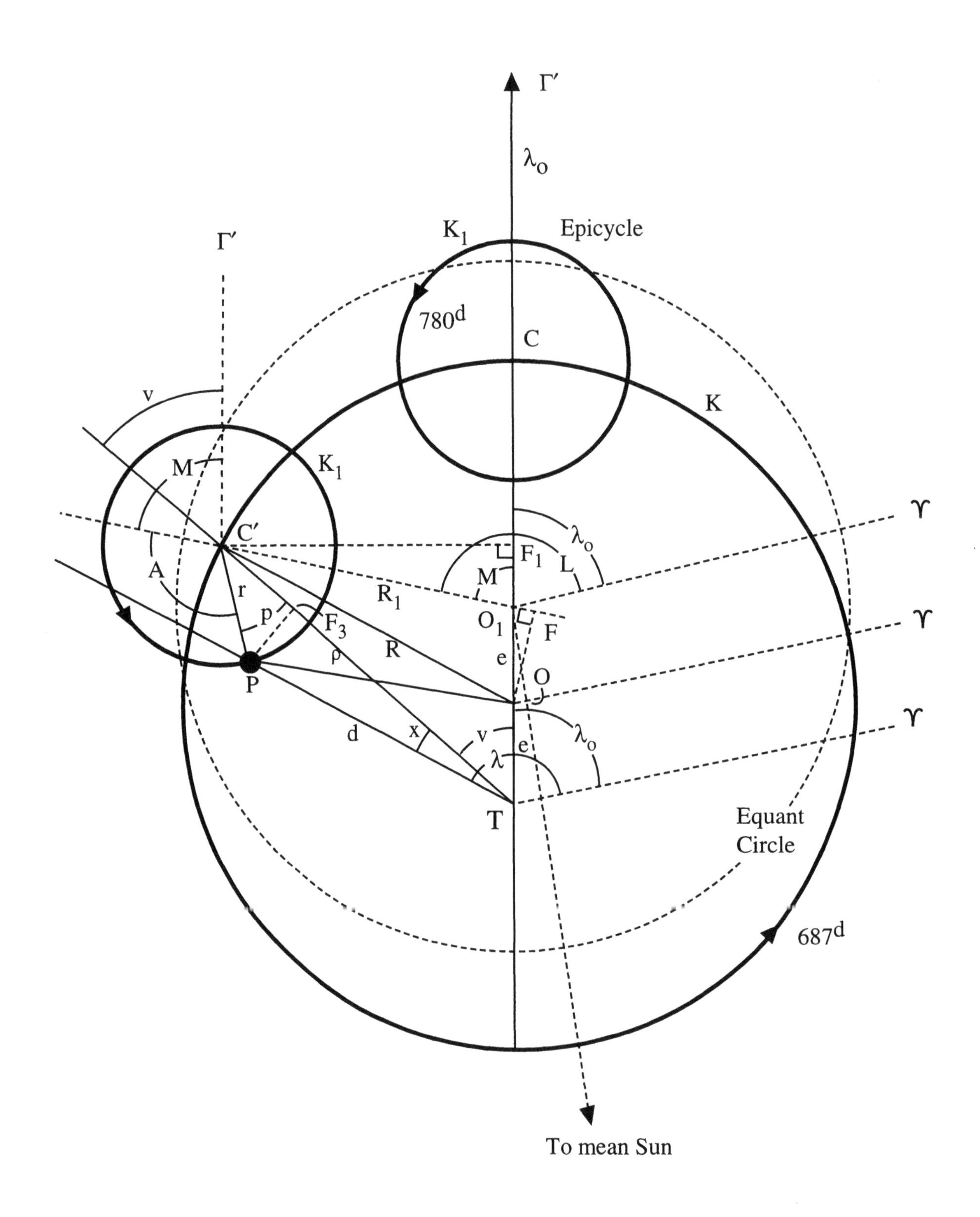

Figure IV.10 — Ptolemy's model of planetary motion for a superior planet (Mars).

Ptolemy's Reason for Introducing an Equant Point in All Planetary Orbits

Ptolemy found from observation and computation that the difference between mean and observed longitude was smaller at apogee (and larger at perigee) than could be accounted for by letting the circular epicycle slide uniformly on any fixed eccentric circular deferent. This would indicate (see Fig. IV.10) that the center of the deferent O is nearer the Earth T than the center of uniform motion O_1. As a second modification Ptolemy therefore introduced the equant point O_1, situated on the line of apsides $T\Gamma'$ on the opposite side of O from T. He found best agreement with observation by making $TO = OO_1$ for the superior planets, as we shall see later. The motion of the center of the epicycle is therefore supposed to be in the deferent centered on O, but with uniform angular velocity (corresponding to the planet's zodiacal period) about O_1. Thus the motion of the epicycle's center C on the deferent K, though still along a circle, was no longer uniform, either with respect to its own center O or with respect to the Earth T! This was the first time in Greek astronomy that the principle of uniform circular motion was violated. (Ptolemy's introduction of the prosneusis into the lunar theory also can be thought of as producing a nonuniformity of motion.)

The excellence of the equant concept for small eccentricities is forcefully illustrated by Ward's Principle: Uniform angular motion about one focus of an ellipse of small eccentricity is very nearly equivalent to elliptic areal motion about the other focus of the ellipse. Hence, the enormous importance of Ptolemy's second improvement, the introduction of equantal motion.

Example of Ptolemy's Prediction of a Superior Planet's Celestial Longitude

To predict the position of a planet at a given time, in Figure IV.10 let

$t =$ the epoch of interest,

$t_0 =$ an epoch for which all constants of the problem are given,

$\lambda_0 =$ the longitude of the deferent's apogee at t_0,

$A_0 =$ the mean epicyclic anomaly at t_0,

$\mu =$ the mean daily motion of the epicycle on the deferent,

$\mu_1 =$ the mean daily motion of the planet on the epicycle,

$\Delta t = t - t_0 =$ the time interval from the given epoch to the epoch of interest,

$L =$ the mean longitude at t (measured at the equant point),

$M =$ the mean anomaly at t (measured at the equant point),

$v =$ the true anomaly at t (measured at the Earth),

$R =$ the radius of the deferent $(= 1)$,

$r =$ the radius of the epicycle,

$e =$ the eccentricity of the deferent,

$R_1 =$ the distance from the equant point to the epicycle's center at t,

$\rho =$ the distance from the Earth to the epicycle's center at t,

$x =$ the prosthaphaeresis at t, of P from TC',

$\lambda =$ geocentric celestial longitude at t (in the plane of the orbit) $= \angle \Upsilon TP$.

The problem is to find λ, given t, t_0, L_0, T_e, A_0, μ, μ_1, R, r, and e. By definition, the mean longitude of C about O_1, measured on K', is

$$L = L_0 + \mu \Delta t\,.$$

The epicyclic anomaly, measured on K_1, is

$$A = A_0 + \mu_1 \Delta t\,.$$

From the figure,

$$M = L - \lambda_0\,.$$

From right triangle OFC',

$$R^2 = (e \sin M)^2 + (R_1 + e \cos M)^2 \qquad \text{or} \qquad R_1^2 = \sqrt{R^2 - e^2 \sin^2 M} - e \cos M\,.$$

From right triangle O_1F_1C',

$$\rho \sin v = R_1 \sin M\,.$$

From right triangles TF_1C' and O_1F_1C',

$$\rho \cos v = R_1 \cos M + 2e\,, \qquad \text{and}$$

$$\angle \Gamma' C' P = A + M, \qquad \angle L_1 C' P = A + M - v\,, \qquad p = 180^\circ - (A + M - v)\,.$$

From ΔPTF_3, by definition,

$$\tan x = \frac{PF_3}{TF_3} = \frac{r \sin p}{\rho - r \cos p}\,.$$

By definition,

$$\lambda = \text{true longitude} = \lambda_0 + v + x\,.$$

This finishes the problem of finding the true geocentric longitude of the planet P. To find d, we know x, p, ρ, r (also $\angle TPC' = 180^\circ - (p + x)$). By the sine formula in $\Delta TPC'$,

$$d = r \sin p \csc x\,,$$

from which d. For a check:

$$d = \rho \sin p \csc TPC'\,.$$

Ptolemy's Determination of the Equant Point's Position

At first Ptolemy, like his predecessors, considered that the center O of the deferent coincided with the center O_1 of uniform motion (see Fig. IV.10). If we take the radius R of the deferent equal to unity, the eccentricity of the deferent circle is TO_1 (since, for the time being, we assume that O coincides with O_1). A model like this for Mars, say, is capable of representing the *spacings* of the planet's retrograde arcs around the zodiac with good accuracy. However, the variable *widths* of these arcs cannot be reproduced with accuracy. In this version of the model, Ptolemy found that Mars's theoretical retrograde arcs near the apogee Γ' were too narrow; conversely, the arcs near the perigee Γ were too wide compared to observation. Ptolemy had the ingenious idea of separating the center O of the deferent circle from the center O_1 of uniform motion by moving O closer to the Earth T. This had the effect of bringing the epicycle a bit closer to the Earth at apogee and so increasing the apparent width of the retrograde arc.

In Ptolemy's final model for Mars, Jupiter, and Saturn, as well as Venus, the center O of the deferent is assumed to lie halfway between the Earth and the center O_1 of uniform motion (the "equant point"). Thus $OT = OO_1 = e$. In separating the center of uniform motion from the center of the deferent, Ptolemy introduced a true nonuniformity of motion into his theory. If we make a comparison to the elliptical astronomy of Kepler, we see that Ptolemy's instincts were good. To put things anachronistically, Ptolemy's equant law is a very good approximation to Kepler's law of areas. Ptolemy's theory (with a center of uniform angular motion separate and distinct from the center of the deferent) made possible a geometrical planetary theory with quantitative predictive power. Nevertheless, he was criticized in the later Middle Ages for this savage rupture of the principles of Aristotelian physics.

This new theory caused some difficulties for Ptolemy in establishing numerical values for the eccentricity and the longitude of the apogee. Ptolemy chose three oppositions of the planet to the mean Sun. At each of these moments the planet P (Fig. IV.10) would lie on the line of sight $C'T$ between the Earth and the center of the epicycle. The use of mean oppositions (when the planet is approximately in the middle of its retrogradation) allows one to see, as it were, point C' – for the direction of the epicycle's center is marked in the sky by the planet P itself. Given the dates and places of three oppositions, Ptolemy could in principle find e and λ_0 in Figure IV.10.

Direct observation gave the longitudinal arcs separating the oppositions from one another (as viewed from Earth). It was easy also to calculate the difference $\Delta\mathrm{M}$ in mean anomaly between successive oppositions, since $\Delta\mathrm{M}$ is simply proportional to the time. Let C_1', C_2', C_3' denote the positions of the epicycle's center at the three oppositions. Ptolemy knew the angles between them as viewed from the Earth ($C_1'TC_2', C_2'TC_3'$) and he knew the angles between them as viewed from the equant ($C_1'O_1C_2', C_2'O_1C_3'$). The problem then was to determine where the equant was – i.e., to determine e and λ_0.

The simpler version of this problem, in which O and O_1 coincide, is not very difficult to solve. Ptolemy's more general problem (with $TO = \frac{1}{2}TO_1$) is extremely difficult. Ptolemy adopted an ingenious method of iterations. First he solved the simple problem (O coinciding with O_1), then he calculated the corrections produced by separating O from O_1. This problem of determining the eccentricity and the longitude of the apogee is perhaps the most brilliant piece of applied mathematics in the *Almagest.*

For Mercury Ptolemy believed that the best agreement with observations was obtained by letting the deferent's center O' retrograde on a circle of radius $\rho = \frac{1}{2}e$. This circle was centered at O, toward the apogee A on the line of apsides at a distance e from the Earth T_e (see Fig. IV.11). The equant point D is placed on the line of apsides halfway between the Earth T and the center of distances O. Thus $TD = \rho = \frac{1}{2}e$, where ρ is the radius of the circle described by the center of the deferent K. (Note the similarity of the gyratory motion of the deferent to the evection in the Moon's orbit.)

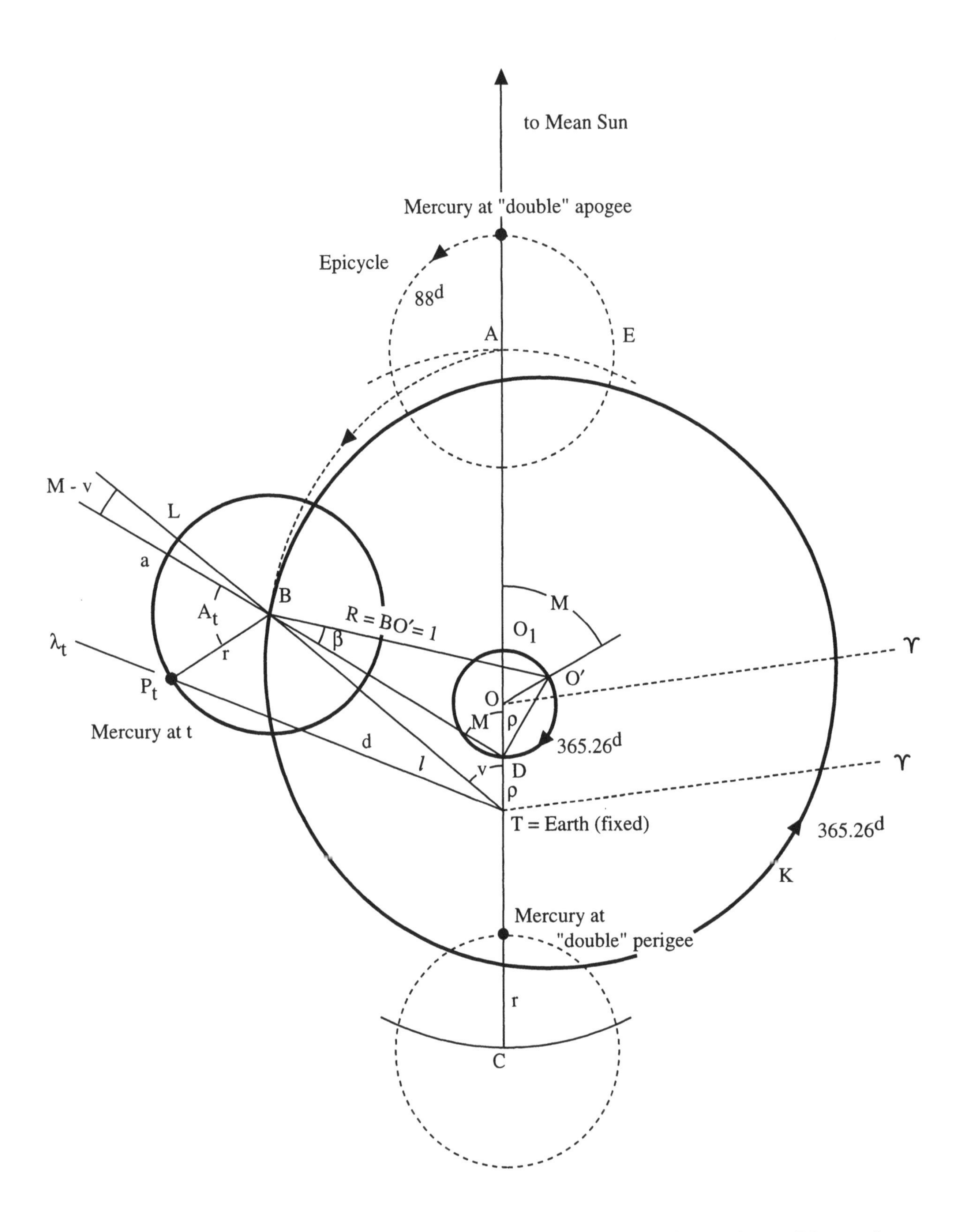

Figure IV.11 — Ptolemy's model of planetary motion for an inferior planet (Mercury).

Ptolemy's Theory of the Celestial Latitudes of the Planets

The Superior Planets

Although the representation of the celestial latitudes of the planets by Ptolemy's system is exceedingly complicated, it never met with much success mainly because *(a)* the planes of the orbits passed (erroneously) through the center of the Earth and *(b)* the distances, although approximated by epicyclic constructions that we now know are very close to elliptic orbits, were not precisely modeled. For each of the three superior planets the plane of the deferent is assumed to pass through the Earth and to be inclined at a constant angle to the plane of the ecliptic. Ptolemy calls this "the inclination (*engklisis*) of the eccentric" (or deferent). According to Ptolemy, the inclinations of the deferents are $2°30'$ for Saturn, $1°30'$ for Jupiter, and $1°$ for Mars. (The modern values of i, measured from the ecliptic to the respective orbital planes, are $2°30'$, $1°18'$, and $1°51'$.) For Mars the line of apsides was assumed to be almost perpendicular to its line of nodes (passing through the Earth), so that Mars's line of apsides nearly coincided with the line joining the points of the planet's greatest north and south celestial latitude, which obtains when the planet is also in epicyclic perigee. For Saturn the point of greatest celestial latitude was about 50° east, and for Jupiter about 20° west, of the line of nodes through the Earth. The epicycles were inclined to the deferents so that their planes wobbled but were always roughly parallel to that of the ecliptic. Ptolemy's reason for this assumption was that observation had shown that the latitude was always greatest at the apogee and perigee of the (eccentric) deferent when the planet was also at the perigee of its epicycle. (Dreyer [1905] remarks: "As the epicycle of an outer planet was nothing but the Earth's annual orbit round the Sun transferred to the planet in question, it was of course quite right that the epicycle should be parallel to the ecliptic.") Observation by Ptolemy and others had shown that the deviations in latitude of the three superior planets were always to the north of the ecliptic.

Ptolemy's mechanism for explaining these celestial latitude deviations of a superior planet (Fig. IV.12) is most simply explained in the case of Mars, where the line of nodes ☊☋ of the deferent plane on the plane of the ecliptic is assumed to be 90° from the line of apsides $\Gamma\Gamma'$. O is the center of the deferent and T, the Earth. The center o of the epicycle $adbc$ moves eastward on the deferent, starting from the line of apsides $\Gamma\Gamma'$. As o moves around the deferent, the plane of the epicycle remains almost (but, as we shall see, not quite) parallel to the plane of the ecliptic. As the epicycle's center moves eastward, the epicycle's line of nodes cd remains always parallel to the line of nodes ☊☋ of the deferent on the plane of the ecliptic. Thus in Figure IV.12, cd would become successively c_1d_1, c_2d_2, c_3d_3, etc. Similarly (in a modern view), as the epicycle moves from o to o_1, o_2, o_3, etc., the diameter ab, being always parallel to the plane of the ecliptic, would take on successive positions a_1b_1, a_2b_2, a_3b_3, etc.

This idea of the movement of a line in space keeping always parallel to itself was, however, entirely foreign to the thinking of the ancients. To them, the diameter ab would always lie in a plane perpendicular to the deferent and passing through O, the center of the deferent. By the motion of the epicycle, ab would become successively $a_1'b_1'$, d_2c_2, b_3a_3, so that its inclination i would gradually decrease from i_o at o through i_1 at o_1 to zero at the descending node o_2. Thus ab, not cd, would become d_2c_2. To effect this gradual tilting of the radially situated epicyclic diameter, Ptolemy introduced a small auxiliary circle perpendicular to the deferent, whose center was in the plane of the deferent and whose radius (equal to $r \sin i_e$)

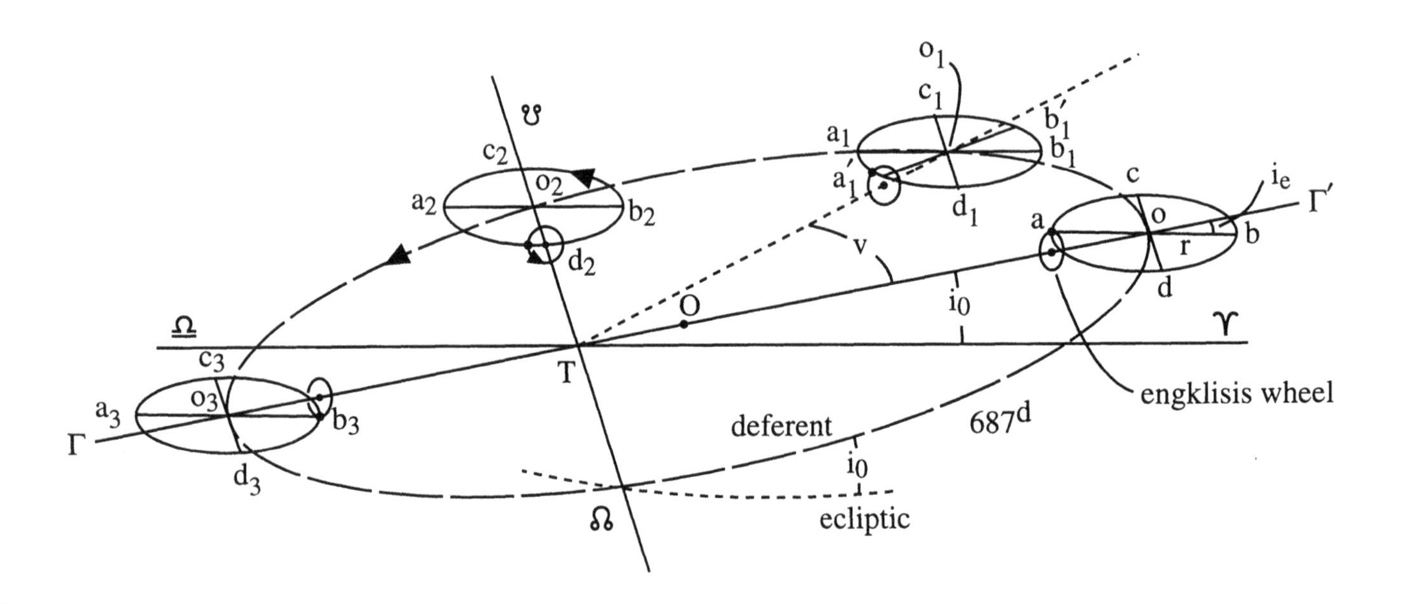

Figure IV.12 — Ptolemy's celestial latitude mechanism for a superior planet (Mars).

was in the plane perpendicular to the plane of the deferent.

As this small circle revolved (in a period equal to the period of revolution of the epicycle around the deferent), it caused the perigee of the epicycle to move up and down with respect to the plane of the deferent. This was Ptolemy's inclination (*engklisis*) of the epicycle. So when the perigee of the epicycle is at a, it is out of the plane of the deferent by its maximum amount; but at d_2 the perigee of the epicycle lies in the plane of the deferent. Now if i_e (the maximum inclination of the epicycle) were equal to i_o (the inclination of the deferent), then the effect of this mechanism would be to keep the plane of the epicycle very nearly parallel to the plane of the ecliptic. However, this is not the case with Ptolemy's latitude theory in the *Almagest,* since he puts i_e equal to $4°30'$ for Saturn, $2°30'$ for Jupiter, and $2°15'$ for Mars.

Along with uniform forward motion of the epicycle the connection between the epicycle and the engklisis wheel shifted gradually, to one side or the other in the plane of the epicycle, through an amplitude equal to the radius of the engklisis wheel. This last fluctuation was adopted to avoid interference with the celestial longitudes. Dreyer (1905) asks us to "imagine a stud on the circumference of the small circle and let it slide in a slot on the edge of the epicycle."

Regarding the motion of the epicyclic diameter *cd*, Ptolemy merely remarks that it always keeps parallel to the plane of the ecliptic. Obviously he thought it sufficiently clear how it could do so. In Figure IV.12, if v is the true anomaly measured from the line of apsides (zero at Γ', with period = 1 sidereal period), the approximate result of the engklisis for Mars is a tilt of the perigee of the epicycle in the amount $2.25° \cos v$.

We have described the latitude theory of the superior planets as presented in the *Almagest.* In his later *Planetary Hypotheses*, Ptolemy adopted a simpler system in which $i_e = i_o$ for each planet. Thus the plane of each epicycle did remain rigorously parallel to the ecliptic plane.

The Inferior Planets

For the motion in latitude of the inferior planets, which is even more complicated than that of Mars, the foregoing description must be replaced by that of a still more elaborate mechanism. In the case of the superior planets, Ptolemy had assumed that the diameter *cd* of the epicycle which was perpendicular to the line from the Earth T to the center of the epicycle o remained always parallel to the plane of the ecliptic. For the inferior planets, Ptolemy assumed that this diameter *cd* of the epicycle is also subject to a second rocking motion that he calls *loxosis*, the "slant" of the epicycle. Figure IV.13 shows the latitude theory for Venus. In addition to two tilts of the epicycle about a pair of perpendicular epicyclic diameters, a third tilt, the *obliquation*, of the deferent itself is added. The reason for the fluctuation of the deferent about its line of nodes on the ecliptic is as follows. Ptolemy had noticed from observations by himself and others that the latitude deviations for Venus were mostly toward the north and for Mercury always toward the south. Since for Venus the rule was violated only when the epicyclic center was near the deferent's line of nodes, Ptolemy concluded that the plane of the deferent must be fluctuating about its line of nodes on the ecliptic in a period of 1 planetary sidereal year, with an amplitude of $10'$ and null values at the nodes.

The mechanism explaining the celestial latitude deviation is identical for both inferior planets, except for different constants and opposite algebraic signs. The case for Venus is summarized in Figure IV.13, in which the line of apsides is assumed to be perpendicular to the line of nodes (also assumed to be the case for Mercury).

In Figure IV.13, let

ρ = radius of deferential hypocycle, which is the circle producing the $10'$ maximum obliquation,

R = radius of the deferent = 1.00,

r = epicyclic radius,

T = the Earth,

O = the mean position of the Sun's annual orbit (the "center of the ecliptic"),

$o = A$ = the epicyclic center when at apogee,

$o_4 = P$ = the epicyclic center when at perigee,

☊☋ = the line of nodes of the deferent on the ecliptic,

EE' = a line in the ecliptic plane orthogonal to ☊☋,

ab = the instantaneous line of apsides of the epicycle (always radial from T),

cd = the epicyclic diameter perpendicular to ab,

v = the true anomaly from the line of apsides (period = 1 sidereal year),

$\angle NTA = 90°$, assumed, for both inner planets.

The models for the three motions affecting the celestial latitude of the inferior planets are as follows:

1. The *obliquation* is a tilting of the plane of the deferent about its line of nodes. For example, for Venus ATP rocks about ☊☋ with an amplitude of $10'$, having a maximum when o is at A ($v = 0°$). Thus the deferent's inclination i is $10' \cos v$. (For Mercury $i = -45' \cos v$.) The effect is to make the center of Venus's epicycle always to be on or above the plane of the ecliptic.

2. The *engklisis* is an inclination or pitching of the epicyclic diameter ab about the perpendicular diameter cd. Due to this motion a moves up and down with amplitude λ_0. The value of the engklisis is zero at A and P and a maximum when o is at ☊ or ☋ ($v = 90°$). Thus the motion of the point a moves as $\lambda_0 \sin v = 2.5° \sin v$ (for Venus).

3. The *loxosis* is a slanting or rolling of the plane of the epicycle about ab, the instantaneous line of epicyclic apsides. Thus c moves up and down with amplitude ϵ_0. The magnitude of the loxosis is maximum at A and P and zero when o is at ☊ or ☋ ($v = 90°$). Thus the epicycle's sideways slant (or roll) is $\epsilon_o \cos v = 3.5° \cos v$, for the point c.

For a superior planet the engklisis correction was then entirely similar to (2) except for the amplitude and a phase change of 90°. The engklisis of a superior planet gives the plane of the epicycle a maximum tilt at the deferent's apsides, while for an inferior planet the plane always coincides with that of the deferent at its apsides.

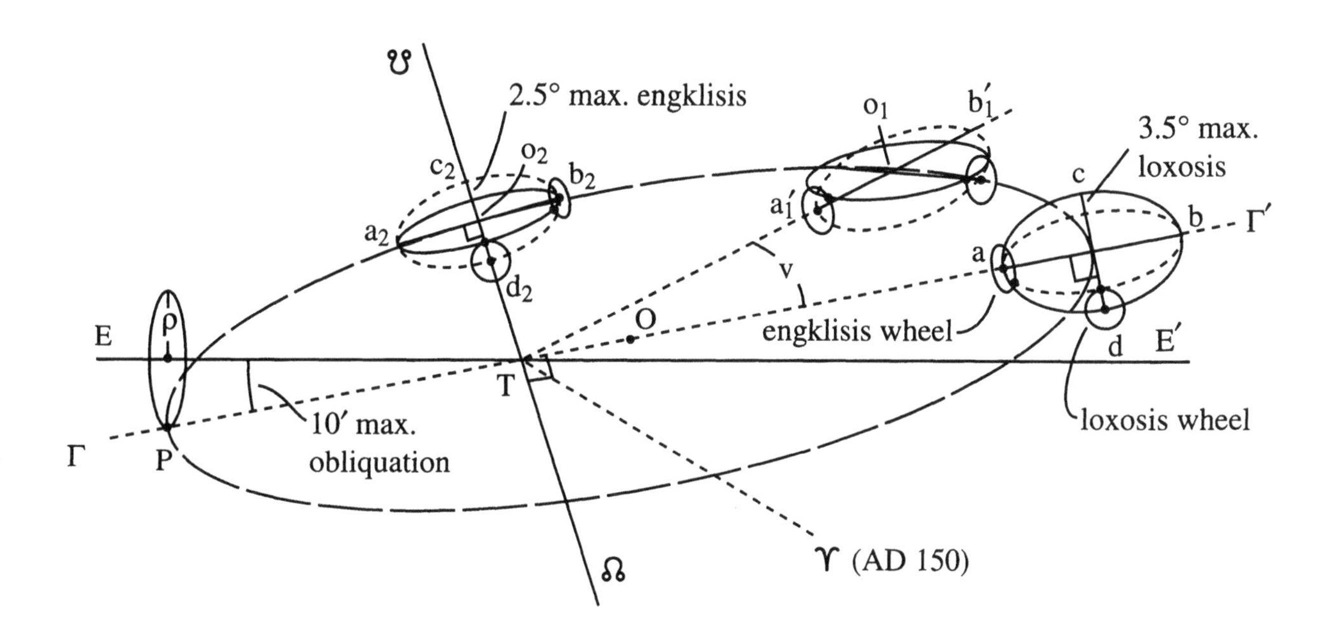

Figure IV.13 — Ptolemy's celestial latitude mechanism for an inferior planet (Venus).

The latitude theories of Venus and Mercury described here are those of the *Almagest.* Later, in his *Planetary Hypotheses*, Ptolemy employs no obliquation of the deferent planes. Both Venus and Mercury are given a constant inclination of 10′. The complex double rocking motion of the epicycle is also suppressed, with Mercury's epicycle given a constant inclination of 6°30′ to the plane of the deferent; Venus's inclination was 3°30′.

CHAPTER V

Copernicus (1473–1543)

However world-shaking and all encompassing have been the consequences of what has been called the Copernican Revolution – the gradual acceptance of the heliocentric idea of planetary cosmology and the abandonment in general of an anthropocentric viewpoint in nearly all human relations – what interests us here is rather to gain some idea of Copernicus *as an astronomer*, not as a cosmologist or a philosopher. After all, knowing something of his intellectual and altruistic character, one would expect him to be the first to minimize the value of mere assertion without proof of the heliocentric viewpoint.

The idea of a central position of the Sun in the planetary system was undoubtedly discussed during the high noon of the Renaissance, the very time when Copernicus studied in Italy. The work of Aristarchus, its greatest proponent in antiquity, was well known to Copernicus but not used by him. Suggestions of a moving (fluttering) Earth were explicit in the works of Cusa, Regiomontanus, and Novarra, whose works Copernicus certainly knew. What made Copernicus great was his purely astronomical activity and the patience which permitted him to work out, during nearly fifty years of study and computation, a specific heliocentric system of planetary motion that was as accurate (to about $10'$) as were the observations at his disposal. This, more than the heliocentric viewpoint as such, marks him as an astronomer. In the following pages, after a brief outline of his many varied activities, we shall try to describe some of the main astronomical processes leading to the final version of his system, as explained in his great work *On the Revolutions of the Celestial Spheres* (hereafter referred to simply as *Revolutions*).

Nicolaus Copernicus was born in 1473 as the youngest of four children of a prosperous immigrant Polish merchant and a rich bourgeois woman, Barbara Waczenrode, in the Hanseatic city of Torun, which had been four years under Polish suzerainty. Barbara's family, including Nicolaus's uncles on the maternal side, belonged to the opulent, typically German business aristocracy of the time. Much controversy has raged about whether Copernicus himself was a German or a Pole. It is true that he spoke and wrote German and never wrote Polish. It is fair to say that he was a "German-speaking Pole" who was only vaguely conscious of the distant overlordship of the Polish Crown. Copernicus himself considered his native country to be Ermeland, then a part of East Prussia, not Poland. His culture was distinctly that of a German humanist.

His father died when he was only ten years old, but Barbara's brother Lucas Waczenrode, a powerful bishop, immediately saw to it that Nicolaus continued his education. At age eighteen Nicolaus registered as a student of liberal arts in the Natio Germanorum of the University of Cracow, then the capital of Poland. There he studied law and also, perhaps mostly, Aristotelian astronomy. He used as textbooks Sacrobosco's *On the Spheres*, as well as Peurbach's *New Theory of Planetary Motion* based on Ptolemy's work. The astronomical instruction at Cracow was at that time better than any in Italy.

At age twenty-two his uncle named him a canon, which included a generous, lifetime annual income. This sinecure was, however, heavily weighted with duties, rules, and regulations. Thus only by special permission from the chapter could he absent himself for study abroad, and in later years he was often required to take care of various mundane affairs. He would occasionally have to visit and supervise the activities of the church farms, to collect rents, to advise and judge in legal matters among the peasants, and to take part in the higher decisions, even military ones, concerning the chapter. Nevertheless, the position allowed him ample time for study and contemplation.

In 1496 he obtained a leave of absence to study law at the Natio Germanorum of the University of Bologna, then the leading law school in Europe. In Bologna he also studied astrology under Domenico Maria da Novarra, to whom he became almost an assistant. For instance, they observed an occultation of Aldebaran by the Moon together. He also studied Greek and became an avid humanist, studying Greek philosophy and reading Ptolemy's *Almagest* in the original.

In 1500 he visited Rome and lectured with success on mathematics. He made important connections among the ecclesiastical hierarchy and discussed freely the great thoughts of the Renaissance then current in intellectual circles of the capital. The following year he registered as a medical student in the University of Padua, then the leading medical school in Europe. He studied medicine and law for two more years and finally returned home in 1503, when he became a secretary and special physician to his ailing uncle.

He now settled down to a quiet, diligent life of helping his uncle, succoring the sick (he was now well versed in medicine), and studying astronomy. In about 1506, he started the first serious work on his heliocentric system of the planets.

In 1510 Copernicus left his uncle's palace (with its many interruptions) and moved to nearby Frauenburg, where he had more peace for contemplation, study, and calculation. When friends importuned him about publishing something of his astronomical research on a heliocentric system, in 1510 he wrote and circulated a few copies of a pamphlet called *Commentariolus* presenting the main (early) features of his system. At this time Copernicus already visualized "a more ample work," probably his *Revolutions*. Copernicus started his detailed work on this great book in 1515, but it was not to appear until twenty-eight years later. In 1517 he was invited to join a commission for calendar reform as an astronomical expert. He declined to give any advice, holding that the time for a permanent settlement of the calendar was premature, in view of the inferior accuracy with which the motions of both the Moon and the Sun were known.

In spite of a voluminous literature by and about Copernicus, it seems difficult to form an intimate picture of his personality traits, probably because so few of his oral pronouncements have been documented. His later technical writings are always clear and to the point and reveal his excellent mind and training. He has been called "weak, timid and reactionary," but maybe this should have been "patient, nonaggressive, and loyal." That he was worldly-wise as well as a devout Catholic cannot be denied. He knew all about Luther's contemporary heresies, which spread with remarkable speed to Ermeland and Prussia. But he chose not to enter into these controversies. Perhaps one comes closest to a personal picture of the man from the following remarks by or about him:

1. His contemporary reputation as an astronomer is contained in a letter from Rheticus

to Schoener: "This man whose work I am now treating is in every field of knowledge and in mastery of astronomy not inferior to Regiomontanus. I rather compare him to Ptolemy, not because I consider Regiomontanus inferior to Ptolemy, but because he shares with Ptolemy the good fortune of completing, with the aid of divine kindness, the reconstruction of astronomy which he began, while Regiomontanus – alas, cruel fate – departed this life before he had time to erect his columns."

2. At one time Rheticus asked Copernicus why he did not observe often and more accurately. Why was he satisfied with results of $10'$ accuracy, obtained by using his wooden triquetrum, when contemporaries like Walther and Schoener, using instruments with steel circles now obtainable in Nürnberg, obtained results to an accuracy of $4'$–$6'$? "If I could bring my computations to agree with the truth to within $10'$, I should be as elated as Pythagoras must have been after he discovered his famous principle," answered Copernicus. Although he was capable of doing observational astronomy, he did not aspire to producing a complete overhaul of the science. He was primarily a philosopher who used his observations only to check certain constants of his system.

3. Copernicus knew that there were errors of $10'$ or more in some of the fixed-star positions in the catalogues of Hipparchus, Menelaus, and Ptolemy; he stated, "Today we have no great observational astronomers on whose results we can place greater reliability than on those of the ancients. Therefore I will rather be satisfied with stating results, the truth of which I can guarantee, than to parade learning and ingenuity by doubtful assertions of accuracy."

4. That Copernicus was greatly esteemed by his colleagues in the chapter at Frauenburg appears in this letter of 1542: "Since even in the days of his good health Copernicus has always loved seclusion, there will probably be only a few friends to succor him now. Yet we are all in debt to him, because of the purity and sincerity of his being, and his wide erudition."

By 1535 Copernicus had prepared new and much more accurate planetary tables to form the basis for computation of almanacs. He wished to publish a general almanac for 1536 and arranged for it to be sent to Vienna for publication. However, it was never published, possibly because in transmitting the manuscript his agent had indicated that Copernicus had for several years been convinced that a certain movement in the sky ought to be attributed to the Earth and that this movement could not be felt. Copernicus was comfortable with the idea of publishing tabular values of planetary positions but much less so with the prospect of provoking his colleagues by publication of new theories. Thus in 1536 Cardinal Nicholas Schönbeg offered to have copies made of Copernicus's works at his own expense, but this offer was not accepted.

Finally, in 1539 Rheticus persuaded Copernicus to let his great work be published. Rheticus suggested that he himself write out and publish as rapidly as possible a description of the new cosmology without giving the name of its author but referring to him always as either "der Hr. Doktor" or "mein Lehrer." Copernicus, probably reluctantly, agreed that this was a sufficiently cautious procedure. Anyway, Rheticus now wrote an excellent summary of the first six books of *Revolutions* (there were originally eight), this being as much as he had time to study in detail under the constant, solicitous supervision of Copernicus. This outline was called *Narratio Prima* and was published close by at Danzig in 1540, with another batch of copies appearing at Basel in 1541. Although the book presupposes considerable astronomical insight on the part of the reader, it is a masterly geometrical but otherwise almost nonmathematical account of the larger work. It has good diagrams and is still the best semipopular exposition of Copernicus's main concepts.

While *Narratio Prima*, written in 1539 by Rheticus, is a true outline of *Revolutions* in its nearly final state, *Commentariolus*, written in 1510 by Copernicus, is not. Both accounts recognize that Copernicus never let the center of a planetary orbit fall exactly in the center of the Sun, but rather always somewhat eccentric to it. Now, this can be effected in two ways: (1) As explained in *Commentariolus*, the center of the circular deferent of the planet (i.e., the orbit itself except for the epicyclic mechanism taking care of the elliptic inequalities of planetary motion) lies on the circumference of a small circular epicycle whose center is the center of the Earth's annual orbit and whose radius is the eccentricity of the deferent. (2) As in *Narratio Prima*, the center of the circular deferent lies eccentric to the annual mean position of the Sun, thus eliminating one circle for each planet.

Nevertheless, *Commentariolus* has been erroneously considered by some authors to be an early description of a version of *Revolutions* and to have been written by Copernicus in 1530. But even in the beginning, i.e., in the 1514 edition of *Commentariolus*, Copernicus did not let the center of any planetary orbit fall exactly in the center of the Sun. At a later time, perhaps prior to 1530, certainly in *Revolutions*, he switched to the second scheme.

Another characteristic difference between the two accounts is that in *Commentariolus* Copernicus considered, with Ptolemy, the celestial longitudes of the aphelia of the planetary orbits to be constant, while at a later stage (in *Narratio Prima*, certainly in *Revolutions*) he assumed them to be secularly increasing (though at a much smaller rate than is actually the case).

Both of these apparently conflicting viewpoints lend themselves to easy explanation when it is realized, first, that a circular direct eccentric deferent can always replace another circular direct concentric deferent plus a circular retrograde epicycle of the same period; and, second, that a secular motion of the aphelion can be adequately described by a uniform motion on a circular direct deferent plus another uniform motion on a circular retrograde epicycle, the epicycle being slightly longer in period if the line of apsides is advancing, shorter if regressing. Thus the differences between the descriptions of planetary motions in *Commentariolus* and in *Revolutions*, while conceptually conspicuous, are mathematically absent and may well reflect a simplification over time in Copernicus's mathematical description of his system.

By June 1541 Rheticus had persuaded Copernicus to actually publish the great work itself, with which he had been occupied for more than a quarter of a century. Copernicus now quit stalling and immediately made some final revisions of his manuscript with a view to publication. In 1542 he wrote his famous preface addressed to Pope Paul III, who appeared to be one of the most liberal leaders of Renaissance thought. In this preface Copernicus very briefly and qualitatively described his planetary system, with the Sun at the center of the world, and the Earth as one of the moving planets. He also expressed the hope that his dedication of the work to His Holiness would help avert ridicule of his system by the lay public.

Many speculations have been made concerning whether there were any precursors of Copernicus and, if so, whether he was influenced by them. During the years Copernicus spent at the universities in Italy, educated persons certainly freely discussed the various Greek theories of the planetary system. Besides the work of Martianus Capella, who in about A.D. 550 advocated Heracleides's view of the heliocentric orbits of Mercury and Venus, there was the translation (A.D. 1031) of a history of India by al-Biruni, who favorably described the completely heliocentric system of Aristarchus, in which he said Indian sages believed. The ideas of Capella are said to have been extended to the orbits of Mars and Jupiter by Duns Scotus (1266–1308) and may have been known to Copernicus. Even if al-Biruni's work may not have been generally known, the possibility that the heliocentric idea was prominently "in the air" during the Renaissance cannot be ruled out. According to J. Naiden (private communication), it was mentioned by at least twelve philosophers and scholars from Plato to Copernicus.

According to Dreyer (1905), Copernicus did not make the mistake, so common today, of thinking that Philolaus the Pythagorean (c. 450 B.C.) had advocated a motion for the Earth about the Sun. Rather, like all Pythagoreans, he thought of the motion of the Earth as describing an oblique path about a central fire, which was much closer than the Sun and about which the Sun also moved in its yearly orbit. It was instead Capella whom Copernicus credited with having first suggested to him that Mercury and Venus move in orbits about the Sun rather than between the Earth and the Sun, as assumed by Ptolemy. So we see that Copernicus was quite ready to give credit wherever it belonged.

But why, then, does Copernicus not mention Aristarchus of Samos (3rd century B.C.) at all in his preface, nor give him credit for being the first to have a heliocentric viewpoint? There are several opinions on this matter. Although we have one extant book by Aristarchus, we have only secondary sources regarding his heliocentric viewpoint. The best authority is an allusion to it by his contemporary Archimedes. According to the *Encyclopaedia Britannica,* Copernicus could not have known about Aristarchus's idea because Archimedes's books were not published in Latin until 1544. This is too categorical a statement, since Copernicus was himself a Greek scholar and might conceivably have seen the Greek manuscript before its translation and publication, but I can accept that possibly he did not rely on the astronomical work of Archimedes.

However, Aristarchus's viewpoint is also described in a treatise on the lunar orbit by the astronomer Plutarch (c. A.D. 400), whose work Copernicus admits having read. Plutarch mentions that Aristarchus was almost accused of impiety for having set "the Hearth of the Universe" in motion. That the Earth was set in motion at the same time seems to have been a less serious consequence! This shows that it was not the Pythagorean motion about the central fire that was contemplated by Aristarchus, for Hestia, the central fire, was the Hearth of the Universe, and the Earth moved with it, as well as around it. This is one place where Copernicus may have seen that Aristarchus's system was a truly heliocentric one in which the Hearth of the Universe and with it the Earth moved about the Sun. In another place Plutarch states that Aristarchus considered his system only a hypothesis useful for explaining motions in the sky, but that Seleucus, his student, afterward maintained that the Earth's motion about the Sun was a real physical motion.

In view of these facts then, why did Copernicus not give Aristarchus any credit at all, while he singled out three Pythagoreans, Hicetas, Heracleides, and Ecphantus, who merely believed in the Earth's rotation on its axis or (for Heracleides) in two rotations, one about its axis and one obliquely about the central fire, which was believed to be not much farther away than, say, 10,000 kilometers? Before we can answer this question we should note one more fact. In the original manuscript of *Revolutions*, immediately following the chapter explaining the motions of the Earth, there are two and a half pages saying that Aristarchus is of the same view as Philolaus, but these have been heavily crossed out by Copernicus himself. Why did he cross this out? There may be at least two reasons: (1) he had by then realized that Philolaus's view was in fact not heliocentric, and (2) he did not want to bring in Aristarchus at all.

I see at least two good reasons for the second view. The details of Aristarchus's explanation are entirely missing from any of the records. Thus Copernicus would have been justified in considering Aristarchus's hypothesis as a mere empty assertion, a hypothesis, without any proof and not worthy of scientific adoption. This was in contradistinction to Copernicus's theory, for which he had shown by observation and computation that it fit the motions in the sky to about the same accuracy with which they could be observed. Furthermore, to focus on Aristarchus's ideas would be to call attention to someone who in about 250 B.C. nearly got in trouble with the authorities for ascribing spatial motion to the solid Earth – the very thing Copernicus was going to do with his *Revolutions*! Although Leo X, Clement VII, and other high authorities seemed to listen with open minds to Copernican ideas, and Paul III apparently gracefully received his dedication, Copernicus undoubtedly knew that he was skating on thin ice and thus refrained from publishing the book for twenty-seven years.

In 1539 Rheticus took Copernicus's manuscript to Wittenberg and tried unsuccessfully to have it published. Although he was personally welcomed and admired there for his mathematical ability and was soon to be elected dean of the university, he quickly ferreted out that the views of the leaders of Protestant thought (Melanchthon and the elector of Saxony) were adverse to the publication of Copernicus's work. As a result of his great contributions to trigonometry (for instance, a ten-place table of trigonometric functions), Rheticus left Wittenberg in 1542 to take up the professorship of mathematics at the University of Leipzig. On the way he stopped in Nürnberg and arranged for the publication of Copernicus's manuscript with the publisher Petreius, who had earlier admired Rheticus's *Narratio Prima* and even expressed the hope that the larger work might soon be published. Rheticus himself supervised publication of the first part of the work but then had to leave town. To finish the job Petreius turned over proofreading to his friend Andreas Osiander, an eminent Lutheran pastor, who was also an excellent mathematician. Osiander modified Copernicus's preface by adding a statement that the work was a mere mathematical device for predicting planetary positions and therefore did not need to be considered a *physical* system of cosmogony. Rheticus immediately flew into a rage over this chickenhearted reservation. However, it is said that Osiander conferred with Copernicus about the change in the preface and obtained his full approval. Much controversy has flared over this point. It seems that if Copernicus only half approved, he could easily and should definitely have said so in a letter to Osiander. But if he did send such a letter, it has been lost, so that we have no real evidence that he disapproved. Considering his general lifestyle, perhaps even this much self-assertion would have been distasteful to the gentle master. Furthermore, he suffered a stroke about this time. Considering his probable subconscious lifelong fears of persecution (heretics were then frequently tortured or burned at the stake), it seems reasonable to believe that Osiander's preface provided Copernicus with a very welcome escape from the necessity of having to give too positive a statement of his personal viewpoints. In reading his words throughout *Revolutions*, one cannot doubt that he believed in the possibility, even the probability, of his system being a physical reality. Moreover, in reading his biography, one becomes equally convinced that as an admirer of Pythagoras he would have been content to have his knowledge diffused to only a dozen or so interested and devoted friends, capable and eager to understand him in detail. Perhaps he was not so much timid as aristocratically disinterested in a wide diffusion of knowledge.

In *Revolutions* Copernicus constantly makes numerous and positive statements and adduces all possible evidence affirming the real motion of the Earth in space, as well as its axial rotation. All this did not "prove" his system but only made it appear very probable and slightly better than Ptolemy's. Moreover, no one was more conscious than he of the increasingly glaring astronomical shortcomings of his own system, which became more inextricably deficient with his latest complicated modifications (such as the mind-boggling "obliquations" and "deviations" of the planet orbits). No wonder if this brilliant recluse finally grew tired of his whole life's effort and only retained the joy of the conviction that his planetary tables were somewhat better than the Alfonsine tables then in vogue.

Beneath many modern efforts to discern Copernicus's own true attitude toward the publication of his work there seems to be one pervasive viewpoint, namely, that Copernicus *must* have desired the publication of his revolutionary ideas, since that meant progress. But he was by no means of a revolutionary temperament. It seems more probable that he really and truly did not wish to bother and upset his colleagues; that he aristocratically wished to "preserve his secrets" as to how his somewhat superior ephemerides were computed; and that he was perfectly happy in his comfortable, if somewhat obscure, ecclesiastical position, where he enjoyed a reputation as a devout Roman Catholic, a fine and helpful physician, a good administrator, and an astronomer of the very first rank. He most certainly did not wish to be a martyr for anything. As he himself stated, "only the initiated and the capable deserve knowledge." That he subscribed to what were during the Middle Ages considered "shallow," if not "subversive," concepts such as "social progress" or "democratic education" seems far-fetched. These concepts were not driving forces of human thinking until later centuries. Furthermore, he probably knew which way the wind was blowing in the ecclesiastical world and was in his later years aware of the gradual change toward increasing intellectual intolerance accompanying the advent of the Counter-Reformation. The Inquisition was reinvigorated in 1542 by Pope Paul III, to whom Copernicus had dedicated his main work. The ecclesiastical reaction that after his death reached its first important official expression at the Council of Trent (1545–63) had already begun during his lifetime and grew steadily in rancor until, after the burning of Bruno (1600) and around the time of the persecution of Galileo (1616), it condemned both Copernicus's *Revolutions* (1615) and Kepler's *Epitome of the Copernican Astronomy* (1618) to the Index of Prohibited Books.

Copernicus's Main Astronomical Contributions

1. Explanation of the diurnal rotation of the celestial sphere as a consequence of the **axial rotation of the Earth**.

2. Explanation of the Sun's annual apparent motion in the ecliptic and of the stations and retrograde motions of the planets near opposition as a consequence of the **Earth's annual orbital revolution** about the mean position of the Sun (which mean position was considered the center of the world).

3. Explanation of **planetary motions from a heliocentric standpoint**, using (in *Commentariolus*) only 34 circles (as against Ptolemy's 57). However, he **abandoned equant motion**, a concept so very useful in simulating elliptic motion!

4. Clear statement of the **modern view to explain retrograde motions** of the planets.

5. Determination of **relative distances of the planets** from observations of their aspects.

6. Adoption (in *Revolutions*) of eccentric circular deferents for the planets, thus **decreasing the number of circles by five**, one for each planet (as described in *Narratio Prima*).

The Copernican System of the Sun, Moon, and Planets

A common misconception, in fact one of the incentives for the present attempt at elucidation, is the often repeated statement that Copernicus simply considered the Sun to be at the center of circular (concentric) planetary orbits, and the Earth to be at the center of the Moon's orbit. This gives only a very crude picture of the real situation, since in fact his system was far more elaborate.

This simplistic view probably derives from the assumption that his main technical ideas can be gleaned from the simple preface that he wrote to his great work in 1542 and that Pope Paul III seemed to gracefully accept. The following portion of the preface shows why people who read only the preface felt satisfied and never fathomed the astronomical difficulties confronting the technical worker in the field. One would jump to the conclusion that the aesthetic illustrations given in the preface were a true description of the system.

> How I came to dare to conceive the motion of the Earth, contrary to the received opinions of the mathematicians, and to the impression of the senses, is probably what Your Holiness will expect to hear. In determining the motions of the Sun, Moon, and planets, the mathematicians do not all use the same principles in their proofs of seeming revolutions and motions. Some use concentric circles, others use circles with centers slightly to one side of the Earth, and still others such eccentric primary circles with secondary circles, whose centers slide on the circumferences of the primary ones. A great many complicated motions can be devised from these systems, but no one has been able to establish a system in which the calculated motions agree fully with the observed phenomena. Those who use eccentric circles have come out tolerably well, but that only by violating the holy principle of uniformity of motion. I therefore took pains to read again the works of all the philosophers on whom I could lay hand, to seek out whether any of them had ever supposed that the motions of the spheres were other than those demanded by the mathematical schools. I first found in Cicero that Hicetas had realized that the Earth moved. Afterward I found in Plutarch the following quotation: Philolaus the Pythagorean says that the Earth moves around the central fire on an oblique circle like the Sun and Moon. Heracleides of Pontus and Ecphantus the Pythagorean also make the Earth to move, not indeed through space but by rotating about its own center, as a wheel on an axle, from west to

> east. Taking advantage of this I too began to think of the mobility of the Earth; and though the opinion seemed absurd, yet knowing now that others before me had been granted freedom to imagine such circles as they chose to explain the phenomena of the stars, I considered that I might be allowed to try whether by assuming some motion of the Earth, sounder explanations than theirs might be so discovered.

Then he goes on with his book. Some of the arguments are prejudiced or aesthetic; for instance, he says that the Universe is spherical because such a figure is the most perfect. Many of his arguments are, however, the same as we have today. For instance, the Earth is spherical because as we pass equal distances in a north-south direction, the stars in the middle of the sky change altitudes by proportional amounts.

In the first part of *Revolutions* (book I, chapter 10) he describes his world system as follows:

> Above all lies the sphere of the fixed stars, containing itself and all things, and for that very reason immovable; it is the frame of the Universe, to which the motion and position of all other bodies are referred. Though some men think it to move, we assign a reason why it only appears to do so; namely, in our theory of the movement of the Earth. Of the moving bodies first comes Saturn, which completes its circuit in thirty years. After that, Jupiter, moving in a twelve-year revolution. Then Mars, which revolves biannually. Fourth in order, an annual cycle takes place, in which is contained the Earth, with the lunar orbit. In the fifth place Venus is carried around in nine months. Then Mercury holds the sixth place, circulating in the space of eighty-eight days. In the middle of all dwells the Sun. Who indeed in this most beautiful temple would place the torch in any other or better place than one whence it can illuminate the whole at the same time? Not ineptly some call it the lamp of the Universe, others its mind, and others its ruler.

Copernicus not only postulated motions for the Earth but seemed to prove them by strict geometrical constructions. His results were for the most part no more accurate than those of Ptolemy, but in one sense they were inferior, for he abandoned the principle of the equant in favor of uniform motions. It is really remarkable that the idea of the Earth's motion in space was not used much earlier. Ptolemy had already found that, though the "invisible hoops" representing the planet orbits were tilted at small angles to the Sun's orbit, the "coils" on the hoops, responsible for the retrograde motions of the planets, were all approximately parallel to the plane of the *Sun's* orbit. Furthermore, the coils appeared smaller, and the backward motions slower and longer lasting, the greater the distance of the planet. Even at the time of Ptolemy any unprejudiced person should have seen that these coils simply represented a reflection on the background of the sky of a heliocentric orbit followed by the Earth over one year. This insight escaped Ptolemy, but fourteen centuries later it came to Copernicus with sufficient clarity to be mathematically proved by him! He could even calculate the exact time when any planet would appear stationary, and start or finish its forward and backward motions in the sky.

As a first inkling of the complications Copernicus met with in his detailed astronomical work, we may note that in the first version of it, described in *Commentariolus*, there were 34 circles involved: 3 for the Earth, 4 for the Moon, 7 for Mercury, and 5 for each of the remaining planets (Venus, Mars, Jupiter, and Saturn). In his final work Copernicus reduced the number of circles by 5, one for each planet. Unlike Ptolemy, whose aim was to keep the planes of the planetary epicycles parallel to that of the ecliptic, Copernicus aimed to keep them parallel to the (inclined) planes of their deferents. However, all the lines of nodes of the deferents passed (erroneously) though the *mean* annual position of the Sun – not through the *current* position of the Sun. This device alone, quite apart from introducing errors in distances, gave rise to unmanageable discrepancies in the heliocentric latitudes.

In an attempt to remedy the errors of the celestial latitudes, Copernicus introduced small periodic fluctuations called *obliquations* in the inclinations of the planetary deferents (except for the Earth). For the orbits of the interior planets he also had to introduce periodic rocking motions about their lines of apsides on the deferents, similar to Ptolemy's loxosis of the epicycles. The orbits of Mercury and Venus also had a fluctuation of the planes of the epicycles, the so-called *deviation*, which was a further periodic tilting of these already rocking orbits. Thus, as Dreyer (1905) points out, the Copernican system was if anything *more* complicated than the Ptolemaic system, even if its description involved a somewhat smaller number of circles. The motion of an inferior planet was as complicated as that of a person circling an island in a boat that was simultaneously rolling and yawing on an ocean strongly affected by complex tides.

Figure V.1 (not to scale) shows the projection on the plane of the ecliptic of some of the 29 circles, which in most cases are at slight tilts about their individual lines of nodes. Thus the figure is a diagram of the Copernican system in "curtate" form. Let

S = the actual Sun.

$C = c_3$ = the A.D. 1500 position of the Earth's orbital center.

$c_1, \dots, c_6$ = A.D. 1500 positions of the centers of the deferents of the six planets: Mercury, Venus, Earth, Mars, Jupiter, Saturn, respectively. The motions are eastward (i.e., counterclockwise as seen from the north side of the ecliptic) on all circles, except for the first epicycle of the Moon and the auxiliary circle (with center c_7) defining the secular change of the Earth's orbital eccentricity.

Υ = the direction of the vernal equinox of A.D. 1500.

$\Gamma_\oplus \Gamma'_\oplus$ = the direction of the line of apsides of the Earth's orbit, A.D. 1500.

$\angle \Upsilon C \Gamma_\oplus = 96°$ = the heliocentric longitude of the perihelion of the Earth's orbit, A.D. 1500.

Note: The lines of nodes of all deferents pass through C, not through S, as would have been a closer representation of reality. All planetary deferents (except that of the Earth) show obliquation; i.e., their inclinations oscillate about a mean value with simple harmonic motion of small amplitude about their lines of nodes. In addition, the orbits of the inferior planets suffer loxosis (rocking) about their instantaneous lines of apsides and also deviation, an additional obliquation-like motion about a uniformly rotating line of nodes lying always in the instantaneous obliquation plane. The line of nodes advances in that plane with a period which ensures a start (i.e., a zero value while passing from south to north) whenever a certain configuration of Sun, Earth, and planet attains. Finally, Mercury oscillates radially with simple harmonic motion, and the center of its deferent describes the small circle centered at c_8. The average obliquation plane was considered the orbital inclination plane. The plane

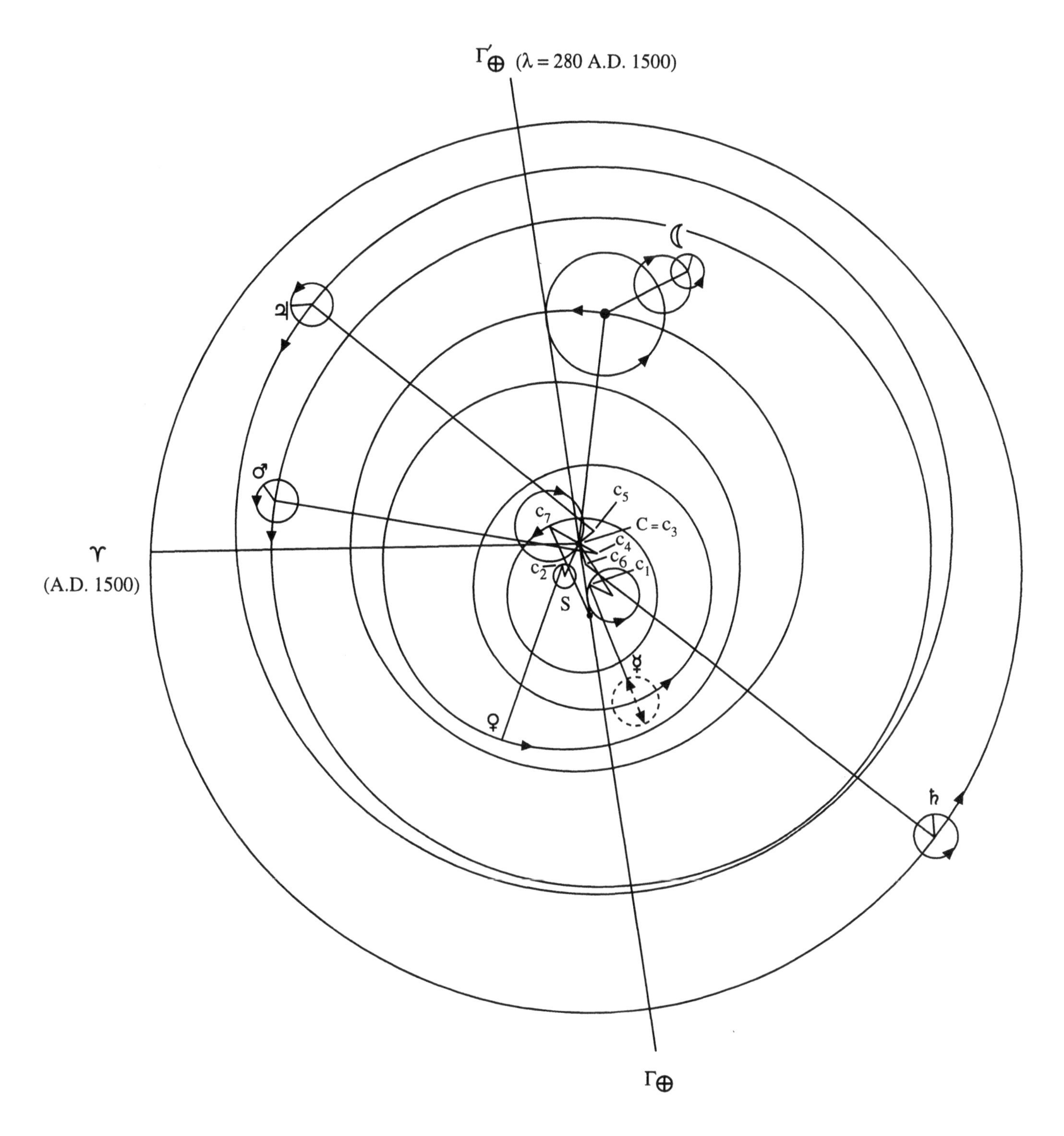

Figure V.1 — The Copernican system of the Sun, Moon, and planets. A curtate view projected on the plane of the ecliptic as seen from the north (not to scale).

thus defined was wrong in two respects: (1) it passed through the mean position of the Sun, not through its current position; and (2) the numerical value of its inclination was therefore wrong, though fairly close to modern values.

Copernicus's Solar Theory

Copernicus's solar theory, or the theory of the Earth's annual revolution about the Sun, is essentially an adaptation to the heliocentric viewpoint of Hipparchus's eccentric theory of the Sun's annual motion. A technical improvement was, however, added by Copernicus; namely, to determine orbital elements he used solar observations about halfway between the equinoxes and the summer solstice rather than observations taken around the time of summer solstice. This improvement depends upon the greater rate of change of solar declination on 5 May ($13'$/day) and 5 August ($16'$/day) than near 21 June and 21 December ($\frac{1}{2}'$/day). The moment of the Sun's passage through a point of the ecliptic 45° from the solstices can thus be determined much more readily and accurately than can the time of the Sun's passage through the solstice itself.

Elementary Considerations

In Figure V.2 the Earth's annual orbit about the Sun is represented by the eccentric circle PHA, center C, and is described with uniform velocity by the Earth T. As the Earth will be at H on 21 March the Sun S will be seen in the direction of the vernal equinox ♈. The Earth will be at F, K, B at the summer solstice, the autumnal equinox, and the winter solstice, respectively, and the Sun will be seen successively in the directions FS, KS, and BS. As the Earth moves uniformly through the semicircle AKP, the Sun will apparently describe the semicircle PHA with continuously varying apparent eastward velocity. The mean daily motion will be about 0.99°/day. The apparent daily motion of the Sun will vary from about 0.97° when seen at S_1 to 1.01° when seen at S_6. Let us illustrate the nonuniform apparent angular motion of the Sun in the ecliptic, resulting from the uniform circular motion of the Earth about C, which is eccentric to the position of the Sun by the distance CS. We choose six positions of the Earth in its orbit: A, E, D, K, B_1, and P, such that $\angle ACE = \angle DCK = \angle B_1CP = 20°$, say. From these positions in space the Sun will be seen in the directions S_1, S_2, S_3, S_4, S_5, and S_6, respectively. The corresponding unequal, eastward apparent angular motions of the Sun will be $\angle S_1SS_2$, $\angle S_3SS_4$, and $\angle S_5SS_6$. A rough measurement of these angles from the figure results in the values 13°, 19°, and 29°, respectively, for the eccentricity CS/AC chosen in the illustration.

Simple Method of Finding the Earth's Orbit

In Figure V.3 is shown a direct elementary method based on Hipparchus's work of finding the elements of the Earth's circular eccentric orbit, as well as the Earth's distance from the Sun on any day of the year. This method requires observations of the two equinoxes and of one solstice. The Earth's orbit is the circular eccentric PHA, whose center C is a point in the ecliptic plane. To find the eccentricity $e = CS$, the longitude of perihelion ω, and the distance ST of the Earth from the Sun at any point T, proceed as follows:

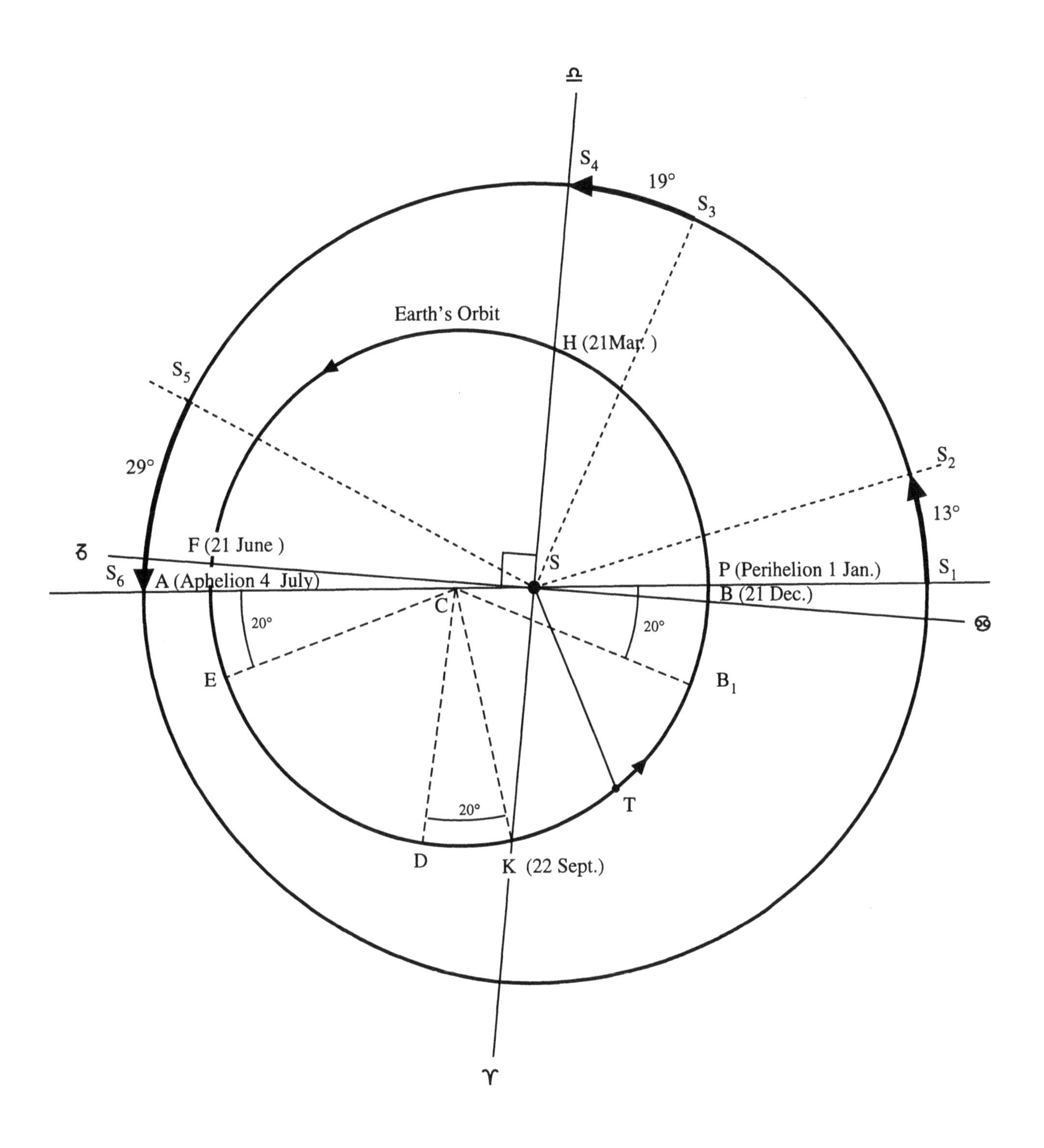

Figure V.2 — Copernicus's solar theory.

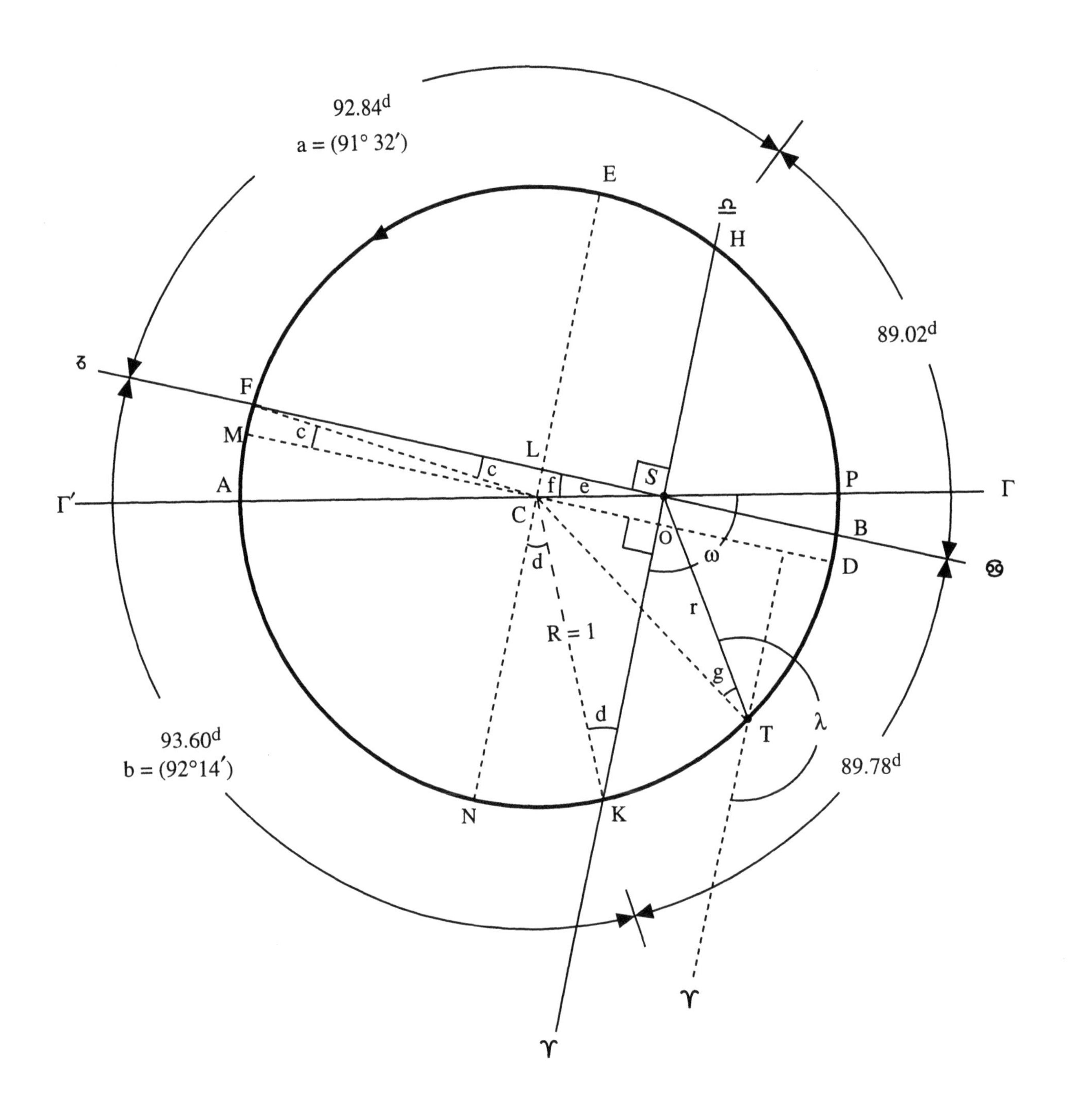

Figure V.3 — Copernicus's elementary theory of the Earth's orbit and of the Sun's position in the ecliptic plane.

Through C draw $NE \parallel KH \parallel T\Upsilon$. Draw $DM \parallel BF \perp S\Upsilon$. Draw CF, ST, CK, and CT as shown. Let the radius R of the circle be unity.

From observations of the Sun's longitude λ at the equinoxes K, H, and solstices B, F, the following arcs are known: $HF = 91°32' = a$; $FK = 92°14' = b$.

To find the position of S and the eccentricity $e = \frac{CS}{CP}$, proceed as follows:

1. $\widehat{FM} = \widehat{HM} - \widehat{HF} = \frac{1}{2}\,(\widehat{HF} + \widehat{FK}) - \widehat{HF} = \frac{1}{2}\,(a + b) - a = \frac{1}{2}(b - a) = c.$
2. $\widehat{NK} = \widehat{KM} - 90° = \frac{1}{2}\,(a + b) - 90° = d.$
3. From right ΔCFL find $CL = R \sin c = R \sin \frac{1}{2}(b - a)$.
4. From right ΔCOK find $CO = R \sin d = R \sin\,[\frac{1}{2}\,(a + b) - 90°] = R \cos\,[\frac{1}{2}\,(a + b)]$.
5. Since $LS = CO$, we can now find our first objective, $e = \sqrt{CL^2 + LS^2}$.

By inspection of the figure, we find the longitude of perihelion

$$\omega = \angle \Upsilon SP = 90° + \angle BSP = 90° + f,$$

where

$$f = \angle LSC = \sin^{-1}\frac{CL}{e} = \tan^{-1}\frac{CL}{LS}.$$

Now proceed to find the distance ST of the Earth from the Sun on any date when the geocentric celestial longitude of the Sun is λ. Thus $\angle \Upsilon TS = \lambda$ is given. By inspection,

$$\angle CST = \angle CSO + \angle OST = \angle PSH + (\lambda - 180°) = (180° - \omega) + (\lambda - 180°) - \lambda = \omega.$$

Also, $CT = R$ and $CS = e$. Thus in ΔCST three parts are given. Find g from

$$g = \sin^{-1}\,[(e/R) \sin CST].$$

Then

$$\angle SCT = 180° - (\angle CST + g) \qquad \text{and} \qquad ST = R\,\frac{\sin SCT}{\sin CST},$$

which solves the problem.

Copernicus's (Improved) Method for Finding Eccentricity and Aphelion of the Earth's Orbit

In regard to the above method of finding e and ω, Copernicus correctly remarks that the direct determination of the moment of occurrence of a solstice from altitude measurements, which forms the basis for the method, is extremely inaccurate because of the very slow change of the Sun's declination near the time of solstice. That the method practically fails in actual use if the observational accuracy is only $10'$ can be seen from the fact that the Sun's declination changes by only $5'$ during the five-day interval on either side of the solstice! Even with an accuracy of $1'$, the moment of the solstice cannot be determined to closer than four days, the period during which the declination remains within $1'$ of its stationary value.

In the case of the equinoxes, the moment of their occurrence can be very accurately determined from observations taken for some days before and after the moment of equinox, when the solar declination changes almost linearly by about 0.4°/day. Copernicus chose to use observations of the Sun's declination at a point of the ecliptic far enough removed from the solstice to ensure an appreciable rate of change of declination to permit high accuracy, and yet far enough removed from the equinox to permit an accurate determination of the position of the orbit's center. He therefore chose the moments corresponding to a longitude $\lambda_\odot = 45°$, as computed with the then known inclination of the ecliptic of 23°52′. Thus he used observations around 5 May at a declination of $\delta = \sin^{-1}(\frac{\sqrt{2}}{2}\sin 23°52')$, or about +16°. The observations had to be corrected for refraction or, if this was unreliable, results from a second set of observations (when the Sun was at the corresponding point of the ecliptic halfway between summer solstice and autumnal equinox, i.e., about 5 August) could be averaged with the May results.

Copernicus's method is outlined below. In Figure V.4, let

S = the Sun at geocentric longitude $\lambda_\odot$,

C = the center of the Earth's orbit,

A = the aphelion of the Earth's orbit (heliocentric longitude λ_A),

E_1, E_2, E_3 = the positions of the Earth in its orbit at the vernal equinox, about halfway between the vernal equinox and the summer solstice, and at the autumnal equinox,

r = the radius of the Earth's orbit,

$e = SC$ = the eccentricity of the Earth's orbit,

τ_1, τ_2, τ_3 = the moments when the Earth is at E_1, E_2, E_3. These quantities are found by interpolation of the times when the Sun's declination equals 0°, $\sin^{-1}(\frac{\sqrt{2}}{2}\sin 23°52')$, and 0°, respectively.

$\odot$ = the mean daily sidereal motion of the Earth in heliocentric longitude, or the constant of angular motion about C,

a, b, c, = the angles of the $\Delta E_1E_2E_3$,

α = the true heliocentric motion of the Sun from the time of the vernal equinox to the epoch E_2. This value could be taken as 45°, or in a general case (as in the figure) it can be found by interpolation of $\lambda_\odot$ (for the moment considered among the observed geocentric longitudes of the Sun). Thus α is given by the observations.

The problem is the following: Given τ_1, τ_2, τ_3, and α, find e and λ_A, where $\odot = 360°/365.26$ days $\approx 0.99°$/day. From the given data,

$$2a = (\tau_2 - \tau_1)/\odot, \qquad 2b = (\tau_3 - \tau_2)/\odot, \qquad 2c = (\tau_1 - \tau_3)/\odot.$$

From the figure,

$$\angle E_2E_3E_1 = a, \qquad \angle E_2E_1E_3 = b, \qquad \angle E_3E_2E_1 = c = 180° - (a+b).$$

Let

$$\angle CE_2S = y, \qquad E_1E_2 = n,\ SE_2 = q, \qquad \angle CSE_2 = x.$$

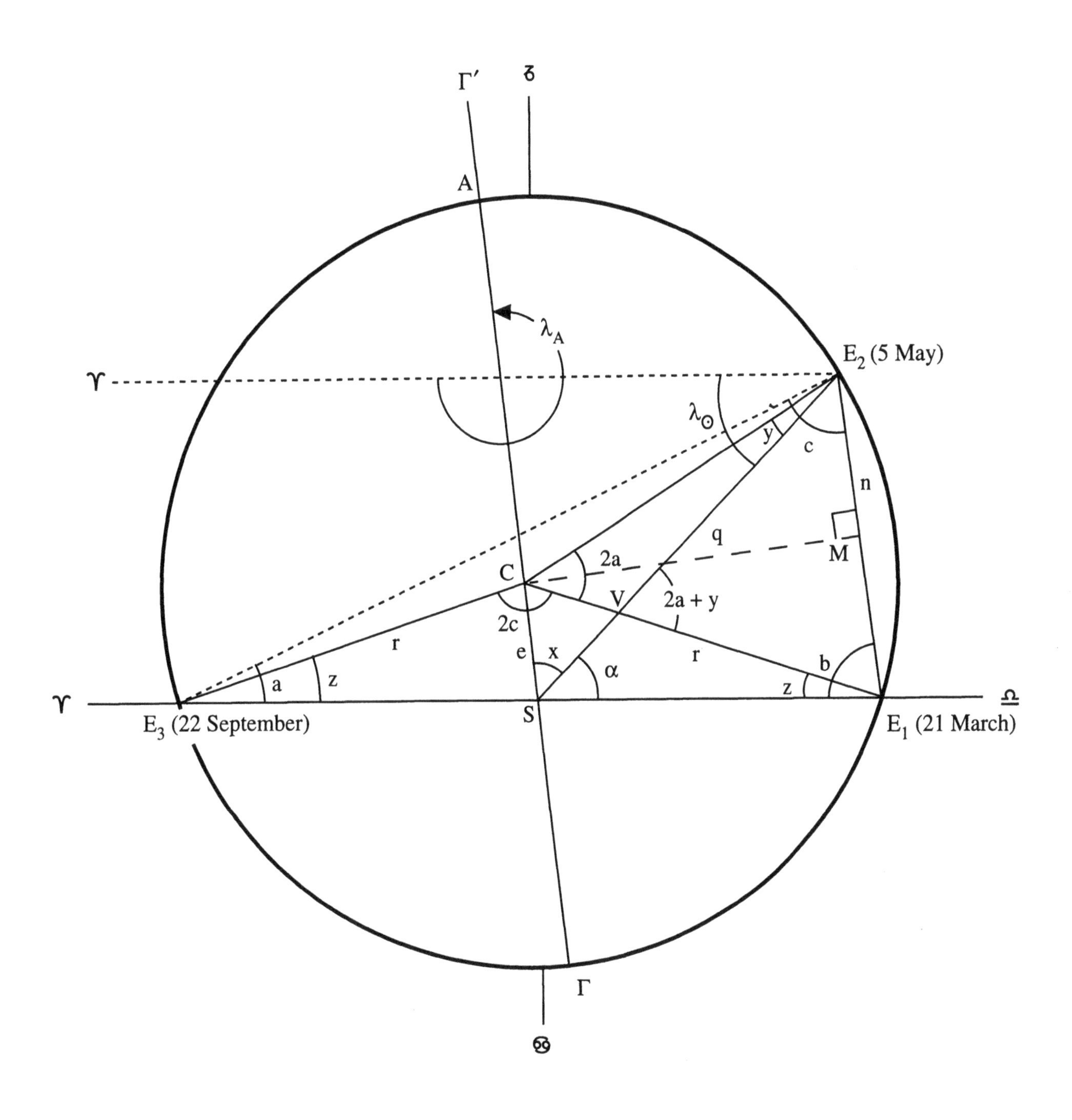

Figure V.4 — Copernicus's determination of the Earth's orbital aphelion and eccentricity.

From ΔE_1CE_3,

$$180° = 2c + 2z \qquad \text{or} \qquad z = 90° - c,$$

from which z. From ΔE_1VS,

$$\alpha + z = 2a + y, \qquad \text{or}$$

$$\begin{aligned} y &= \alpha + z - 2a \\ &= \alpha + (180° - c - 2a) - 90° \\ &= \alpha + (b - a) - 90°, \end{aligned}$$

from which y. From right ΔE_2MC, $n = 2r \sin a$, and from ΔE_1SE_2, $q/n = \sin b/\sin \alpha$, so that $q = 2r \sin a \sin b/\sin \alpha$, from which q. From ΔE_2CS, $e \sin x = r \sin y$, and $e \cos x = q - r \cos y$, from which e and x. Then $\lambda_A = 180 + \alpha + x$, which solves the problem. The resulting values were $\lambda_A = 96°40'$ and $e = 0.0323$.

Copernicus's Orbit of the Earth

By his accurate method described above, Copernicus found that the heliocentric longitude of the Earth's aphelion was 96°40′ (for the epoch A.D. 1500). He then compared this result with ancient values for the geocentric longitude of the apogee of the Sun's orbit (Hipparchus had determined, and Ptolemy later adopted, the value 65°30′). His conclusion, following Thābit ibn Qurrah (9th century), was that the heliocentric longitude of the Earth's aphelion was increasing. But he thought erroneously that, since Ptolemy adopted Hipparchus's value, there had been no change over the three centuries between Hipparchus and Ptolemy. He thought the value might have been decreasing at a still earlier epoch. To explain all this he adopted the arrangement shown schematically (though not to scale) in Figure V.5.

The Sun S is at the center of the world. Around it a point A moves with uniform direct motion in a circle of radius SA with a period of 53,000 years. This period corresponds to that of the advance of the line of apsides of the Earth's orbit. In addition the center B of the Earth's orbit moves with retrograde uniform motion in a period of 3434 years on a circular epicycle of radius AB centered at A. This latter period corresponds to the variation of the obliquity of the ecliptic. Due to the motion in the smaller circle, the eccentricity of the Earth's orbit (the length of SB) also changes with a 3434-year period. The Earth itself moves with uniform, direct motion in its one-year circular orbit of radius BT.

In his *Narratio Prima*, Rheticus gave an astrological significance to the changes of the Earth's orbit. When the center B of the orbit was at *1* (c. 100 B.C.), causing the eccentricity BS to be a maximum, the Roman republic was starting to degenerate into an empire, which then declined and vanished. When the eccentricity was a mean value as at *2* (c. A.D. 700), Mohammedanism emerged and gave rise to another great empire, which, alas, had not yet disappeared, but which, he hoped, would disappear when the eccentricity reached a minimum, as at *3*, in the seventeenth century. Finally, when the eccentricity reached its next mean value as at *4* (c. A.D. 2500), the Second Coming of Christ might be expected, for the Earth's orbit would then be in the same position as it was at the creation of the world some 6000 years earlier! Nothing of this is mentioned by Copernicus himself.

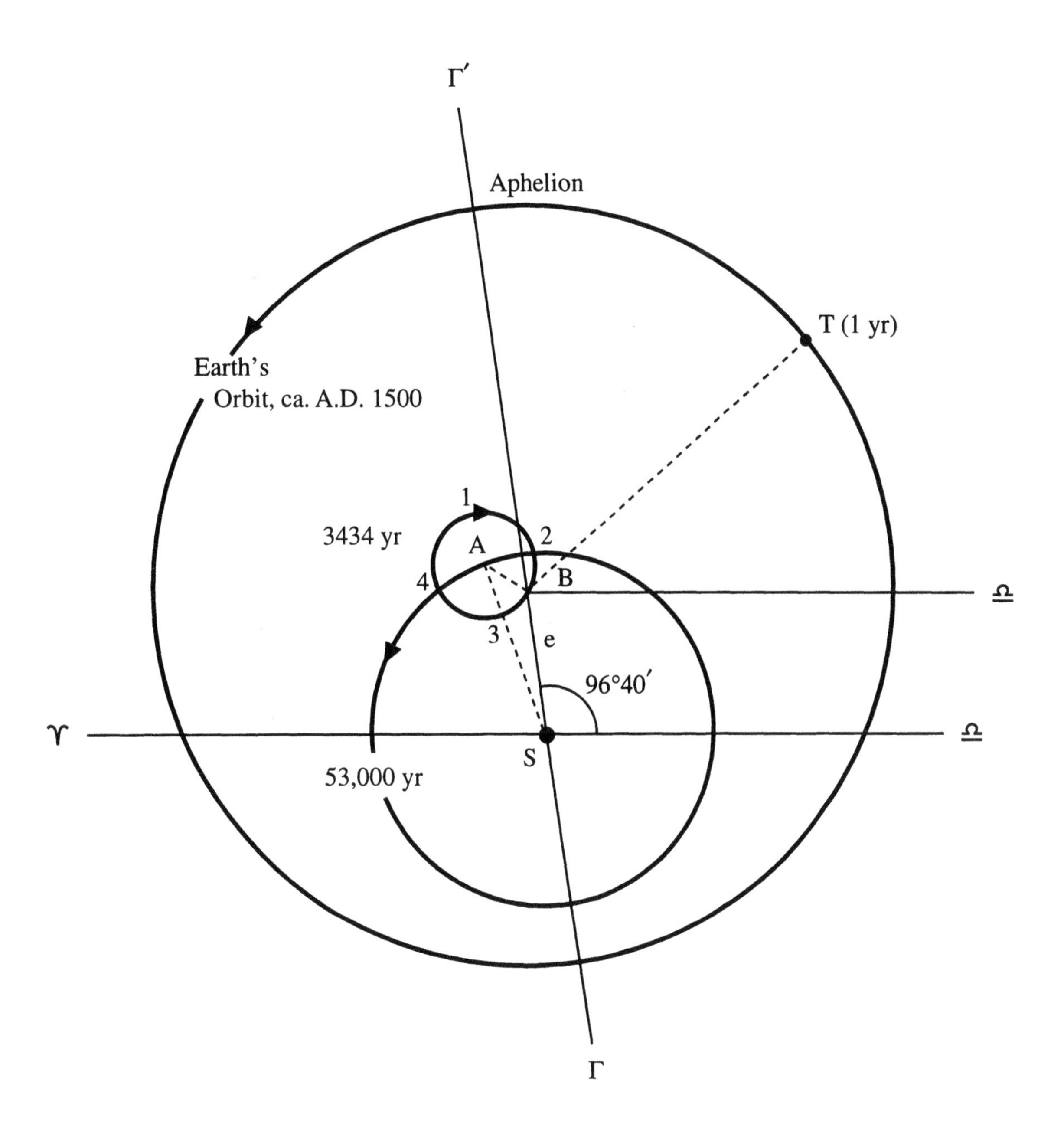

Figure V.5 — Copernicus's orbit of the Earth.

Notes on Some Elementary Methods of Finding the Relative Distances in the Planetary System

By adding epicycle upon epicycle the geocentric systems of the Greeks could be made to agree fairly well with the angular motions of the planets in longitude, with an accuracy about equal to that of contemporary observations. This is of course why these systems were adopted and why they held the stage for two millennia. In fact, they may be said to have been scientifically acceptable when judged by the modern criterion of satisfactory agreement with observations. Relative distances to the planets, however, were a different story. With the exception of a few special cases (the Moon, the inferior planets), relative distances could only be unreliably estimated from brightness variations based on eye estimates combined with erroneous assumptions about the orbital motions. But with the Copernican assumption of a heliocentric arrangement for the planetary orbits, it became possible for the first time to find reliable relative distances of the planets. Before proceeding further with the details of this Copernican achievement, we shall describe basic triangulation methods of finding distances in the solar system from observations of geocentric longitudes and latitudes.

The Distance to the Moon Found from Observations by Triangulation. In Figure V.6, let A and B be two observatories on, or very nearly on, the same line of longitude and very far apart. Let

M = the Moon,

O = the center of the Earth,

P = the terrestrial North Pole,

QE = the terrestrial equator,

A = an observatory in the Northern Hemisphere (e.g., Greenwich),

B = an observatory in the Southern Hemisphere (e.g., Cape of Good Hope),

$AM = d$ = topocentric distance of the Moon from A,

$OM = R$ = geocentric distance of the Moon,

$z = \angle ZAM$ = observed zenith distance of M at A,

$z' = \angle Z'BM$ = simultaneously observed zenith distance of M at B,

ϕ = geocentric latitude of A,

ϕ' = geocentric latitude of B,

c = chord connecting A and B,

$\gamma = \angle AOB$,

$\beta = \angle BAM$,

$\beta' = \angle ABM$,

$\delta = 180° - (\beta + \beta')$,

r = radius of the Earth, considered spherical,

α = base angle in equilateral ΔAOB.

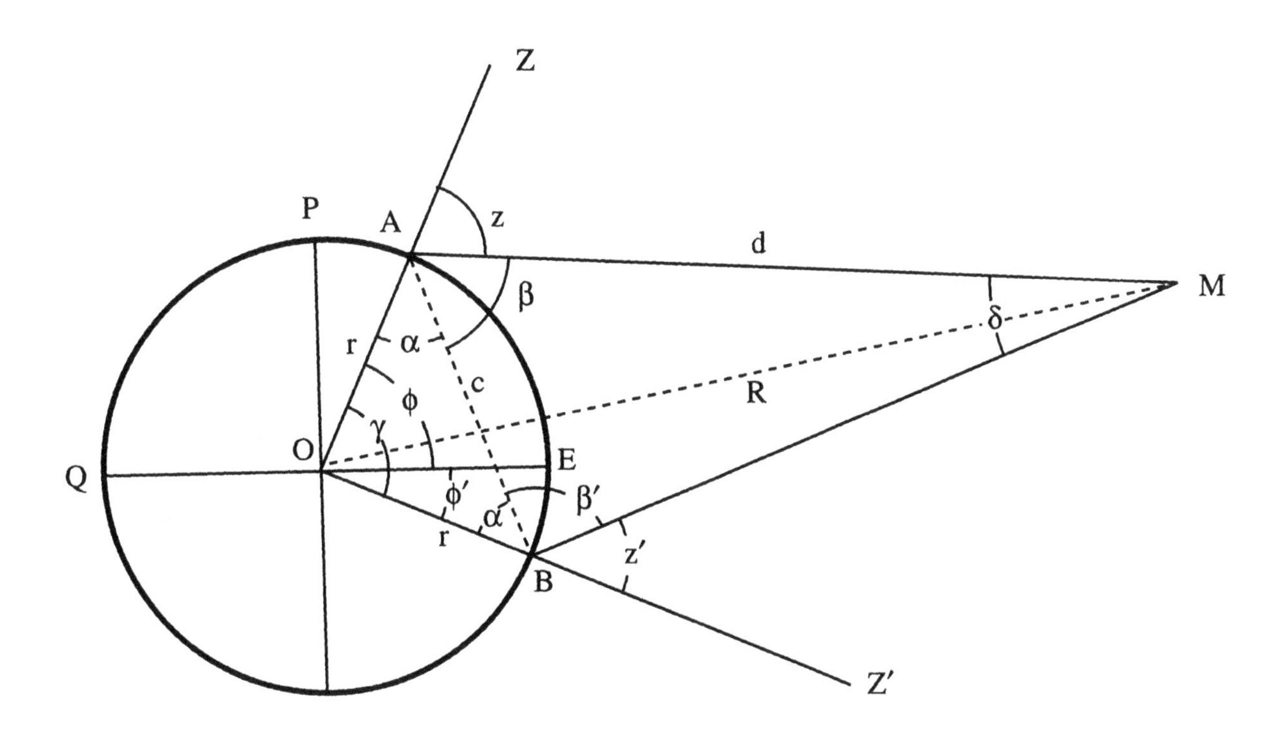

Figure V.6 — Finding the distance to the Moon by triangulation.

Now, let the zenith distances of the Moon be observed simultaneously at the moment of its transits at A and B and reduced to a common meridian. The problem may now be stated as follows: Given z, z', r, ϕ, ϕ'; find R. From the definitions given above it follows that

$$\gamma = \phi + \phi',$$

$$\alpha = 90^\circ - (\gamma/2),$$

$$c = 2r \sin \alpha,$$

$$\beta = 180^\circ - (\alpha + z),$$

$$\beta' = 180^\circ - (\alpha + z').$$

From ΔABM,

$$d = c \sin \beta' \csc \delta = 2r \sin \alpha \sin \beta' \csc \delta.$$

From ΔAMO,

$$R^2 = r^2 + d^2 - 2rd \cos (\alpha + \beta),$$

which solves the problem.

In accurate work $OA \neq OB$, $\gamma \neq \phi + \phi'$, and the times of transits at A and B differ. But OA, OB, ϕ, and ϕ' can be accurately found from a knowledge of the Earth's size and shape, and γ can be calculated by spherical trigonometry. However, the Moon's motion during the interval between the two transits, as well as the components of its motion affecting the size of the base angles in ΔABM, must also be allowed for; and all quantities must be reduced to the same instant, say, one of the transits. Thus only the principle of the method is simple.

Aristarchus's Method of Finding the Approximate Distance of an Inferior Planet from the Sun. In Figure V.7 let the orbits of Venus and Earth be circular and concentric in the Sun. The radius of Earth's orbit SE is taken as unity. The distance x of Venus from the Sun is to be found.

Method: Find by observation (and interpolation, if necessary) the greatest elongation of Venus from the Sun. Result: 47°.

From the figure: $x = 1 \sin 47^\circ = 0.72$.

Copernicus's Method of Finding the Approximate Distance of a Superior Planet from the Sun. In Figure V.8 let the orbits of Jupiter and Earth be circular and concentric in the Sun S. The radius SE_2 of Earth's orbit is taken as unity. The distance x of Jupiter J_2 from the Sun is to be found.

Let E_1 and J_1 be the positions of Earth and Jupiter, respectively, at the time of an opposition of Jupiter with the Sun, and let E_2 and J_2 be their respective positions at the following eastern quadrature. Neglecting the eccentric positions and mutual inclination of the orbits, SE_1J_1 is a straight line, and SE_2J_2 is a plane right triangle. The method is as follows:

Observe the date of an opposition (e.g., 16 February 1956).

Observe the date of the following eastern quadrature (e.g., 13 May 1956).

The difference is 87 days.

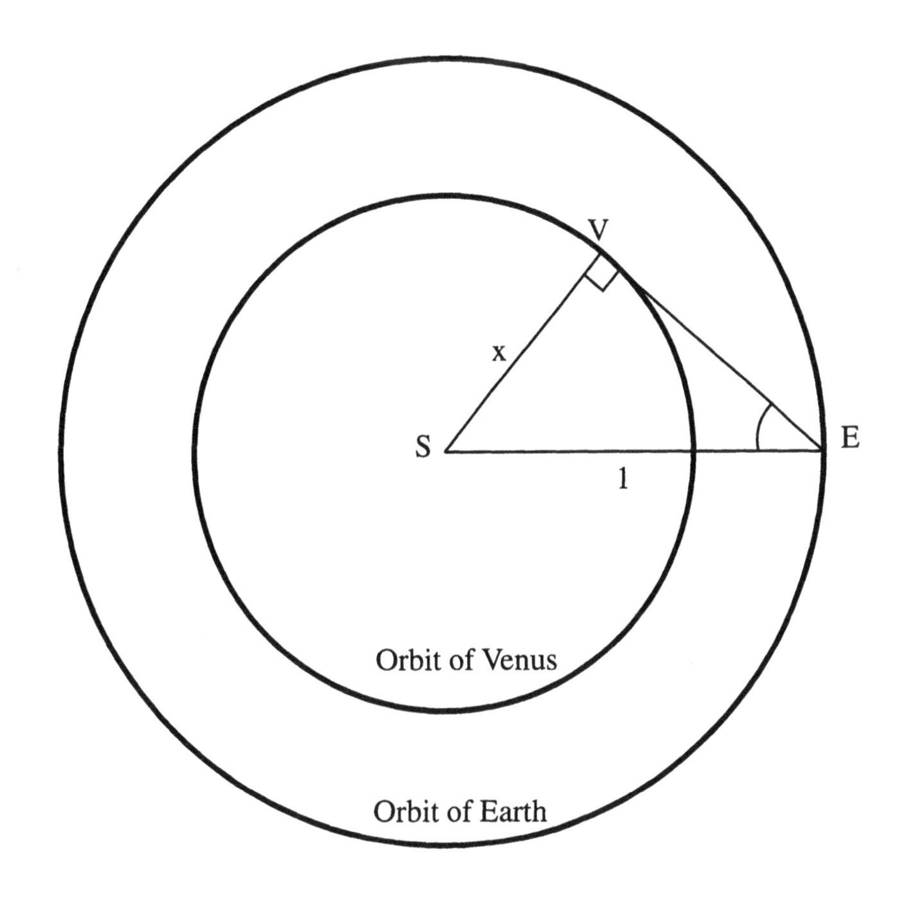

Figure V.7 — Aristarchus's method for finding the approximate distance of Venus.

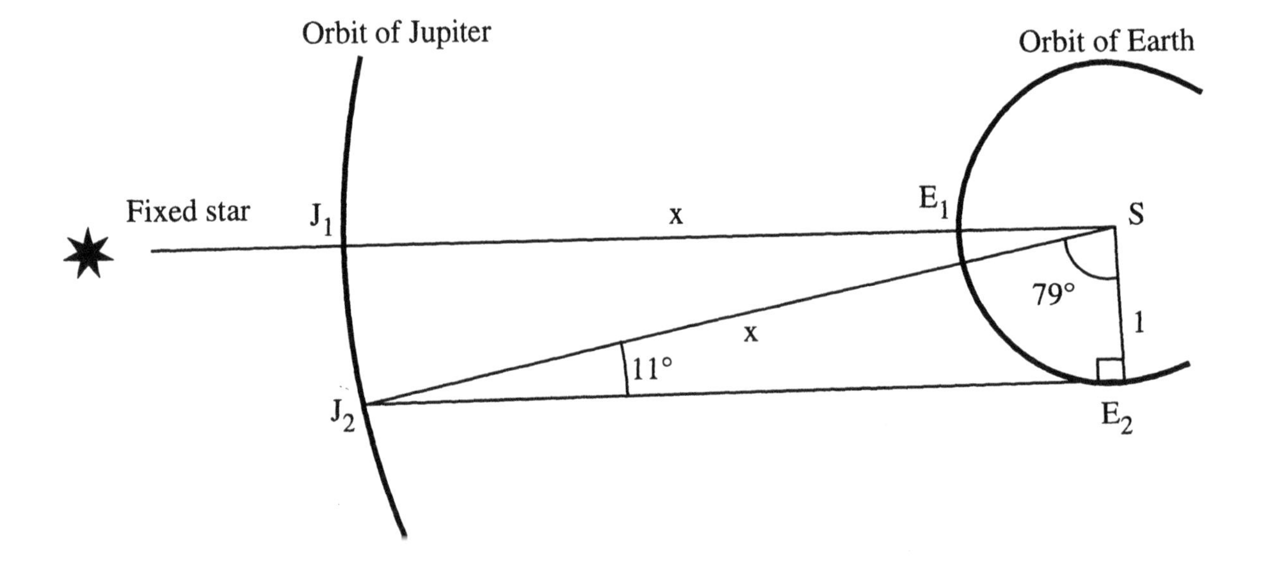

Figure V.8 — Copernicus's method for finding the approximate distance of a superior planet from the Sun.

The angular gain of SE_1 on the direction of a fixed star S_* is about 1°/day. But the sidereal period of Jupiter is about 12 years, so the angular gain of SJ_1 on S_* is about $\frac{1}{12}^\circ$/day. Therefore the angular gain of SE_1 on $SJ_1 = \frac{11}{12}^\circ$/day. Hence, $\angle J_2SE_2 = 87$ days$(\frac{11}{12}^\circ/\text{day})$ $= 79.8°$. From ΔSE_2J_2, $x = \csc 10.2° = 5.6$, approximately. Thus Jupiter's orbit is 5.6 times larger than the Earth's.

By more elaborate methods Copernicus found the relative mean distances given in Table V.1 of the planets from the Sun. As a result Copernicus was able to find the relative distances of all the planets not only at their chief aspects but at any position in their orbits. Since all the sidereal periods were well known, even in antiquity, had he been of a more speculative mind, he might thus easily have discovered Kepler's harmonic law.

Table V.1 — Relative Mean Distances
of the Planets from the Sun

Planet	Copernicus	Modern Value
Mercury	0.3763	0.3871
Venus	0.7193	0.7233
Earth	1.0000	1.0000
Mars	1.5198	1.5237
Jupiter	5.2192	5.2028
Saturn	9.1743	9.5389

The Precession of the Equinoxes and the Trepidation

Following the general order of the topics treated in *Revolutions*, we find a section on the precession of the equinoxes and the trepidation. Prior to Copernicus these slow changes were believed to be caused by a slow motion of the ecliptic along the fixed equator. It is one of Copernicus's achievements to be the first to explain both motions by a slow change of the Earth's axis of rotation, so that the poles of the rotation intersect the celestial sphere at continually different points. The corresponding motion of the equator on the celestial sphere results in the sliding of the equinoxes westward on the fixed ecliptic.

Instead of the slow, almost secular, progressive change we know as precession, some ancient astronomers believed in another, somewhat faster, *periodic* fluctuation of the equinoxes and solstices, resulting from an oscillation of the inclination of the ecliptic. This movement was called the *trepidation.* Medieval astronomers (and Copernicus) believed in a combination of the two motions.

In order to appreciate the situation we will describe the two motions separately.

The Precession of the Equinoxes

Hipparchus's discovery and approximate evaluation of this movement was described in Chapter III.

The Trepidation

Hipparchus's discovery of precession as a uniform increase in the celestial longitudes of the fixed stars, of about 2° in 150 years, was adopted by Ptolemy. But other astronomers (e.g., Proklus, A.D. 410–85) misunderstood the very nature of Hipparchus's explanation, namely, a uniform sliding of the celestial equator westward on the ecliptic, with always the same angle between the two great circles. They thought instead of a uniform, presently westward motion, always at a rate of 1° in 80 years, that reached a limit of 8° and then suddenly reversed in direction. The absurdity of a sudden reversal was somewhat mitigated by Thābit ibn Qurrah (9th century), who substituted a simple harmonic variation of amplitude $10\frac{3}{4}^{\circ}$ and period 4096.9 years, called the *trepidation*.

Ibn Qurrah's theory of the trepidation, which was adapted by Copernicus, is shown in Figure V.9. It depends upon the introduction of a ninth celestial sphere (called the primum mobile) with fixed equator A_9 (pole p_9), fixed ecliptic E_9 (pole K_9), and a perfectly uniform westward diurnal motion. Within this sphere the eighth, or "fixed-star," sphere (pole p_8) is suspended on two diametrically opposite pivots that are fixed on the ninth sphere at $♈_0$ and $♎_0$, while a movable ecliptic of the eighth sphere is suspended on two opposite movable pivots P and P' of this sphere. P and P' move uniformly counterclockwise (as seen from "the outside") on the circumference of a pair of equal, oppositely situated small circles with centers $♈_0$ and $♎_0$ and radius 4°18′33″, their centers being fixed on the ninth sphere (thus the equators A_8 and A_9, and the poles p_8 and p_9, always coincide). The movable ecliptic E_8 of the eighth sphere coincides with the fixed ecliptic E_9 of the ninth sphere only twice in each trepidational period, i.e., when the pivots of the eighth sphere lie on the fixed ecliptic of the ninth sphere, as at P_0 and P'_0. At that time the summer solstice is in its mean position $V_9 = V_{8_0}$. The inclination of both ecliptics in this position is $\varepsilon_0 = 23°33'30''$, its maximum value. Whenever the ecliptic of the eighth sphere is tangent to the small circles, as at P_2, P'_2, the equinoxes $♈_2$ and $♎_2$ and the corresponding summer solstice V_{8_2} of this sphere (the fixed stars) are the farthest westward (by $10\frac{3}{4}^{\circ}$) of their respective means. The summer solstice occurs in Gemini (as at V_{8_2}) and the inclination is now a minimum value ε_2. Half a trepidational period later, the opposite tangency occurs. The equinoxes are now farthest east (by $10\frac{3}{4}^{\circ}$) on the equator at $♈_1$ and $♎_1$, the summer solstice V_{8_1} occurs in Cancer, and the inclination is again the same minimum value.

Thus there was not precession in the sense of a constant, continual, westward progression of the equinoxes on the equator but a periodic fluctuation of amplitude $10\frac{3}{4}^{\circ}$ along the equator. Later astronomers, Copernicus included, combined a Hipparchus-type precession with the nonexistent trepidation. This required a tenth sphere acting as primum mobile, which communicated a perfectly uniform diurnal motion to all the others. Inside it was suspended a ninth sphere on pivots through an equinox moving constantly westward on the fixed ecliptic and equator of the tenth sphere. This produced the constant, westward, uniform motion now called precession. The eighth sphere was pivoted on the moving equinox of the ninth (i.e., on ibn Qurrah's mean equinox), but its moving ecliptic was pivoted on opposite points located on the small circles centered on the mean equinox. The motion thus produced was as in ibn Qurrah's theory explained above. Al-Sufi and a few other Arab astronomers did not believe in the trepidation, while others retained the theory with variously assigned amplitudes and periods.

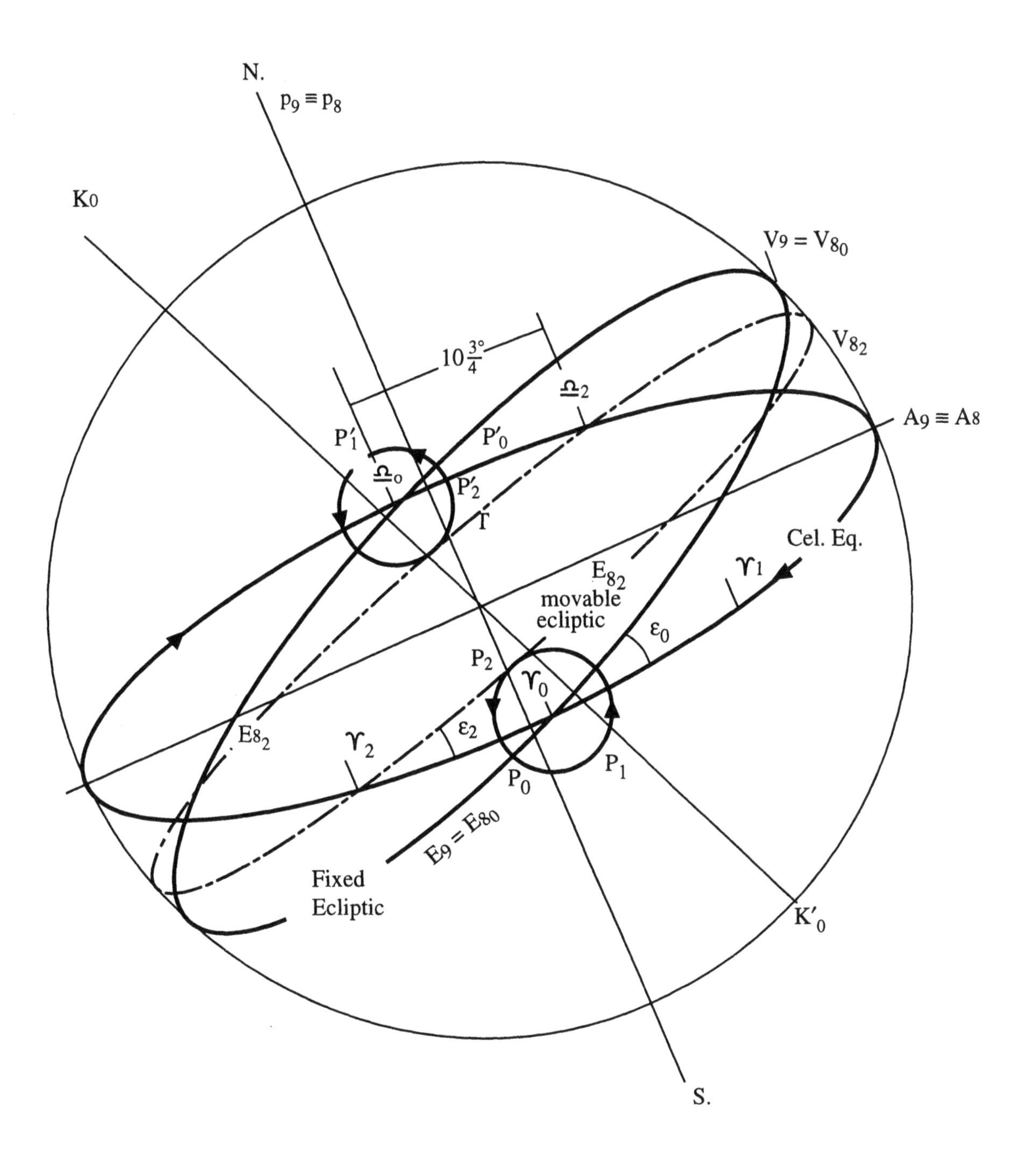

Figure V.9 — The "trepidation" according to Thābit ibn Qurrah.

Copernicus's Treatment of Precession and Trepidation

In order to find the various values of the constant of precession resulting from observations taken at different epochs, from the time of the ancients to his own determination of the position of Spica, Copernicus made estimates based upon the available observations of this star, with the results listed in Table V.2.

Table V.2 — Estimates of Precession at Various Epochs

Observer and Epoch	Rate of Precession in Celestial Longitude	
1. Timocharis, 293 B.C. Hipparchus, 127 B.C.	1° in 72 years	50″/yr
2. Hipparchus, 127 B.C. Menelaus, A.D. 98	1° in 100 years	36
3. Menelaus, A.D. 98 Ptolemy, A.D. 138	1° in 86 years	42
4. Ptolemy, A.D. 138 Albategnius, A.D. 880	1° in 66 years	54
5. Albategnius, A.D. 880 Copernicus, A.D. 1525	1° in 70 years	51

From these data Copernicus took a minimum to have occurred about 64 B.C. and a maximum about A.D. 794. Noting that the precession was smaller in the centuries before Ptolemy than afterward, and also that apparently there were corresponding irregular changes in the inclination of the ecliptic, he then explained these two (erroneous) results with a mechanism that graduately shifted the direction of the Earth's axis of rotation.

The observations suggested to Copernicus, first, a variation of 24′ in the inclination of the ecliptic described by a sinusoidal variation of period 3434 years, amplitude 12′, and minimum value 23°28′ in 64 B.C. Second, there was a deviation in the position of the equinoxes relative to that given by a constant precession. This fluctuation was sinusoidal in longitude with amplitude 28′ and period 1717 years and was centered on the point of the solstitial colure determined by projection of the first motion. The second motion was projected at approximately right angles to the first, i.e., on the parallel of latitude through the star. The actual pole traced out a distorted figure eight, resulting from the two simultaneous movements at right angles, one having half the period of the other. The actual displacement of the equinox was the amount of the second motion projected upon the ecliptic. Thus the maximum excursion of the instantaneous solstitial colure, and hence of the equinoxes, was the projected value $28' \sec 66\frac{1}{2}^\circ = 70'$.

Copernicus's theory of precession and trepidation is illustrated in Figure V.10, in which

O = the center of the celestial sphere, i.e., the observer's eye,

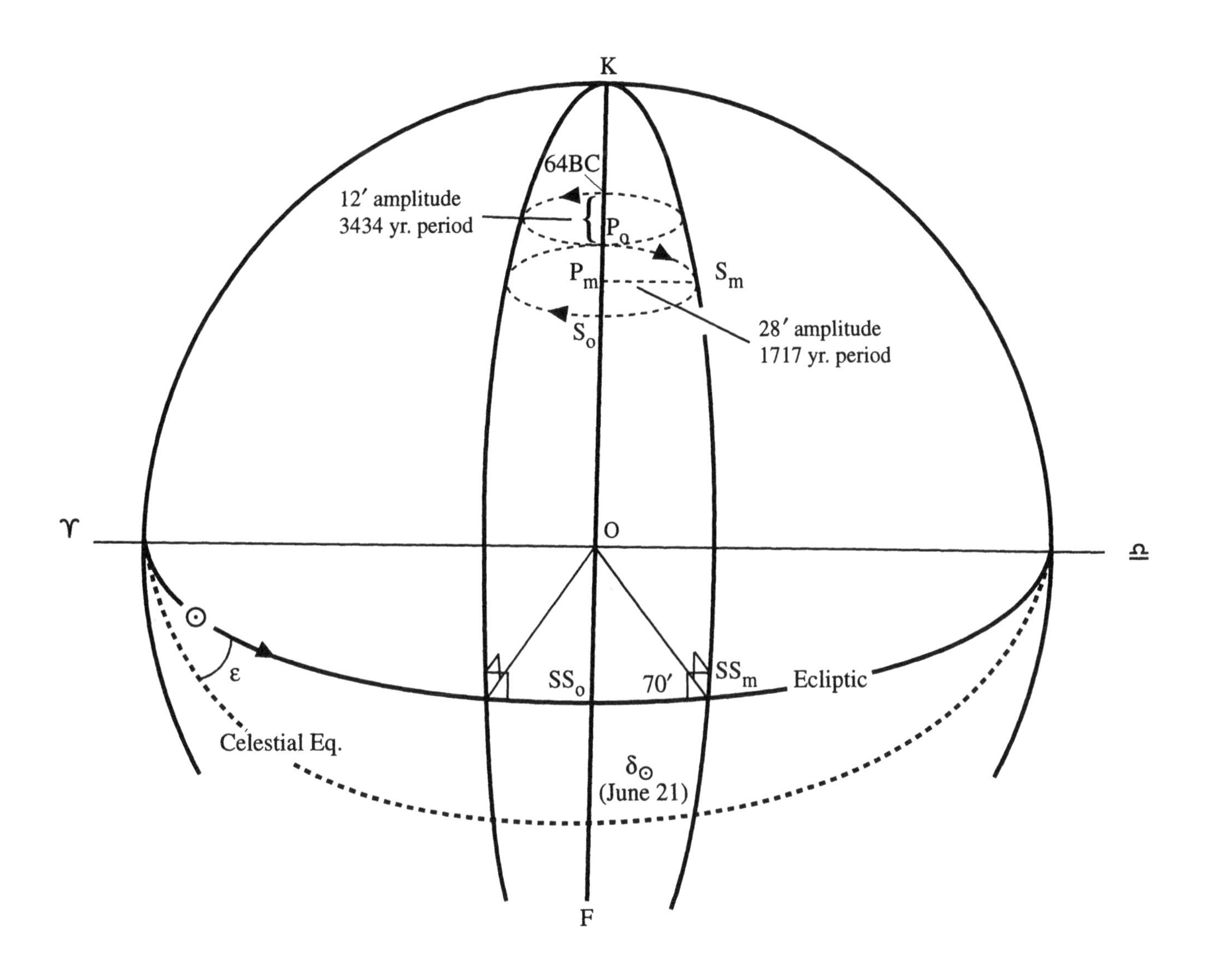

Figure V.10 — Copernicus's explanation of precession and trepidation (view of the celestial sphere as seen from the north and from the "outside").

Υ = vernal equinox,

Ω = autumnal equinox,

ε = inclination of the ecliptic,

K = pole of the ecliptic,

P_0 = mean position of the north celestial pole, with inclination $i = KP_0 = 23°40'$,

SS_0 = position of the summer solstice,

KP_0F = solstitial colure,

S_0 = initial position of the north celestial pole at the ecliptic's maximum inclination KS_0,

S_m = position of the pole at the first trepidational maximum,

P_m = projection of S_m upon the mean solstitial colure, amplitude = 28′, $i = KP_m = SF_m$,

SS_m = projection of S_m upon the ecliptic at maximum trepidation of 70′,

$\odot$ = position of the Sun about 1 April.

Copernicus thus believed in a uniform precession overlaid by a small trepidation or, as he called it, an "oscillation." Because of this complicated motion of the celestial poles, he counted celestial longitudes from the position of a "fixed" mark, namely, the "first" star of Aries (α Arietis). As measured from the moving equinox, the celestial longitude λ of α Ari was slowly increasing at a variable rate. At any moment the longitude difference of this star from the instantaneous vernal equinox was given by the formula

$$\lambda = 5°32' + 50.2''T + 70' \sin(0.02096°T + 13.2°) \qquad \text{(period = 1717 yr)},$$

where T is in tropical years since A.D. 1. Hence, if λ_* denotes the celestial longitude of an object, reckoned from α Ari at time T, and λ_E the value reckoned from the instantaneous vernal equinox, then $\lambda_E = \lambda_* + \lambda$. The last term in the formula for λ was considered to be spurious by Tycho and was omitted by him. The coefficient of the second term is remarkably close to its modern value of 50.26″.

For the inclination of the ecliptic Copernicus found the formula

$$\varepsilon = 23°40' + 12' \cos(0.1048°T + 6.75°) \qquad \text{(period = 3434 yr)}.$$

This formula gives much too high values throughout historical times. Even Copernicus's own approximate solar observations should have revealed its inadequacy.

In book III of *Revolutions* Copernicus considers the motion of the eighth (fixed-star sphere) by describing the combined progressive and oscillatory movements of a fixed star on the ecliptic. He compares his theory with all the known observations of Spica, $\delta_{(1500)} = -8°40'$, a star close enough to the ecliptic for illustration. He studies the varying values of the constant of precession from a table similar to that above and concludes that we are in the second revolution of a cycle that started in 64 B.C.

The formula describing the progress of the celestial longitude of Spica and, conversely, the drift of the equinox causing the nonuniform increase in longitude (proper motion was then unknown) is derived by Zeller (1943) in the notes to his German translation of Rheticus's *Narratio Prima*. By taking the derivative of the above expression for λ, one finds an expression for the rate of steady precession plus trepidation:

$$\frac{d\lambda}{dt} = 50.2'' + 15.4'' \cos [0.2096° \ (T + 64)],$$

where t is the number of years since A.D. 1.

Copernicus's Lunar Theory

In his lunar theory Copernicus explains the first inequality (roughly the elliptical terms) in exactly the same manner as did Ptolemy, with no eccentrics and employing two epicycles I and II (Fig. V.11); and just like Ptolemy he determines their dimensions from eclipse observations. His model requires (1) a retrograde epicycle I sliding with uniform direct motion on a deferent D, concentric with the Earth T and in a period of 1 mean sidereal month, and (2) a second, smaller, direct epicycle II centered on I and sliding with retrograde uniform motion in a period of 1 mean anomalistic month, counted from apogee. (The difference between these two periods, as in the case of Ptolemy's construction, ensures the correct mean advance of the line of apsides by 3.3° per sidereal month.) The Moon itself (e in Fig. V.11) is situated on the circumference of the smaller epicycle II and has a direct uniform motion in a period of $\frac{1}{2}$ mean synodic month. Copernicus retained Ptolemy's values of the two inequalities, 5° and 7°40′, giving radii of 0.1097 and 0.0237 for the first and second epicycles (taking that of the deferent as unity). The second epicycle (amplitude 1°20′) fairly well accounts for the "evection."

By the above arrangement the enormous changes in the distance of the Moon under Ptolemy's construction were considerably diminished. The greatest and smallest distances of the Moon from the Earth were found to be $68\frac{1}{3}$ and $52\frac{17}{60}$ semidiameters of the Earth. The corresponding angular diameters of the Moon were 37′34″ and 28′45″, a great improvement (as Copernicus himself remarks) on the theory of Ptolemy, according to which the apparent diameter would be nearly 1° at perigee. (The modern values are 33′32″ and 29′26″ for the Moon's extreme angular diameters.)

Copernicus's Orbit of Venus

As in the case of the Earth, the inferior planets were moving uniformly in perfect circles. However, for Mercury and Venus the orbits were shifting about, so that the apparent motions of these planets were neither circular nor uniform. For both Mercury and Venus the greatest elongations were known to be unequal and variable. For the celestial longitudes of Venus, the following construction, shown in Figure V.12, was adopted:

O = average position of center of Earth's annual orbit,

E = a representative position of the Earth in its orbit (period = 365.26 days),

OE = radius of Earth's orbit = 1 A.U.,

p = instantaneous position of center of Venus's orbit,

P = a representative position of Venus (period = 225 days),

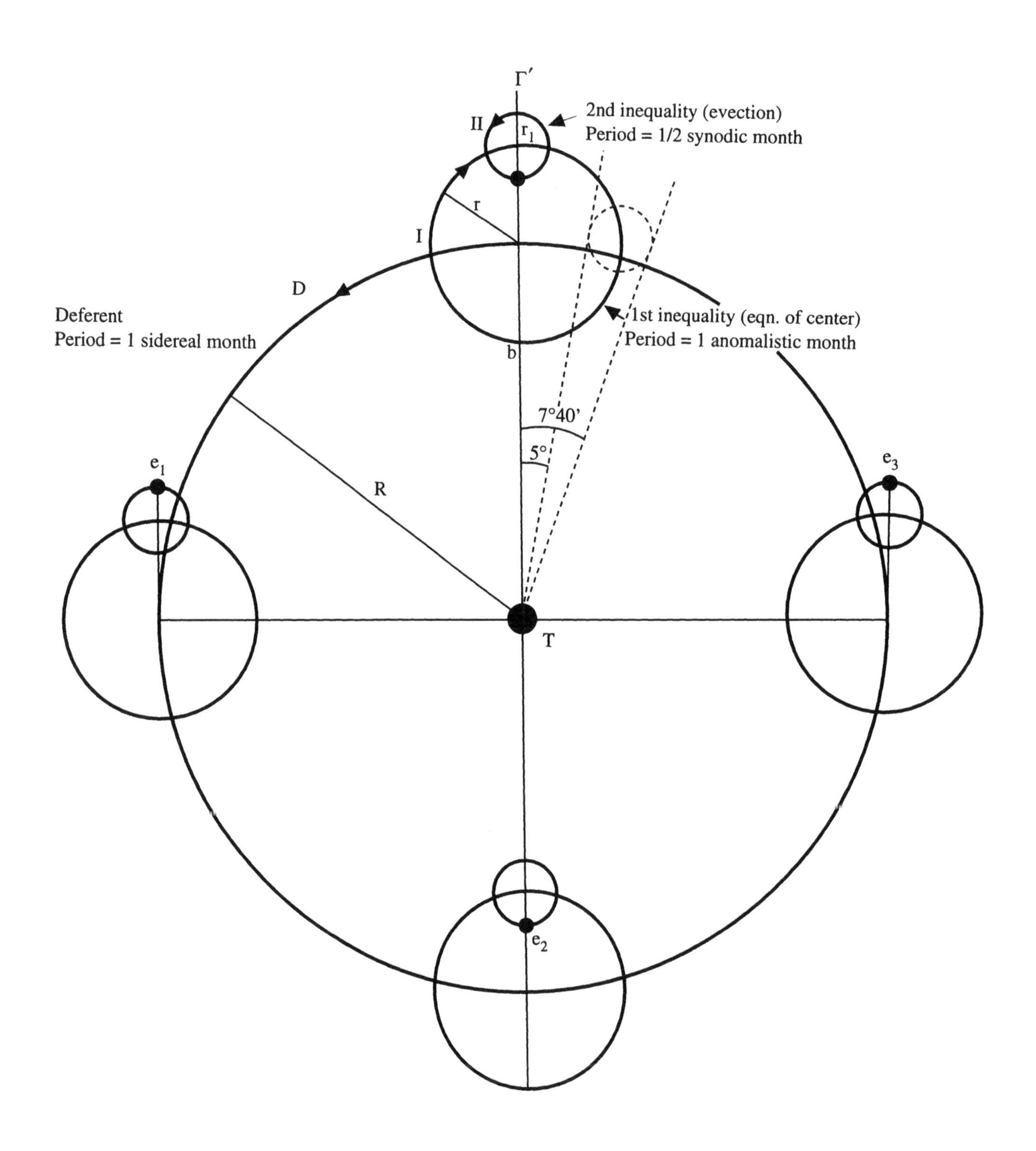

Figure V.11 — Copernicus's lunar theory.

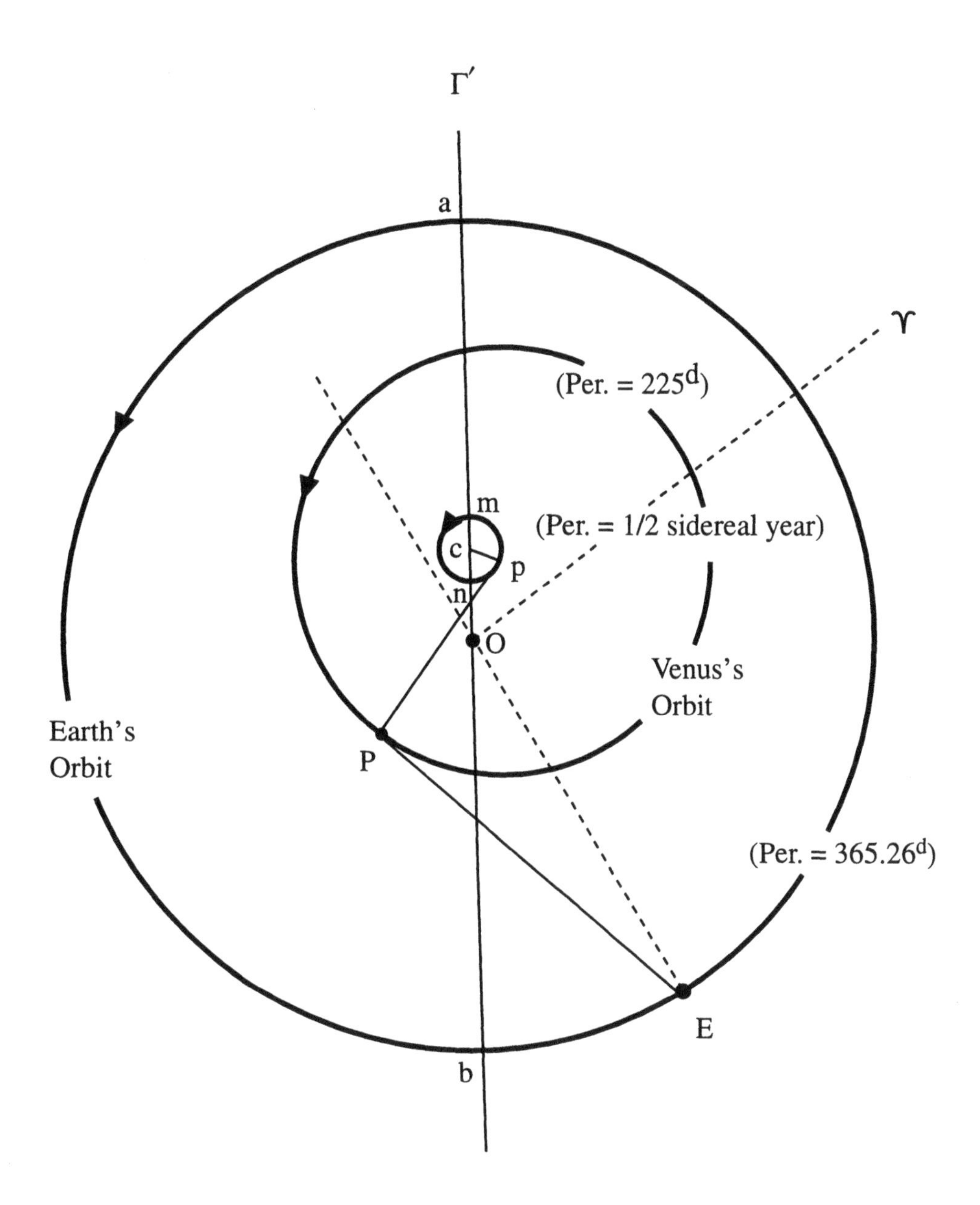

Figure V.12 — Copernicus's orbit of Venus (not to scale).

pP = radius of Venus's orbit = 0.719 A.U.,

c = center of hypocircle npm described by p with uniform motion (period = $\frac{1}{2}$ sidereal year),

$Oc = e = 0.02$, mean eccentricity of center of Venus's orbit,

$cp = \frac{1}{3}e$ (best value found by trials),

ab = mean position of line of apsides of Venus's orbit; longitude of aphelion, $\Gamma' = 55°$.

All motions in the diagram are direct and take place as follows: p coincides with n when E is in either a or b. It follows that the so-called angle of commutation $\angle ncp = 2\,\angle bOE$. It further follows that p coincides with m when the Earth is at right angles to the line of apsides of Venus's orbit.

Because of the small value of the eccentricity of the center and the moderate value of the inclination of Venus's orbit, this arrangement fairly well represents the planet's motion in longitude.

Copernicus's Orbit of Mercury

In Figure V.13 is shown the construction adopted by Copernicus to explain the highly complicated motions of Mercury in heliocentric longitude. With certain phase shifts and changes of dimensions, the first part of this mechanism is identical to that adopted for Venus. The second part of the mechanism comprises a simple harmonic oscillation along P_1P', the radius of the instantaneous orbit. In Figure V.13 we have

O = average position of Sun and center of Earth's annual orbit,

E = a representative position of the Earth (period = 365.26 days),

OE = radius of Earth's orbit = R = 1 A.U.,

P = a representative position of Mercury,

p = instantaneous position of center of Mercury's orbit,

$P_1P' = \rho_1$ = radius of Mercury's epicycle II,

P' = a point on the circle (of center p) that is the center of epicycle II giving radial simple harmonic oscillation (amplitude ρ_1, period = $\frac{1}{2}$ sidereal year),

α = displacement of P resulting from al-Ṭusi's construction (P at P' when P' at Γ'),

ρ_1 = amplitude of linear radial displacement,

P_1, P_2 = maxium and minimum positions of Mercury's oscillation along pP', the instantaneous orbital radius (period = $\frac{1}{2}$ sidereal year and double amplitude P_1P_2),

pP' = radius of epicycle I = planet's orbit = 0.38 A.U. (orbital period = 88 days),

c = center of hypocircle described uniformly by p (period = $\frac{1}{2}$ sidereal year),

$Oc = e = 0.20$, average eccentricity of center of Mercury's orbit,

$Op = e'$ = instantaneous eccentricity of center of Mercury's orbit,

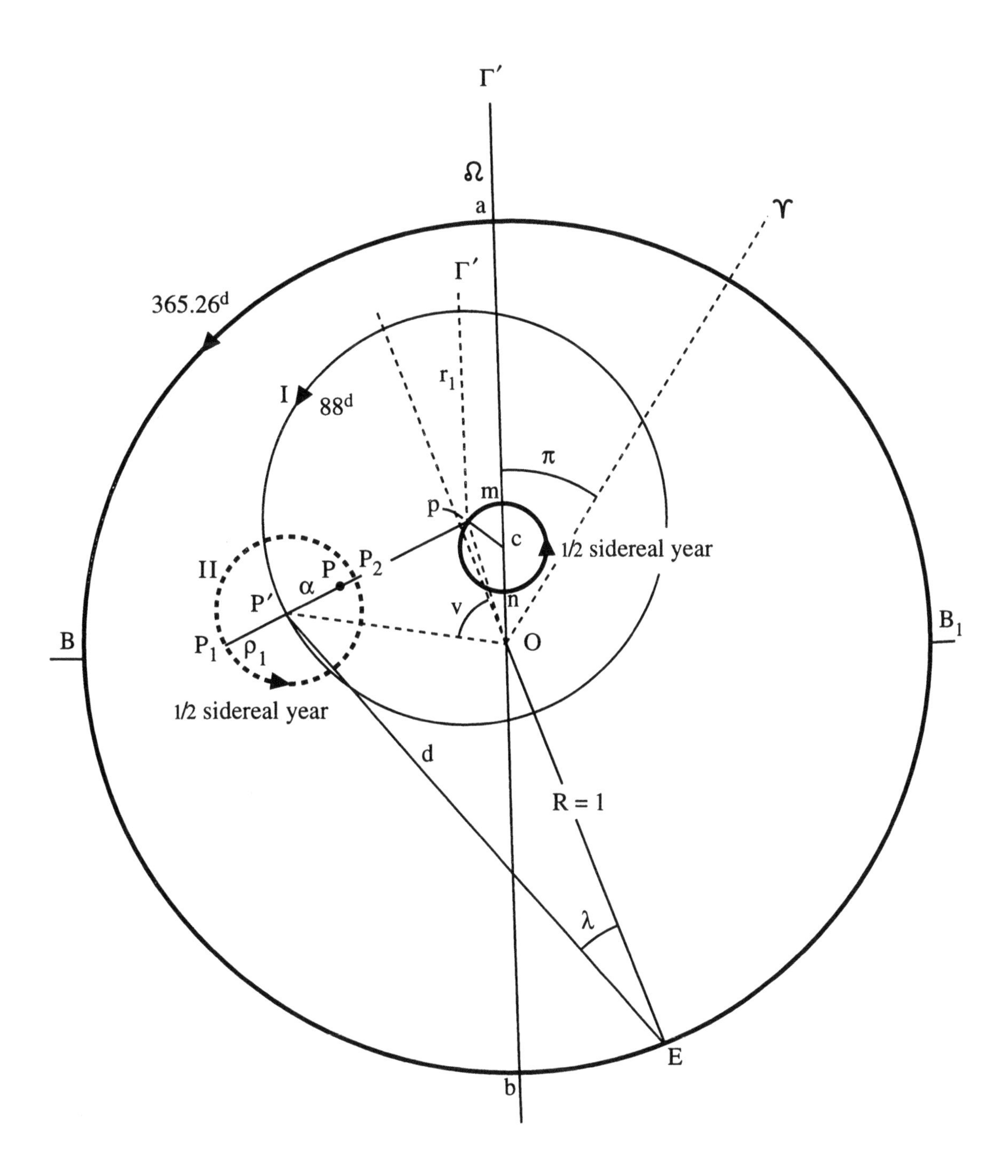

Figure V.13 — Copernicus's orbit of Mercury (not to scale).

$cp = \rho = \frac{1}{3}e$,

ab = mean position of the line of apsides of Mercury's orbit, also taken as line of nodes of Mercury's orbit,

Υ= vernal equinox,

π = longitude of aphelion Γ' and of ascending node Ω.

All motions are direct and take place such that p coincides with m when E is at a or b. It follows that p coincides with n when E is at B or B_1, at right angles to the line of apsides ab, and that $\angle mcp = 2\ \angle bOE$. Also, Mercury is in P_2 when p coincides with m (i.e., when the Earth is in either a or b); and Mercury is in P_1 when p coincides with n (i.e., when the Earth is perpendicular to ab, at B or B_1).

The linear simple harmonic motion P_1 to P_2 is conceived as a uniform circular motion of a point on a hypocircle of radius $\frac{1}{2}\ \rho_1$, as explained by al-Ṭusi's theorem.

Copernicus's Orbit of a Superior Planet

Copernicus's heliocentric model was a great advance toward the true explanation of the phenomena of the superior planets. However, he introduced a nearly fatal "simplification" when he abandoned the equant principle of Ptolemy, so valuable in simulating elliptic motion. This was done, he explains in the preface to *Revolutions*, to preserve "the holy principle of uniform circular motion," so ingrown in the consciousness of all astronomers even until long after Kepler's work. On the other hand, Copernicus was the first to determine the advance of the line of apsides of the planetary orbits. From a comparison of oppositions determined by Ptolemy with three observed by himself, he estimated that the amounts of advance were 1° in 100 years for Saturn, 1° in 300 years for Jupiter, and 1° in 130 years for Mars.[1] Thus Copernicus considered that the instantaneous longitude of aphelion Π is given by $\Pi = \Pi_0 + \pi t$, where t denotes the time in years measured from an epoch when the longitude of aphelion is Π_0 and π denotes the annual rate of advance.

For each of the superior planets Copernicus considers direct uniform motion on a circular epicycle of center E and radius r_1, moving directly and uniformly on a circular deferent of center D and radius r (Fig. V.14). The Earth T is shown on a circular orbit of center O and radius R_O. The point a is the aphelion of the planet orbit. Given elements include r, r_1, the position e of the eccentric D on the line of apsides Oa, and periods on the deferent and epicycle, both of which are equal to one sidereal period of the planet (neglecting the advance of the line of apsides). Let P and T be any general positions of the planet and the Earth at time t.

To locate the planet (i.e., to find its geocentric longitude λ), let M = the mean anomaly $= \angle aDE = \angle DEP$, μ = the rate of motion in mean longitude L, and π = the rate of mean motion in true longitude ℓ. From these definitions and from the figure we have $\ell = \ell_0 + \pi t$; $L = L_0 + \mu t$; $M = L - \Pi$; where L_0 is the mean longitude at $t = 0$.

In ΔEDO,

$$\sigma \sin x = e \sin M \qquad \text{and} \qquad \sigma \cos x = r + e \cos M.$$

from which σ and x. In ΔEPO,

[1]Modern values for Saturn, Jupiter, and Mars are 1° in 115 years, 100 years, and 130 years, respectively.

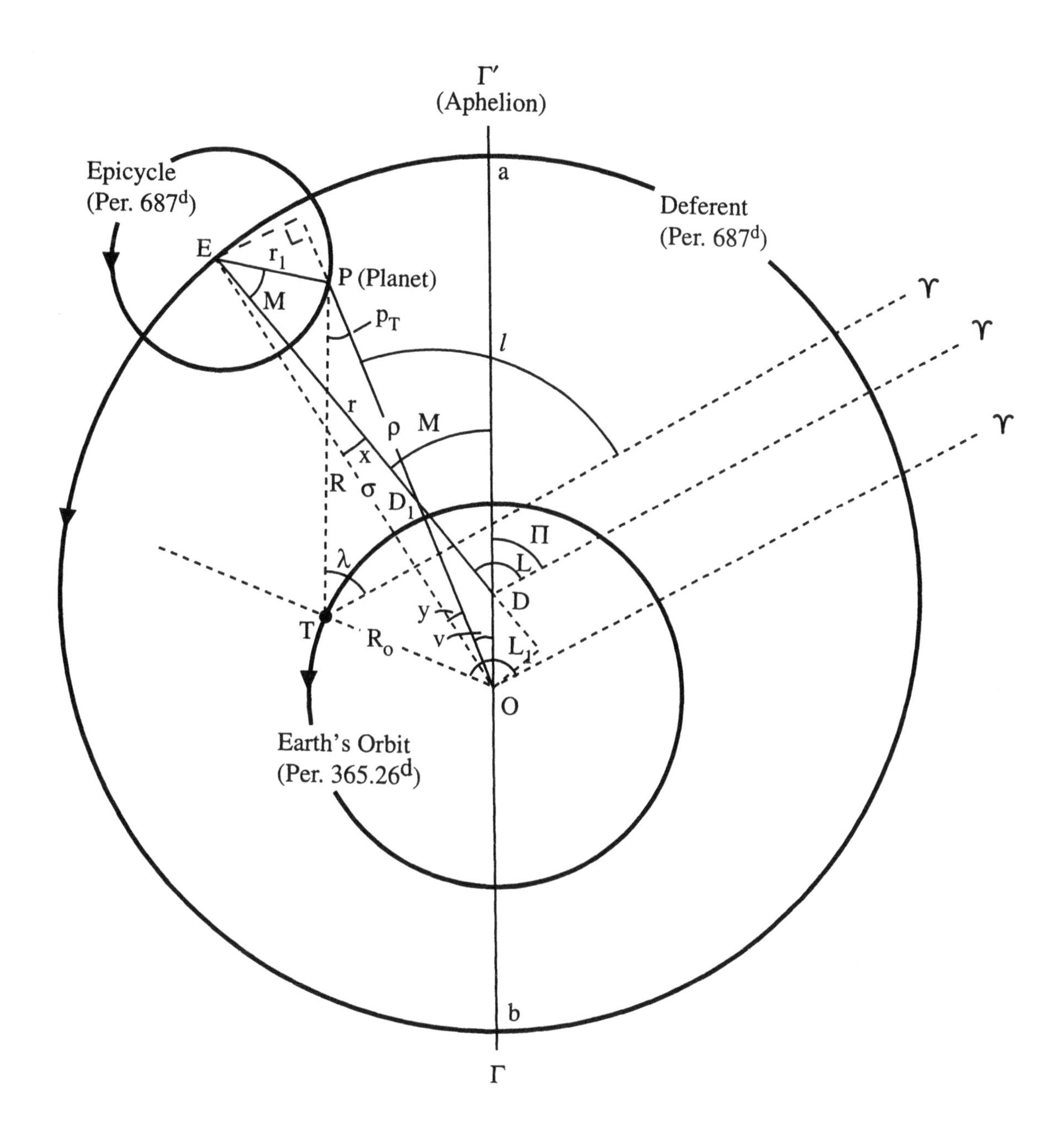

Figure V.14 — Copernicus's orbit of a superior planet (Mars—not to scale).

$$\rho \sin y = r_1 \sin (M + x) \qquad \text{and} \qquad \rho \cos y = \sigma - r_1 \cos (M + x),$$

from which ρ and y. From ΔED_1O and ΔD_1DO,

$$M - v = x + y, \qquad \text{or} \qquad v = M - (x + y).$$

Also, $\ell = \mathrm{v} + \Pi =$ the true heliocentric longitude.

This solves the problems of locating the planet in heliocentric longitude. To find its geocentric position and distance, Copernicus introduced the angle p_T, the "parallax of the Earth in its orbit." Let L_1 be the true heliocentric longitude of the Earth at time t (given, since Earth's orbit is considered known). Then the geocentric longitude λ and the geocentric distance R are to be found. In Figure V.14,

$$\angle TOP = L_1 - \ell,$$

$$\lambda = \angle \Upsilon TP,$$

$$\ell = \angle \Upsilon OP = \angle \Upsilon D_2 P,$$

where Υ is the vernal equinox. In ΔPTD_2,

$$\ell = \lambda + p_T.$$

In ΔPTO,

$$R \sin p_T = R_0 \sin (L_1 - \ell) \qquad \text{and} \qquad R \cos p_T = \rho - R_0 \cos (L_1 - \ell);$$

from which R and p_T. R is given in terms of R_0, the known heliocentric distance of the Earth, and $\lambda = \ell - p_T$, solving the problem.

Since Copernicus's system permitted only uniform motion in perfect circles, he realized that if e and Π were given, a few well-distributed observations should yield all the linear elements, i.e., r and r_1. As a start he derived the values of e and Π from three oppositions of each planet observed or determined by Ptolemy. He then checked these values against three oppositions observed by himself, and this led to his discovery of the advances of the lines of apsides of the three outer-planet orbits. He also found that the best representation of the motion for each planet was obtained by adopting a value of $\frac{1}{3}e$ for the epicyclic radius r_1. Throughout his career he tested this hypothesis by occasionally making observations of his own, but he seems to have used only about 27 observations as a check on all the work in *Revolutions*!

Knowing the angle p_T and the elements e, Π, r, and r_1, one can obviously find the maximum value p_{Tmax} of the angle subtended at P by the radius R_0 of the Earth's orbit, i.e., the apparent size of the Earth's orbit as seen from the planet. In the case of Mars the angle p_{Tmax} was later studied in great detail by Tycho, who discovered from persistent observations that, even for Mars in the same position, p_{Tmax} depended upon the Earth's heliocentric longitude L_1 (Chapter VI). Since the geometry was fully taken into account, it appeared to Tycho that the Earth's orbit size depended on the position of the Earth along its orbit. This was perhaps the first observational evidence that the Earth's orbit is not circular. The same effect also showed up for the cases of Jupiter and Saturn, but to a much smaller extent. The true interpretation of Tycho's observational results was given by Kepler, who showed that the Earth's orbit, considered as an eccentric circle, needed an equant point – the very thing that Copernicus persistently denied (Chapter VII).

Copernicus's Theory of the Celestial Latitudes of the Superior Planets

In describing Copernicus's ingenious, though unsuccessful, account of the heliocentric and geocentric latitudes of the planets, it is well to review the simplest case of a fixed orbit with known elements. Given the position of the ascending node ☊, the inclination i, and the heliocentric longitude ℓ, one finds that the heliocentric latitude b is given by $\sin b = \sin i \sin (\ell - ☊)$. The geocentric latitude β is then found from $\sin \beta = \frac{R}{\rho} \sin b$, where R is the instantaneous heliocentric distance of the planet, and ρ is the instantaneous geocentric distance. Disregarding eccentricity and noncircularity, the maximum and minimum values of the geocentric latitude occur when the planet is in opposition and conjunction at 90° from its node. Conversely, latitudes and longitudes can be observed at any time (except close to solar conjunction) and used to compute the inclination from the above formula. The node can be found by searching for the zero values of β, and therefore of b.

Copernicus's Obliquation

In Copernicus's theory of the superior planets, the geocentric latitude observations of Ptolemy and others were used to find the inclinations. Observed latitudes are in general largest near opposition and smallest near solar conjunction and when 90° from the node. Although observations cannot be made exactly at conjunction, latitude values can be approximately computed from other well-spaced observations. From many observations it appeared that the maximum values for latitudes occurred at about 90° east of the line of nodes, i.e., at heliocentric longitudes $B_0 \approx ☊ + 90°$. However, values of B_0 often differed a few degrees from ☊ ± 90° due to the asymmetry of the orientation of the orbits of the planet and the Earth, or to erroneous values for inclination, line of nodes, or distances. Tycho was well aware of this phenomenon, and Kepler explained it and corrected it as soon as he could supply the correct elements and distances based upon elliptic orbits (see Chapter VII). Copernicus found the following values of B_0 (measured from ♈): for Mars, 117°; for Jupiter, 207°; for Saturn, 217°. These directions were within about 20° of the perpendiculars to the lines of apsides of these planets. It further appeared to him that the inclination of each planet was not constant but exhibited a variation with a period equal to the planet's synodic period.

Following Herz (1887), let

i = the instantaneous inclination,

i_0 = the mean inclination $= \frac{1}{2}(i_{max} + i_{min})$,

i' = the amplitude $= \frac{1}{2}(i_{max} - i_{min})$,

M' = the commutation angle = the angle between the heliocentric directions of the Earth and the mean position of the planet,

ℓ_P = the heliocentric longitude of the planet,

$\ell_\oplus$ = the heliocentric longitude of the Earth.

Then $M' = \ell_P - \ell_\oplus$ is the difference between the heliocentric longitudes of the planet and the Earth. The period of M' is one synodic period of the planet, with a value of zero at opposition and 180° at conjunction. Consequently, $i = i_0 + i' \cos M'$, with a maximum at opposition and a minimum at conjunction. From spherical trigonometry the corresponding heliocentric latitude b is found from $\sin b = \sin i \cos(\ell_P - B_0)$. The corresponding geocentric latitude β is found from $\sin\beta = \frac{R}{\rho} \sin b = \frac{R}{\rho} \cos(\ell_P - B_0) \sin i$, where ρ denotes the distance between the planet and the Sun and R denotes the distance of the planet from the Earth. Finally,

$$\sin\beta = \frac{R}{\rho} \cos(\ell_P - B_0) \sin(i_0 + i' \cos M') . \tag{1}$$

In this formula the cosine factor has a periodicity of one sidereal period of the planet, and the sine factor varies over one synodic period. This formulation for latitude includes the *obliquation*, or periodic oscillation of the planet's orbit about the line of nodes passing (erroneously) through the "center" of the Earth's orbit, called the "center of the ecliptic" by Copernicus. Only the cosine term ever attains negative and zero values. By means of equation (1), i_0 and i' can be found from two observations of β. (This formula cannot be developed into a series because i' is never of higher order than i_0.) Copernicus adopted the values shown in Table V.3.

Table V.3

Planet	i_0	i'
Mars	1°00′	0°51′
Jupiter	1°30′	0°12′
Saturn	2°30′	0°14′

The obliquation factor $\sin(i_0 + i' \cos M')$ contains the planet orbit's mean inclination i_0 to the ecliptic (deferent and epicycle in one plane) about the line of nodes, and the amplitude i' of the "obliquation" term. (According to Zeller (1943), the axis of obliquation used by Copernicus for any planet is the line perpendicular to the line joining the centers of the instantaneous orbits of the Earth and planet. This is only very approximately the line of nodes of the planet orbit. For Mars the difference between the above-mentioned perpendicular and the line of nodes was about 8° in A.D. 1500.)

Thus obliquation is seen to have the following definition: the correction to the geocentric latitude resulting from a sinusoidally varying oscillation (period equal to the sidereal period of the planet, amplitude itself sinusoidal with the synodic period) of the plane of the planet orbit, about a mean inclination of the orbital plane and referred to an erroneous line of nodes passing through the mean position of the Sun (not through the Sun itself)! The correction is zero at the quadratures and maximum at the syzygies, where $i' \cos M' = \pm i'$. It is given by the above formulas along a latitude circle, and its value along a declination circle may be easily found. However, even though the value is a correction to the latitude, not to the declination, Copernicus called it *declinatio* if seen radially from the Earth and *reflexio* if seen tangentially to the planet orbit, or not quite radially. Of course it can be seen tangentially only for the orbits of the inferior planets. To sum up: A pure *declinatio* can be observed only for a superior planet (at opposition) but never for an inferior planet, because of its conjunction with the Sun at the time of its occurrence. Similarly, a pure *reflexio* can be observed only for an inferior planet at greatest elongation. All other observations of celestial latitude are compounded of both *declinatio* and *reflexio.*

A Mechanical Model of the Obliquation

A mechanical aid to the visualization of Copernicus's obliquation movement of the planetary orbit planes may be imagined as follows:

In Figure V.15 let BFC be the ecliptic. Let the center of the celestial sphere of radius r be C_1, and let the ascending and descending nodes of the orbit plane, be B and C. For simplicity take the point $B_0 = 90°$ east of B, in the "average" orbital plane BB_0C. Let the positions of the planetary orbit plane at minimum, average, and maximum inclination be $Bh'C$, BB_0C, and BhC. The vertices of the orbit plane with respect to the ecliptic in these crucial positions are thus h', B_0, and h, all located on the celestial longitude colure FK, 90° east of the ascending node B. Also, the inclinations of the orbit in these different crucial positions are $\angle FC_1h' = i_0 - i'$, $\angle FC_1B_0 = i_0$, and $\angle FC_1h = i_0 + i'$.

Now imagine a small circle (center b, radius bh) described on the sphere (center C_1, radius $C_1b = r$). Let a solid wheel (center b', radius d, handle h) be mounted in a solid plane framework $aa'b'f'f$, fixed perpendicular to C_1b' at b'. At h let the wheel have a pivot, riding in the circular groove of diameter $h'h$.

If now the wheel, starting from h, is turned uniformly in either direction, the plane of the deferent BhC will oscillate with simple harmonic motion about the line BC_1, thus producing at maximum a change of inclination of the orbit to the average inclination i_0, which at any time is $i' = \frac{d}{r}$, since i' is small. If now the period of rotation of the wheel is taken equal to the commutation period of M' measured from B_0h, the value of i will be close to that desired, namely, $i = i_0 + i' \cos M'$.

Some Special Observational Facts

Before describing Copernicus's celestial latitude theory for the inferior planets, it is well to note some observational facts known since the time of Ptolemy but often overlooked – though not by Copernicus, who was a great student and admirer of Ptolemy. We refer to the somewhat cryptic statement by Dreyer (1905): "When the apogee or perigee of the planet is turned towards the Earth, Venus always deviates most to the north and Mercury to the south." We will clarify this statement shortly, but the principle involved was fully recognized by Copernicus, and to comply with it he formulated a complicated theory for the latitude changes undergone by the inferior planets. In addition to an obliquation about the line of nodes similar to that of the superior planets, he introduced two *more* movements into each orbit. He even went so far as to form a theory in which Venus *always* deviated to the north and Mercury to the south.

Some commentators (e.g., Herz 1887) have erroneously held that these latitude deviations of the inferior planets were not a consequence of their real motions but a fallacious conclusion from observational errors by Ptolemy and others. The principle enunciated by Dreyer, however, is a real consequence of the relative orientations, sizes, and shapes of the orbits of the three innermost planets. A modified statement of the principle, more in keeping with our present discussion, is as follows:

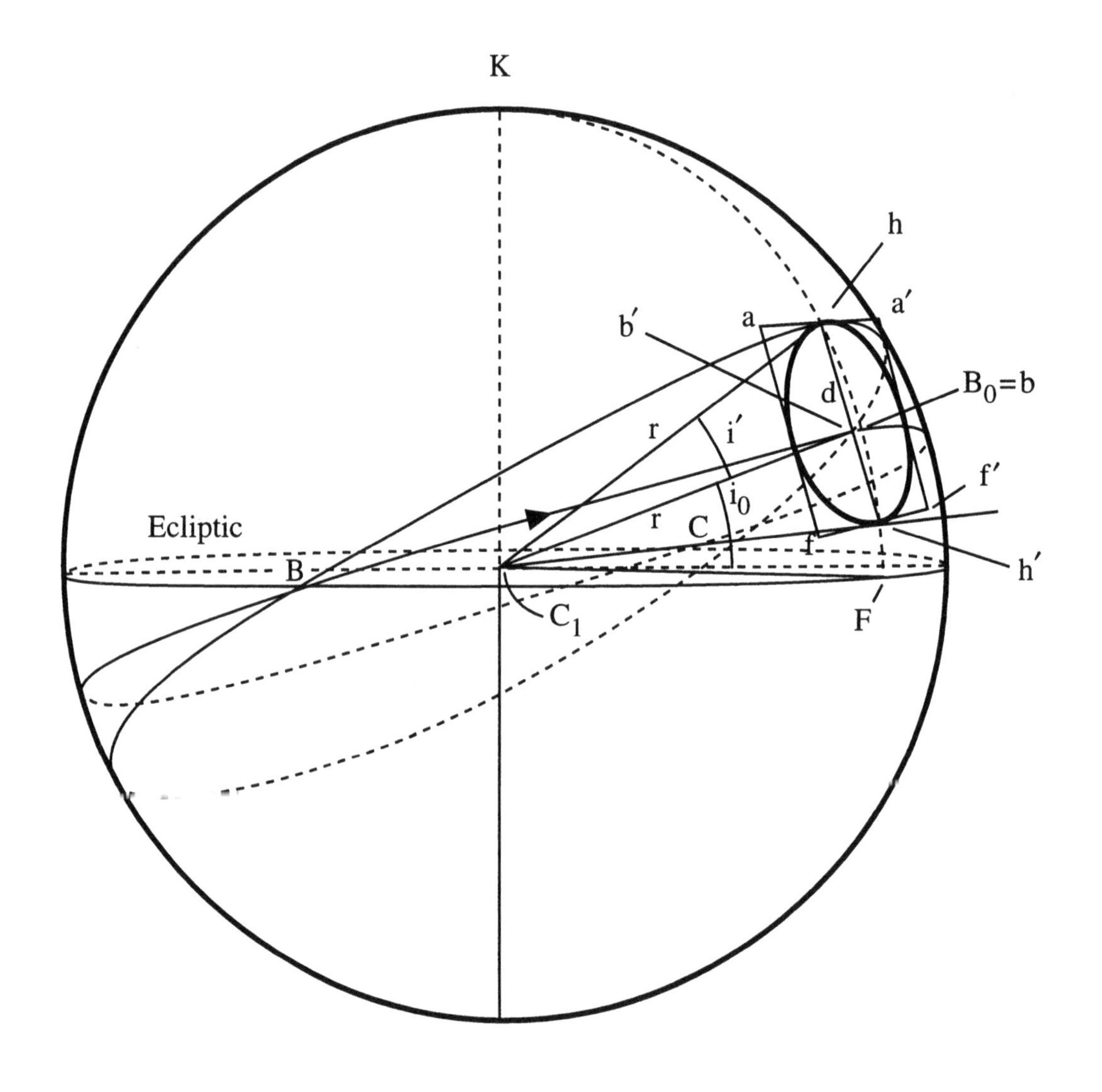

Figure V.15 — Mechanical model of Copernicus's obliquation (seen from the outside and from the north, not to scale).

For Venus: When the heliocentric longitude of the Earth is "not too far" (i.e., 30° to 40°, say) from that of the aphelion or perihelion of the Earth's orbit, deviations of Venus in either heliocentric or geocentric latitude are mostly positive. (The angular distance of the Earth from its own perihelion is of greater importance than its angular distance from the perihelion of Venus's orbit because (1) the Earth's orbital eccentricity is more than twice that of Venus, and (2) Earth's radius vector is about 40% longer.)

For Mercury: When the heliocentric longitude of the Earth "is not too far" from that of the aphelion or perihelion of Mercury's orbit, deviations of Mercury are always negative.

In both cases the latitude deviations are defined as observed values minus those computed from equation (1).

At the time of Hipparchus (and for purposes of illustration even today) the eccentricities, inclinations, lines of nodes, and positions of perigee of the three innermost planets were approximately as given in Table V.4.

Table V.4

	Orbital Eccentricity	Longitude of Ascending Node	Longitude of Perihelion	Orbital Inclination
Mercury	0.20	51°	73°	7°
Venus	0.007	86°	130°	$3\frac{1}{2}°$
Earth	0.016	—	95°	—

Figure V.16 provides a view of the situation described by this table as seen from the north side of the ecliptic plane. For simplicity the orbits, whether circular or elliptic, are referred to C, the average annual position of the Sun, or the Earth's orbital center. The northern part of each orbit (from the ascending node N to the descending node N') is shown by a heavy curve, the southern by a broken curve. A cursory glance at the figure reveals that (1) the perihelia of both inferior planets lie on the north side of the ecliptic plane, and (2) the longitude of the Earth's perihelion is "not too far" from those of Mercury and Venus, 22° and 35°, respectively. The predominantly northern latitude error for Venus obtains mainly because of the smallness of its orbital eccentricity compared with that of the Earth. In the case of Mercury's orbit the principal cause of the planet's southern latitude error is the smallness of the Earth's orbital eccentricity as compared with that of Mercury.

A more detailed study of Figure V.16 permits a qualitative appreciation of the above-mentioned rules. We note that the perifocal radius of an elliptic orbit of eccentricity e is shorter than the semimajor axis a by the distance ae; and that for about a quadrant of anomaly on either side of perihelion, the distance from the Sun in an elliptic orbit is smaller than in a circular orbit of center C and diameter $2a$. In Figure V.16 we neglect the slightly different positions of the centers of the orbits (i.e., from Figure V.1 we let $c_1 = c_2 = c_3 = C = S$). Further, in Figure V.16 let

C = the average position of the Sun,

$EB_1E'B_1'$ = the (circular) Copernican orbit of the Earth,

$P_1B_1''A_1B_1'''$ = the actual (elliptic) orbit of the Earth,

$VB_2V'B_2'$ = the (circular) Copernican orbit of Venus,

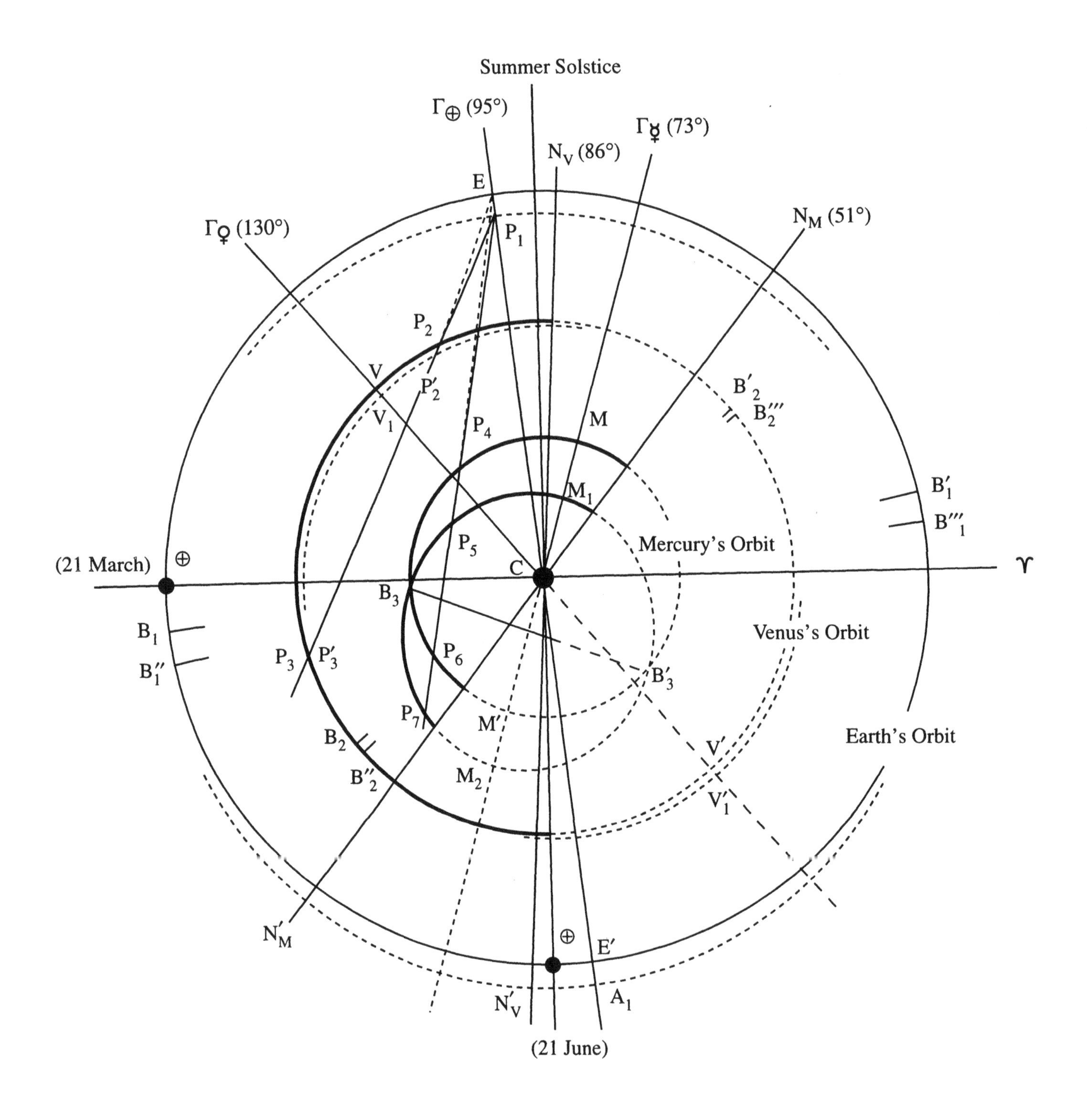

Figure V.16 — Relative positions of the orbits of Earth, Venus, and Mercury. Curtate view, not to scale, about 100 B.C., referred to the Sun and to a convenient equinox. Thin lines in the plane of the ecliptic, heavy lines above (north) the plane of the ecliptic, and broken lines below (south) the plane of the ecliptic.

$V_1 B_2'' V_1' B_2'''$ = the actual (elliptic) orbit of Venus,

MP_4M' = the (circular) Copernican orbit of Mercury, and

$M_1 B_3 M_2 B_3'$ = the actual (elliptic) orbit of Mercury.

The last two orbits are coplanar at an inclination about $N_M N_M'$ of $\sim 7°$; Venus's orbit plane has an inclination about $N_V N_V'$ of $\sim 3\frac{1}{2}°$.

Observations are best performed near an elongation with the planets rising in the morning sky (as is the case in the figure) or setting in the evening sky. For morning observations let P_2 and P_3 be two observed positions of Venus among the fixed stars, and let their observed (reduced) geocentric latitudes be O_2 and O_3. Let the corresponding latitudes (from the Earth at E) as computed under Copernican theory be C_2 and C_3. These values were computed by using $\rho_2 = EP_2$ and $\rho_3 = EP_3$ in equation (1). But the Earth is actually in its elliptical orbit at P_1 and Venus at P_2'. Therefore the correct distances to use in the computation are $\rho_2 = P_1P_2'$ and $\rho_3 = P_1P_3'$, both of which are shorter than the corresponding Copernican values. It follows from equation (1) that the values of C_2' and C_3' computed with these correct distances are larger than C_2 and C_3, respectively. Excepting observational errors, we then have $C_2' - C_2 = O_2 - C_2 > 0$ and $C_3' - C_3 = O_3 - C_3 > 0$.

Similar considerations reveal positive residuals obtained with the Earth at or near A_1 and the planet south of the ecliptic and in the morning sky. The deviations are then toward the south; but calculated with the correct distances for Earth and Venus, they are then smaller than those computed from the wrong distances in the best-fitting circular (Copernican) orbits.

In the case of Mercury's latitudes these considerations also apply, but they are far outweighed by the large eccentricity of Mercury's orbit, acting in the opposite direction. Let P_5 and P_7 be two observed points on the path of Mercury among the fixed stars, their observed (reduced) geocentric latitudes being O_5 and O_7. As before, the Copernican values of the corresponding quantities C_5 and C_7 are computed by equation (1) using $\rho_5 = EP_4$ and $\rho_6 = EP_6$. Due to the elliptical shape, high eccentricity, and location of Mercury's orbit, the true values of the distances to be used, P_1P_5 and P_1P_7, are both longer than the Copernican distances EP_4 and EP_6, respectively. Since the latitudes are both north, the values of the residuals are both negative. Similar considerations reveal that negative residuals are also obtained when the Earth is at or near A_1 and Mercury is south of the ecliptic, in the morning sky. The actual computations are of course more rigorous, but the qualitative results remain the same.

On the basis of these facts, as true in A.D. 1500 as they were in 100 B.C., Copernicus constructed his mind-boggling latitude theory for the two inferior planets as follows.

Copernicus's Theory of the Celestial Latitudes of the Inner Planets, Involving Both Obliquation and Deviation

Because of the better approximation to the true conditions of planetary motion embodied in Copernicus's viewpoint, one might expect a corresponding simplification from Ptolemy's complicated theory of the planetary celestial latitudes. No considerable improvement followed immediately from the adoption of the heliocentric viewpoint, however, mainly because the distances and longitudes, still based on circular deferents without equants, were prohibitively in error. (As we shall see in Chapter VII, Kepler was the first to give a correct

Table V.5 — Sources of Celestial Latitude Deviations in the Systems of Ptolemy and Copernicus

Term		Superior Planet (Assuming $\Gamma\Gamma' \perp$ ☊☋)		Inferior Planet (Assuming $\Gamma\Gamma' \parallel$ ☊☋)	
		Ptolemy	Copernicus	Ptolemy	Copernicus
Epicyle	Loxosis	"rolling"	—	"rolling"	—
	Engklisis	"pitching"	—	"pitching"	—
Deferent	Obliquation	—	"oscillating"	"oscillating" slightly	"oscillating"
	Deviation	—	—	"gyrating"	"oscillating and yawing"

Loxosis ("obliquity"): Tilting of epicyclic plane about $\Gamma\Gamma'$, the line of apsides; "rolling."

Engklisis ("inclination"): Tilting of epicyclic plane about the axis perpendicular to the line of apsides; "pitching."

Obliquation: Tilting of deferential plane about the line of nodes.

Declination: The part of the obliquation seen perpendicular to the line of nodes.

Reflexion: The part of the obliquation seen along the line of nodes.

Deviation: Tilting of deferential plane about a line rotating (advancing) in the instantaneous obliquation plane; "yawing."

explanation of the celestial latitudes, after he had remedied many accumulated defects of the earlier systems.) Copernicus not only found it necessary to retain the *obliquation* of the orbit planes about the lines of nodes (for the inferior planets he considered these coincident with the lines of apsides) but also introduced one of Ptolemy's details, the *loxosis*, in a modified form – and added a further complication of his own, called the *deviation* (see Table V.5).

The obliquation, or periodic fluctuation of the inclination of the orbit plane, was retained, as in the case of the superior planets, but with the modification that the plane oscillated about the line joining the planet's apsides (or nodes), and not, as in the case of the superior planets, about a line placed 90° from B_0, the point of maximum geocentric latitude (north or south). Since this point is not always exactly 90° from the node, the axis of oscillation was only approximately parallel to that of the line of nodes. The period of the obliquation was in all cases $\frac{1}{2}$ sidereal year of the Earth (neglecting the secular advance of the planetary perihelia). The obliquation reached a maximum when the Earth was on the line of apsides of the planet orbit (when the commutation angle M' was zero).

The discussion of the latitude deviations of all five planets is in book 6 of *Revolutions*. The explanation proceeds in terms of trigonometric processes involving several (permissible) approximations, appropriate to the smallness of the angles considered. The constants are determined from a few observations (as usual). The results, rather than being given by formulas, are embodied in tables giving corrections at 3° intervals for the "anomaly of the eccentric circle" (approximately the true anomaly). Directions are given for easily finding values of the corrections for any anomaly. The use of the tables is so described that "the deviations always remain northerly for Venus, southerly for Mercury." The great work of *Revolutions* ends abruptly with these directions, as if Copernicus, in true astronomical fashion, had been content to determine the quantities with which he was dealing and to make available his results, not his theoretical viewpoints. In contrast Rheticus, who must have been an excellent teacher as well as a superb mathematician, in *Narratio Prima* elucidates the subject much further for Schoener, to whom he "narrates" his discoveries received from privileged daily contacts with Copernicus. Yet he gives no diagrams, but rather long, cumbersome (yet clear) descriptions of the various planes, their tilts, oscillations, and rotations. At the end of these descriptions he says that Copernicus himself considered "this complicated mechanism, which in essence is modeled on Ptolemy's work," to be too obscure; and that Copernicus, after finishing the manuscript of *Revolutions*, made additional efforts to clarify the latitudinal movements of the planets, by a study of the oscillations of the poles of the various planes involved, as he had done in the case of the precessional and trepidational movements of the celestial poles. He had also temporarily considered an explanation requiring the orbital planes to be parallel to that of the ecliptic but at some distance from it, Venus to the north and Mercury to the south.

As if to indicate that his latitude theory for the inferior planets was too complicated, Copernicus tried another explanation of the deviation assuming that the center of the planet orbit was always on one side of the plane of the ecliptic and at a distance from it varying in a period of one year. He soon gave up this idea as unsatisfactory and too much of an *ad hoc* assumption.

According to Dreyer (1905):

> It was natural that Ptolemy should find great difficulty in representing the latitudes, since he had to let the line of nodes pass through the Earth instead of through the Sun. But Copernicus also erred, though to a smaller extent, by letting it pass through the center of the Earth's orbit. [This is 5 million kilometers from the Sun, toward Sagittarius.] This displaced the nodes, so that a planet was found to have some latitude when it ought to have none (or been in the ecliptic), and this amount of latitude varied with the place of the Earth in its orbit. For the same reason the greatest north latitude of a planet would turn out different from its greatest south latitude, and the amount of the difference would also seem to vary with the position of the Earth. No wonder that it was necessary to assume oscillations of the orbits.

To a question by Rheticus about the dearth of observations utilized in building the system, Copernicus answered that he would be as delighted as was Pythagoras when he discovered his famous theorem if he could make his planetary theory agree with the observed positions of the planets within $10'$. The accuracy of the system was very far from reaching this modest limit. A fundamental error was taking the center of the Earth's orbit as the center of all motion. Kepler pointed out that this error alone could result in a discrepancy of nearly 5° in the longitude of Mars when at right angles to the line of apsides of its orbit.

Such then are the main features of the Copernican system as presented in *Revolutions*, a masterpiece of constructive astronomy that in its time could only be equaled by Ptolemy's *Almagest.* Although *Revolutions* is far more detailed than appears from the present discussion, a sufficient appreciation of it has been given here to dispel the widely held view of the Copernican system as merely coplanar circles concentric in the Sun.

CHAPTER VI

Tycho Brahe (1546–1601)

In his preface to the *Rudolphine Tables* (1627), Kepler gives credit to his older colleague Tycho for having strongly stressed the necessity of producing extended and continuous series of observations of the Sun, Moon, and planets at *all* times, not merely at a small number of interesting points such as syzygies, quarters, and greatest elongations.

Kepler says of Tycho: "He was a man of outstanding nobility in the Kingdom of Denmark, who chose for himself the restoration of astronomy with his immense intellect, spending his life and the splendid family fortunes on his task. He was a most outstanding man, the proposer of the *Rudolphine Tables*, the recorder of a thousand fixed-star positions, the investigator of the motions of the Sun and the Moon (with an accuracy that even Hipparchus could not approach), the observer of all the planets for 38 years (and continuously for the last 20). He excelled all human expectation in diligence, observation, patience, and reliability. The restoration of astronomy he first conceived in the year 1564."

According to Dreyer's 1890 biography, Tycho already had the grand concept of overhauling the science of astronomy in 1562 as a sixteen-year-old student at the University of Leipzig. And only a little later he acquired his cosmopolitan culture and his rare view of astronomy as an international science. Three years earlier Tycho had enrolled as a student in the University of Copenhagen, the wish of his family being to give him a good education in rhetoric and philosophy as a background for high administrative or ambassadorial service. The boy already read, wrote, and spoke Latin with ease, having studied this cosmopolitan language since the age of seven. Between 1560, when his interest in the science of astronomy was first aroused by his observing a partial solar eclipse at Copenhagen, and 1563, when he observed the "Great Conjunction" of Jupiter and Saturn, he had estimated enough planet positions with his primitive cross-staff to prove that errors up to 5° occasionally existed in the *Alphonsine Tables* then in common use. He also showed that nearly as large errors (4°) were possible in working with the more recent *Prutenic Tables* of 1551. While in Copenhagen Tycho studied mathematics with Pratensis, who became his lifelong friend. He also bought himself a splendid copy of Ptolemy's *Almagest* with a commentary by Theon bound in green velvet. This book has been preserved and contains many annotations in Tycho's hand.

For 10 years (1562–72) Tycho traveled extensively in Germany, Switzerland, and Venice. He studied at famous universities such as Leipzig, Rostock, Wittenberg, Augsburg, and Basel, concentrating on mathematical sciences and astronomy. His broad view of the needs of astronomy was fortified in 1569 by his meeting in Basel with Petrus Ramus, a professor from the Sorbonne in Paris who had eloquently expressed similar viewpoints.

From 1570, when he permanently returned to Denmark, until 1575 Tycho lived with his maternal uncle, Steen Bille, at Knudstrup, some 20 miles east of Helsingborg. A few miles from there was Herrevad Abbey, a Benedictine monastery prior to the Reformation (which had been introduced into Denmark in 1536). The abbey still had a staff of ecclesiastics and was used for Protestant religious services. Steen Bille had been directed by the Crown to "clean up" the place, where "ungodly living" had developed among the residents. Among other things he was "to drive out all learned, superfluous, and useless persons" from the premises. Nevertheless he kept his own alchemical laboratory there and arranged for an observatory where Tycho could set up many of the splendid instruments acquired during his European travels. Tycho was a good alchemist, astrologer, chemist, and therapist, and he concentrated for a while almost exclusively on these subjects. Later that year (1570) Steen Bille died, and Tycho inherited one-half of the property of Knudstrup.

On the evening of 11 November 1572, upon emerging from his chemical laboratory, Tycho saw near the zenith, in Cassiopeia, a blazing new star (a supernova, as we now know), at its peak brighter than Venus and easily visible in the daytime even through thin clouds. This exciting experience rekindled his ambition to overhaul astronomy. He immediately wrote a small treatise on the new star, containing his observations of its position, proving it to be a fixed star that belonged to the "ethereal regions" and not to sublunar space. The book was sent to many European authorities and immediately made him world famous. By 1575, his reputation as a leading astronomer was well established; he made a tour of Europe and visited friends and astronomers in Wittenberg, Augsburg, Basel, and Venice. In Augsburg he attended the coronation of Rudolph II of Bohemia as Holy Roman Emperor and also made acquaintances that much later became valuable when he emigrated to Prague.

Tycho's life was forever changed when on 23 May 1576 King Frederick II of Denmark conferred upon him the island of Hven in fief for life. The next day he was already taking first observations from the new site. The weather was beautifully clear, and on the evening of 24 May Tycho observed a conjunction of Mars and the Moon with one of his portable instruments. In the words of A. Petersen (1924): "Anyone superstitiously inclined could consider this observation an auspicious omen, for these two celestial bodies were to be the ones most assiduously observed at Uraniborg and became the two greatest causes of Tycho's fame, the first because of its being important for Kepler's epoch-making discoveries concerning the shape of the planetary orbits, the second because its motion across the sky was later determined by Tycho with far greater accuracy than had been possible before and led to the redetermination or discovery of all the lunar perturbations visible to the naked eye."

Tycho received not only the 2000-acre island plus the income from its hundred tenants but also a cash sum "for the building of a house," by which Tycho understood the erection of the fabulous Uraniborg castle and observatory. Later Tycho received a truly royal salary of 4000 rigsdalers annually, making him one of the richest men in Denmark.

There were several reasons for the king's magnanimous support of Tycho. First, he probably felt a debt of gratitude toward the Brahe family, since Tycho's uncle and foster father Jörgen Brahe had lost his life in 1565 only days after saving the king from drowning in the icy waters of the palace moat. Second, the astronomer and *landgrave* (count) Wilhelm of Hesse had written a letter to the king pointing out the enormous prestige that would be lost to Denmark if Tycho were to settle anywhere abroad. And third, it may be surmised that Tycho was already a hard bargainer, even with kings and emperors, whom he considered his equals, if not in some respects his inferiors! Before the final arrangements about Hven were made, he had refused the offer of any one of several castles as the price for his staying in Denmark. Yet he was always in debt because of his extravagant expenditures on instruments, mechanical devices, printing press, bindery, paper mill, forge, chemical ovens, laboratories, pharmaceutical gardens, zoo, and buildings. The king bailed him out again and again, but when in 1588 Christian IV came to the throne, the patience of the Crown gradually and inevitably gave out. Immediately after his coronation, Christian IV once more generously assumed a large debt incurred by Tycho. But the latter, instead of retrenching his style of living, kept on in his usual extravagant manner. Enemies at court, who greatly resented Tycho's large salary, gradually caused Christian IV to deprive Tycho of all sources of income that were not deeded to him for life. Although the later years at Uraniborg were undoubtedly unhappy ones for Tycho, perhaps mainly due to himself, his great astronomical work was never allowed to lag.

There is no doubt that Tycho relentlessly drove his subordinates. Not only did his tenants despise him, but Kepler, after praising him to the skies for his astronomical excellence, stated that "no one can work long with Tycho without being subjected to the greatest humiliations." Tycho was despised also by many of his peers. With the loss of income, the question arose of finding another government that could sufficiently appreciate his work to employ him at an adequate salary. Tycho's final choice was the position of imperial mathematician to Emperor Rudolph II of Bohemia, himself an avid alchemist. Tycho, as we have seen, was not only an astronomer but also an alchemist and physician. The emperor's physician, Hagecius, had been a close friend of Tycho's ever since they had met at the coronation of Rudolph. Hagecius was an alchemist and astronomer, as well as a physician. Thus there would be a strong concurrence of interests if Tycho were to settle in Prague. This he did in 1599 after three years of strenuous travel, but he died only two years later.

During his life Tycho succeeded in nothing less than completely overhauling astronomy. As part of this he redetermined with superior accuracy every important astronomical constant then known, with the single exception of the solar parallax.

Tycho's Main Astronomical Contributions

At the end of a description of his 22 major instruments Tycho (1598) includes a section "On that which we have hitherto accomplished." Primarily from this is derived the following list of Tycho's principal accomplishments, as he saw them:

1. He improved the general accuracy of observation by one order of magnitude by
   a. building larger and more stable quadrants, sextants, and octants,
   b. using a system of diagonal divisions on the arcs of his instruments, and
   c. inventing a slit-pinule for sighting a celestial object by an alidade, thereby increasing the accuracy of sighting from about $2'$ to about $15''$–$30''$.
2. He investigated atmospheric refraction at all altitude angles and constructed a more reliable table of refraction. The results were good to $1'$ except at very low altitudes, where the error increased to about $5'$. Earlier tables were good only to about $20'$.
3. He accurately observed the position and estimated the brightness and color of the nova in Cassiopeia of 1572 and proved from its lack of parallax that its distance exceeded that of the Moon.

4. He accurately observed the positions of the Great Comet of 1577 and published them in a book that was sent to several scientists, including Galileo. He showed that the comet at one time was outside and at another time inside the crystalline sphere of Mars, thus disproving the physical nature of that sphere. He also made many accurate observations of six other comets over the following 20 years.
5. He advanced the Tychonic system of the cosmos, in which the Sun moved about a fixed Earth, but all the planets moved about the Sun.
6. He made a catalogue of 777 fixed-star positions with unprecedented accuracy amounting to an average error of $\pm 48''$. This was later extended to 1000 stars with an average error of $24''$ in declination and $25''$ in right ascension.
7. He made 20 years of accurate, continuous observations of the Sun, Moon, and planets, particularly Mars, with an accuracy of about $2'$.
8. For the Moon's orbit he improved the values of the equation of the center and of the evection. He discovered and evaluated the "variation," as well as discovered and estimated the value of the "annual equation." At the time of his death he had work in progress to improve this estimate, "almost discovering" the parallactic inequality.
9. He improved Copernicus's estimated value for the advance of the Sun's apogee with respect to the vernal equinox (Copernicus: $24''$/yr; Tycho: $41''$/yr; modern: $61''$/yr).
10. He determined the length of the tropical year to better than 1-second accuracy.
11. He constructed tables of the Sun's position to an accuracy of perhaps $10''$, certainly $20''$. Earlier tables were good to only $15'$–$20'$.
12. He redetermined the value of the precession of the equinoxes and rejected the trepidation.
13. He discovered and determined the variation of the Moon's orbital inclination.
14. He determined the rate of regression of the nodes of the Moon's orbit.
15. He explained the 19-year variation of the inclination of the lunar orbit using an epicycle of $4^\circ 58'$ radius.

The Tychonic System

In his great work on the comet of 1577, published in 1588, Tycho included a section on the planetary system that he had developed as early as 1573. A sketch of the system is shown in Figure VI.1. The Earth was stationary at the center of the Universe, whose boundary was the fixed-star sphere (with a radius of 1.21 A.U.). The distance of the Moon was about 52 Earth radii (which is the correct order of magnitude) and that of the Sun, 1040 Earth radii (more than 10 times too small). The greatest distance of Saturn was 1.06 A.U. (about 10 times too small).

To account for the diurnal motion of the heavens the firmament of fixed stars rotated uniformly westward about the Earth in a period of 24 sidereal hours, dragging the Sun, Moon, and planets with it. The material world extended to the Moon's orbit, and anything beyond was ethereal, devoid of mass. There were no solid crystalline spheres. Closest to the Earth the Moon completed its eastward orbital motion in 27.3 days. Much farther away the

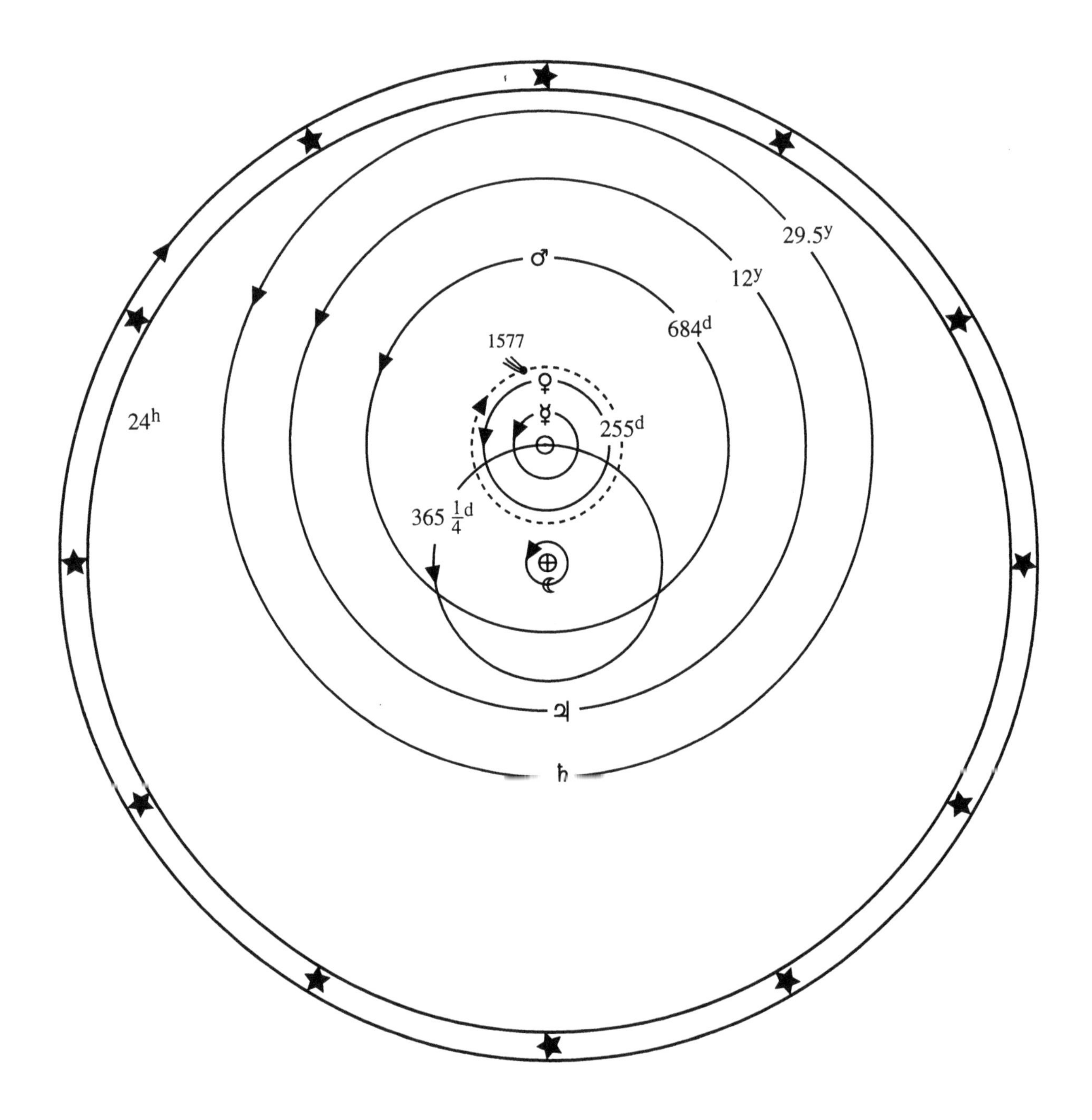

Figure VI.1 — The Tychonic system of the world (not to scale).

Sun moved eastward in an eccentric circle, completing its course in 1 year, serving as an approximate center for all the circular orbits of the planets and for that of the Great Comet of 1577. The planets moved eastward about the Sun in their respective sidereal periods. The comet moved about the Sun in a retrograde circular orbit just outside Venus. However, the orbital motion of the comet was not uniform, an important detail. Still more important was Tycho's suggestion that "the figure of the orbit may not have been exactly circular but somewhat oblong, like the figure commonly called an oval." Its period, if any, was unknown. All the planets moved in circular, direct orbits centered near the Sun (and moving with it). They were in the usual order: Mercury and Venus between the Sun and the Earth; then Mars, Jupiter, and Saturn, in successively larger orbits surrounding both the Sun and the Earth.

Because of the reliability and accuracy of his observations Tycho soon found it necessary to introduce all sorts of epicycles and/or eccentrics in describing the details of his system. Nevertheless his system was a historically important compromise appealing to those uncomfortable with both the Ptolemaic and Copernican systems. The retrograde motion of the outer planets when at opposition was on the whole naturally and well explained by an excess of apparently westward drift of that half of the orbit enveloping the Earth, caused by the Sun's eastward angular drift about the Earth always being in excess of the planet's angular drift about the Sun. All other main features of planetary motions were at least as well explained as in the older systems. The Tychonic system avoided the main objection of the confirmed Ptolemaists, namely the assumption of a moving Earth, and at the same time it also met the chief objection of the confirmed Copernicans, the existence of crystalline spheres. However, Tycho's lunar theory was fully as complicated as anything the ancients had produced. He also left none of the planetary orbits, except perhaps that of Venus, exactly centered on the Sun but referred them to points eccentric to the Sun, just as the Sun's orbit was eccentric to the Earth.

Note on Reymers's System

In 1588 Tycho published his system in a book, *On the Aethereal World*, but only a few copies were distributed to friends and correspondents. The system nevertheless soon became extensively known and several lecturers claimed it as their own invention. In the same year Nicolai Reymers, often known as Ursus, published his *Fundamentals of Astronomy*, the last chapter of which treats of "his new hypothesis" of planetary motion. This system is exactly like Tycho's except for two main differences. First, Mars's orbit about the Sun entirely surrounds the Sun's orbit about the Earth. Second, the diurnal westward motion of the heavens is caused by the Earth's eastward rotation on its axis, as in the Copernican system. The first difference is very important astronomically; the second is solely of cosmological importance.

It seems that the failure of Reymers to ever mention Tycho's name in his book caused Tycho to accuse him of having stolen the idea of the system during his visit to Hven in 1584. A priority feud concerning the two systems ensued. Much can be said on both sides of the controversy. One of Kepler's first assignments when he joined Tycho was to write a pamphlet in defense of Tycho's priority to his system. During the late decades of the 16th century the safeguarding of priorities was a real problem. Even in 1610 Galileo had to announce his telescopic discoveries in secret anagrams. As so often happens when a new intellectual synthesis is felt to be necessary, the main outlines of the solution may occur independently and simultaneously to several persons. Reymers, then court mathematician of Rudolph II, may well have been capable of originating a conceptual solution. After all, his system is cosmologically one step closer to the Copernican view than is Tycho's, in that it admits the as-yet-unproved diurnal rotation of the Earth, a fact subscribed to by several of the best minds of the age. Although Reymers's system is astronomically inferior to that of Tycho, it seems fair to consider it one of the stepping stones to the final acceptance of the Copernican planetary arrangement, even though it is not bold enough to enlist the Earth as a planet. This last step was so difficult that its full acceptance came only about a hundred years later. Neither Galileo's mental bludgeoning nor Kepler's epoch-making discoveries could produce the necessary intellectual revolution that eventually came with Newton's comprehensive synthesis.

Tycho's Solar Theory

Tycho's solar observations illustrate the extreme accuracy and completeness of his observational work. Starting late in 1576, the year Uraniborg was built, he observed the Sun's transit time and meridian altitude regularly every day of the year, if possible; somewhat later, this was done with three or four metal instruments simultaneously. From 1582 on, the great mural quadrant was employed. Circumspect employment of giant instruments with diagonally graduated arcs added a new dimension to observational science and suddenly increased the attainable accuracy by fully one order of magnitude. He soon determined improved values of the times and positions of the equinoxes and solstices. The position of the summer solstice was especially well determined, since it was based upon solar positions halfway between the equinoxes and halfway between positions at declinations near $+16°$, permitting the result to be little influenced by refraction (in contrast with Copernicus's use of midway *southern* declinations).

From his well-determined values for the times and positions of the equinoxes and solstices for the years 1584–88, Tycho derived an accurate ecliptic inclination and, by Hipparchus's method (see Chapter III), the elements of the solar orbit considered as one of uniform motion along an eccentric circle. He found the longitude of the solar apogee to be $90°30'$ (modern value: $90°10'$ for 1586), advancing annually by $45''$ (modern value: $61''$), an improvement on Copernicus's value of $24''$. The eccentricity of the circular solar orbit was found to be 0.03584, an accurate figure for this model. He also determined the length of the tropical year to be $365^d5^h48^m45^s$, only about 1^s too short! This was effected by a comparison of his own equinox determination with that of Bernhard Walther of Nürnberg about a century earlier, after the earlier determination had been rereduced using a better latitude for the place of observation.

From his profusion of accurate solar observations, often involving following the Sun across the sky with large armillaries throughout the day, Tycho constructed a table of the Sun's apparent yearly motion along the ecliptic. The averages of his observations, and therefore his tables, were accurate to $10''$–$20''$! Other Renaissance astronomers such as Regiomontanus, Walther of Nürnberg, the *landgrave* of Kassel, and Hainzel in Augsburg, all using steel instruments, had also been able to make good solar observations. Tycho showed that their errors were generally a small fraction of an arc minute, while the Alfonsine and Copernican tables were in error by $15'$–$20'$.

Tycho's Orbit of Saturn: An Example of an Outer Planet

Tycho observed all the planets with the greatest possible accuracy and continuity, but he did not derive their orbits in detail, although he knew many exact details such as the nodes, inclinations, and positions of apogee. As a result we have only the merest sketch of the kind of construction he favored for the description of their orbits. Figure VI.2 shows the main features of the system of Saturn. The Earth is stationary in the center of the Universe. The Sun ☉ revolves in a slightly eccentric circle about the Earth with an average angular daily eastward drift of about 0.99°/day. The deferent of Saturn with its attendant epicycles is centered near the Sun, and all the orbital machinery of Saturn performs its own yearly gyration about the Earth. On the deferent (period 1 yr) is the center C_1 of an epicycle K_1 that slides uniformly eastward on the deferent with the planet's sidereal period of $29\frac{1}{2}$ years. On this epicycle is the center C_2 of a smaller epicycle K_2 that moves retrogradely with a period of $29\frac{1}{2}$ years. The planet itself is located on K_2, moving retrogradely with a period of half the sidereal period, or $14\frac{3}{4}$ years.

The early death of Tycho (1601) prevented any elaboration of this system, but by then Kepler had already taken over the theoretical work on Mars. By May 1605 he had brilliantly proved that the orbit of that planet is an exact ellipse. Later he made many other advances (Chapter VII), all based on Tycho's treasure of accurate and continuous observations for more than 20 successive years, using the giant instruments that he conceived and acquired.

Tycho's Lunar Theory

Tycho made the first important step forward in the lunar theory since Ptolemy by accurately and diligently observing the Moon's positions continuously round the sky numerous times and not, like the ancients and Copernicus, merely at opposition and other interesting points of its orbit. He redetermined the values of the two largest lunar perturbations, and discovered and determined three more, the five constituting all those visible to the naked eye.

He also discovered the variability of the inclination of the lunar orbit and the nonuniform rate of regression of its nodes, correctly finding at times a difference of $1\frac{3}{4}^\circ$ between the actual and mean positions of the node. When evaluating his lunar observations of 1577 he found that the 5° value for the inclination of the lunar orbit to the ecliptic, adopted since the time of Hipparchus, was too small. As a result of a lifetime of observations, he finally showed that the inclination varies between 4°58′30″ and 5°17′30″ (both values correct to about $\frac{1}{2}'$) in a period of 32.28 days.

To represent the position and motion of the plane of the lunar orbit, Tycho introduced a small epicycle on the celestial sphere for the motion of the pole of the orbit plane (Fig. VI.3). In a period of 32.28 days the orbital pole P_o moves directly and uniformly in a small circular epicycle of radius 9.5′. This epicycle's center C moves retrogressively in a period of 18.6 years on a small circular deferent of radius 5°8′ concentric with the pole of the ecliptic P_E. These two motions together account for the variation of the inclination and the regression of the nodes of the orbit.

To represent the motion of the Moon in celestial longitude Tycho adopted the following scheme, all in the oscillating and regressing plane of the lunar orbit (see Fig. VI.4):

1. The center C_1 of the deferent (of radius $C_1C_1' = 1$) moves directly with a period of $\frac{1}{2}$ synodic month on a small circle of center C_2 and radius $\rho = 0.02174$. The Earth at A

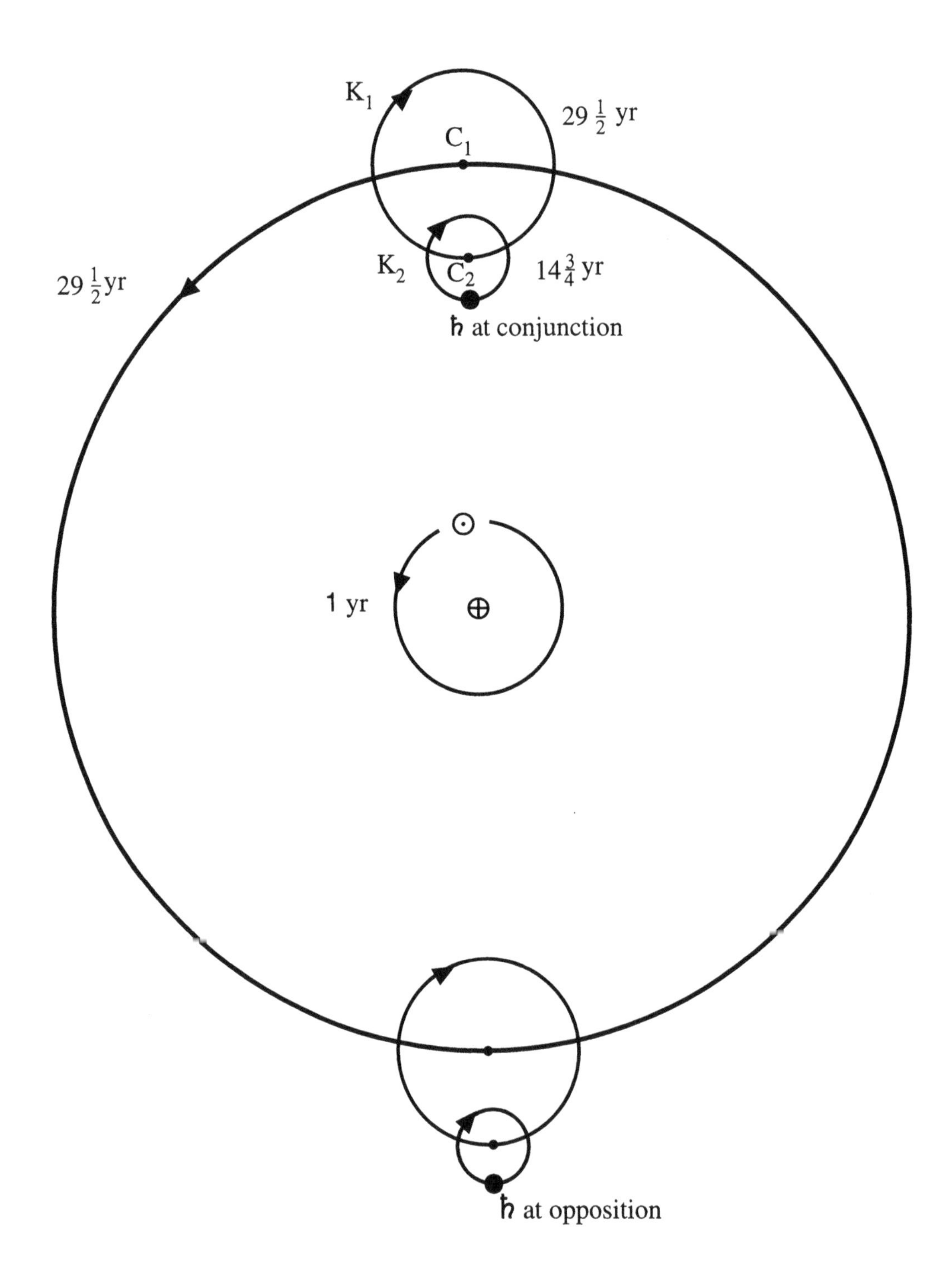

Figure VI.2 — Tycho's orbit of Saturn (not to scale).

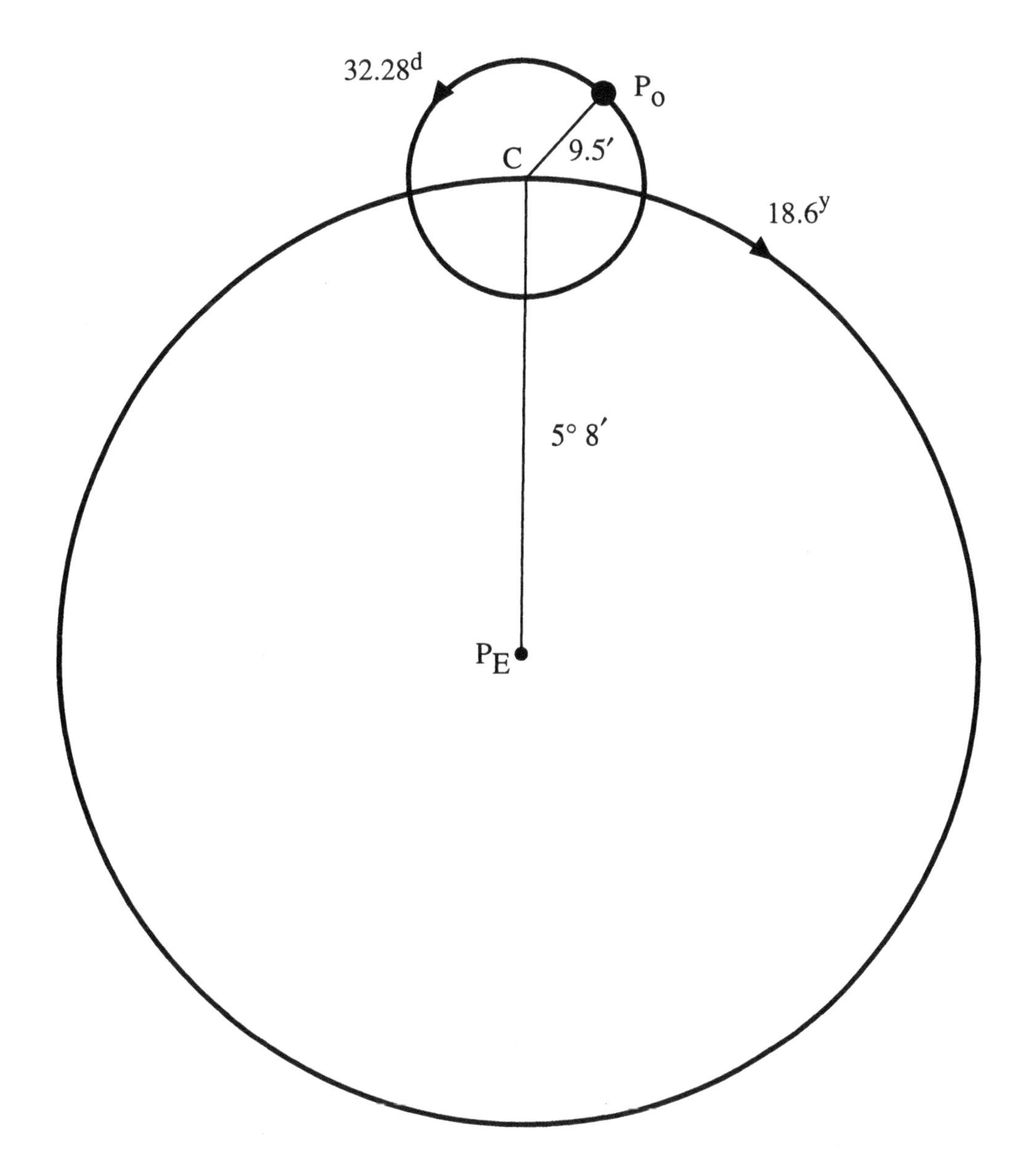

Figure VI.3 — Motion of the pole of the lunar orbit (not to scale).

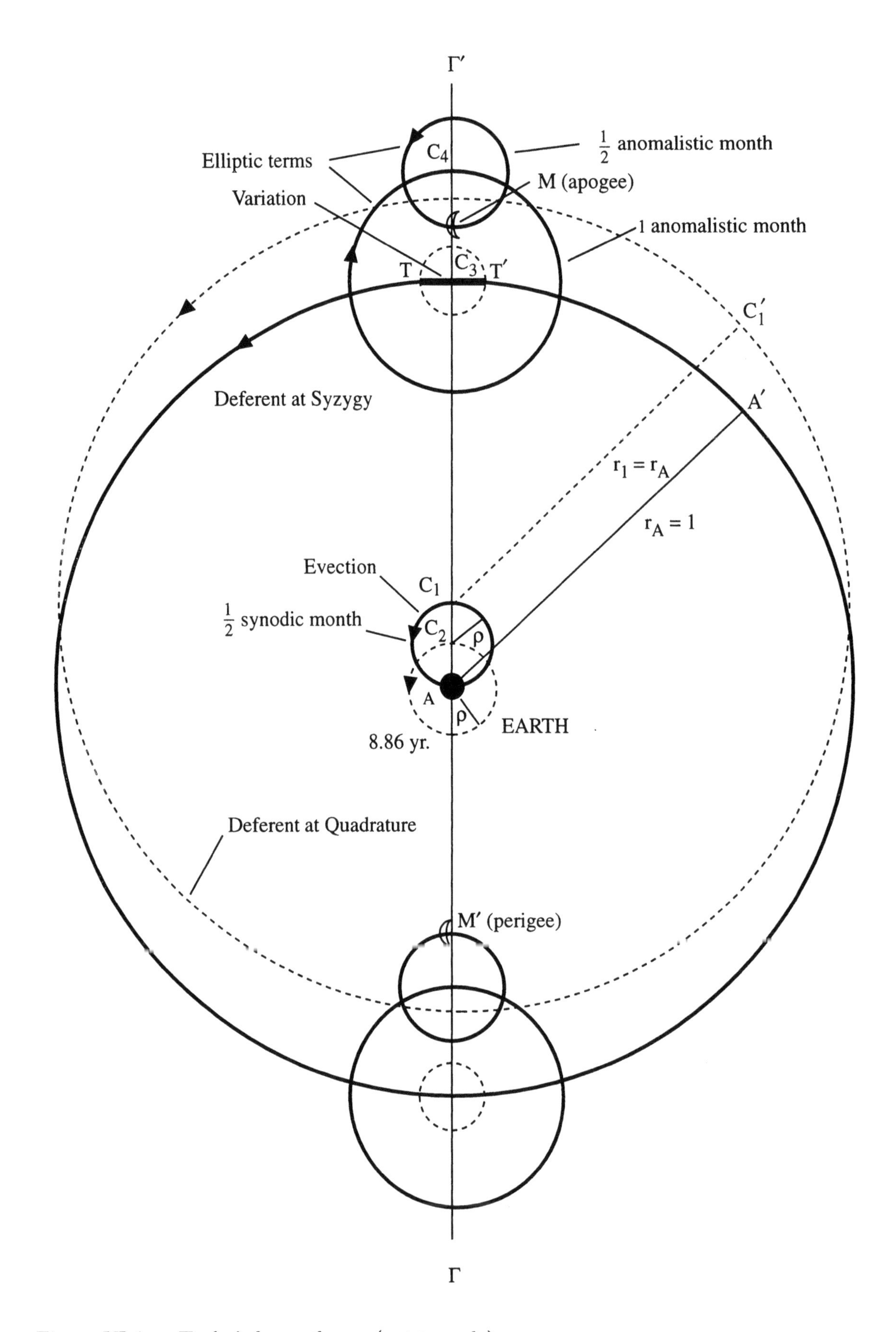

Figure VI.4 — Tycho's lunar theory (not to scale).

and a second circle of the same radius are fixed in space. The center of the deferent C_1 coincides with the Earth at the syzygies and is farthest from it at the quadratures. This arrangement accounts for the second inequality, the *evection*. To account for the advance of the line of apsides, the center C_2 moves directly in a period of 8.86 years on the circle concentric with the Earth and of radius $\rho = 0.02174$.

2. At C_3 (on the shifting deferent) is centered an epicycle (of radius $r_3 = 0.058$) with retrograde motion in a period of 1 anomalistic month, on which is moving the center C_4 of an epicycle (of radius $r_4 = 0.029$) with direct motion in a period of $\frac{1}{2}$ anomalistic month. The Moon M is situated on the circumference of this smaller epicycle, being at perigee when the center C_4 is at apogee on the larger epicycle. These two epicycles together account for the *elliptic* terms in the Moon's motion in geocentric longitude (see eqn. 10).

3. To account for the third inequality, the *variation*, the center C_3 of the larger epicycle, in addition to its progress along the deferent, shifts back and forth by a total distance of $1°21'$ along a tangent TT' to the instantaneous deferent, with simple harmonic motion in a period of $\frac{1}{2}$ synodic month. This last refinement was introduced when Tycho, on his way to Prague, discovered this perturbation. It had been long known that an oscillation along a straight line TT' (called by Tycho a "libration") can be identically represented by two uniform circular motions with an amplitude ratio of 2:1. Thus was the principle of uniform circular motion preserved.

   Let a circle of center C_5 and radius $r = 20.25'$ roll without slipping with a period of $\frac{1}{4}$ synodic month inside a circle of center C_3 and twice the radius. The motion of point P on the rolling circle may be visualized in Figure VI.5 as follows: At the initial instant P coincides with C_3. After $\frac{1}{16}$ synodic period, P will lie at P' on the line segment TT' tangent to the deferent at C_3. After $\frac{1}{8}$ synodic period P will lie at P'', which coincides with T. By al-Ṭusi's Theorem P will have traced out a rectilinear path $PP'P''$ coinciding with TT', beginning at C_3 and ending at T.

   To produce the "variation," Tycho let C_3 coincide with P, starting at syzygy when the variation is zero. At the end of $\frac{1}{8}$ synodic month, C_3 will be at the first octant on the deferent, but in addition it will be displaced eastward by C_3P'', or $40.5'$ along the tangent TT', producing the maximum "variation" at the octant. Thus, the Moon will be ahead in heliocentric celestial longitude by $40.5'$ at the first octant due to this perturbation. Similar considerations show that the variation is zero at the syzygies and quarters, a maximum of $+40.5'$ at the first and fifth octants, and $-40.5'$ at the third and seventh octants. It varies with simple harmonic motion, changing most rapidly when its value is zero.

Also, on his way to Prague, while working at Wittenberg in 1599, Tycho showed the existence of the fourth inequality in longitude, the *parallactic inequality*. The period of this was 1 anomalistic solar year, and it revealed itself in that the Moon's observed place in its orbit was behind the computed one between solar perigee (1 January) and apogee (4 July), and ahead during the other six months of the year. Tycho found that it was inconvenient to represent this effect by another epicycle in his already complicated system, and he turned the further discussion of the Moon's orbit over to his chief assistant, Longomontanus.

Longomontanus refused to further complicate the theory and tried with only moderate success (residuals $\sim 6'$) to explain the effect by claiming that the problem was not in the

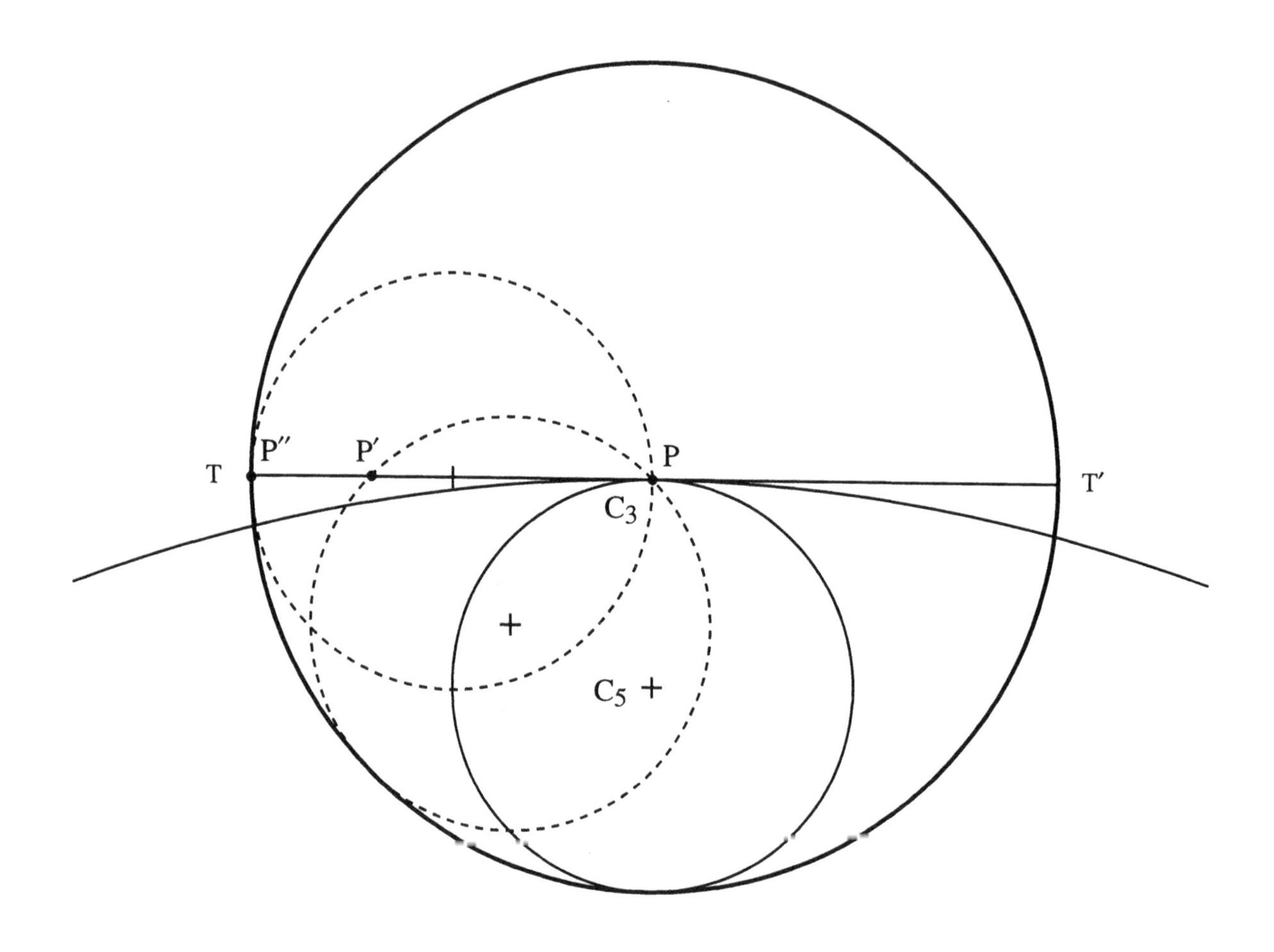

Figure VI.5 — Illustration of al-Ṭusi's Theorem.

Moon's positions but in the calculated mean times. He said that the (decreasing) slope of the equation of time curve was too steep between 11 February and 5 May, and again between 21 July and 2 November, giving values for the length of the mean solar day too short during the parts of the year when the slope is mostly negative, and too large during the rest of the year. Soon after this work was finished Longomontanus left to take up the chair of mathematics at the University of Copenhagen.

Before leaving this description of Tycho's lunar orbit, one improvement over all previous systems should be noted. One of the principal weaknesses of Ptolemy's lunar theory was the disparate relation of the geocentric distances of the Moon at perigee and at apogee. Ptolemy's perigee distance required an angular diameter of the Moon nearly twice that at apogee, an obvious conflict even with naked-eye observations. The correct relation of these distances is on average 1.07:1. Tycho's theory makes it 1.25:1, a distinctly better value than that of Copernicus (1.31:1) or of Ptolemy (1.93:1).

Tycho's Method of Predicting the Moon's Place

Figure VI.6 shows both the initial (apogee) position P of the Moon at a time $\tau = 0$ and a general (later) position P_1 at a time τ. Let the circles k, K', K_1, K_2 have radii ρ, r, r_1, r_2, respectively. Let the average rate of sidereal motion of the Moon be $\mu = 2\pi/27.32$ rad/day and the average rate of anomalistic motion of the Moon be $\mu_1 = 2\pi/27.55$ rad/day. Let M be the mean geocentric celestial longitude of the Moon and M_1 its mean anomaly reckoned from AP. Then $M = \mu\tau$ and $M_1 = \mu_1\tau$, both counted from an apogee at new moon.

At the moment when P arrives at P_1, motion has occurred on all five circles shown by solid lines at their respective rates, thus bringing A to A_1, P to P_1, C to C_1, and D to D_1. Since uniform motions in a circular epicycle and circular deferent of the same period can always be replaced by a single uniform circular eccentric motion, we will simplify the situation by replacing the motions on K' and K_1 by a single uniform circular motion on an eccentric deferent K'' with center E_1, radius $r'' = r$, and eccentricity $AE = CD = C_1D_1 = A_1E_1$.

It follows that D_1 lies on K'' and that the remaining part of the motion can now be discussed as taking place about E_1 rather than A_1. The details are seen in Figure VI.7. Let

$$\angle D_1C_1P_1 = \tau_1 \qquad \text{and} \qquad \angle D'A_1P_1 = \tau_2\,.$$

Then

$$\angle D'C_1P_1 = M_1 + \tau_1 \qquad \text{and} \qquad \angle P_1A_1C' = M' = M_1 - \tau_2\,.$$

Because of the respective rates of motion in the circles considered,

$$\angle C'A_1D' = \angle D_1C_1D' = \angle C'E_1D_1 = M_1 \qquad \text{and} \qquad \angle C_1D_1P_1 = 2M_1\,.$$

Now an inspection of $\Delta A_1C_1P_1$ in Figure VI.7 shows that

$$\cos\tau_1 = r_1 - r_2\cos 2M_1\,. \tag{1}$$

Also,

$$r_2 = \tfrac{1}{2}r_1 = 0.029 \tag{1a}$$

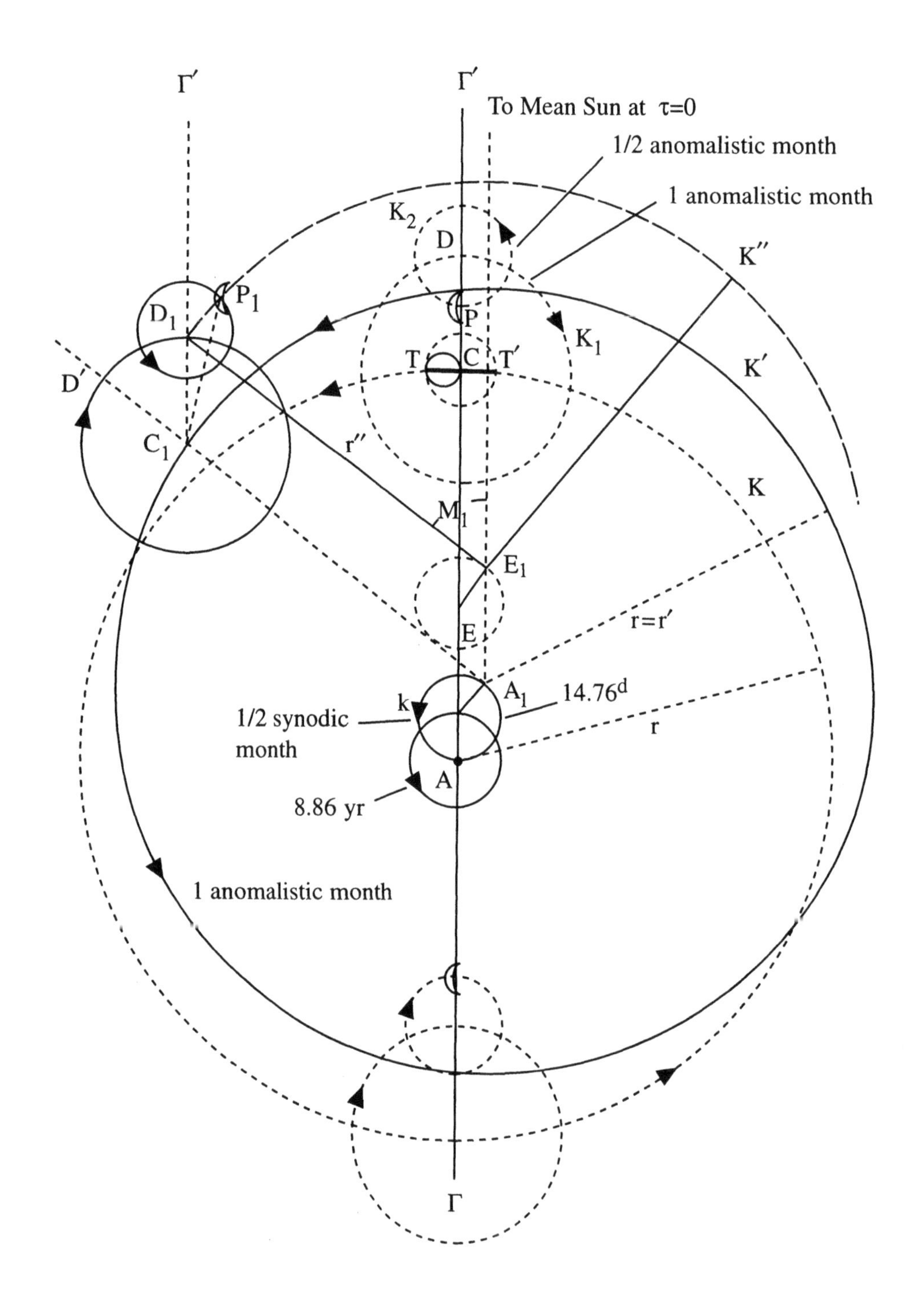

Figure VI.6 — Predicting the Moon's position in Tycho's theory.

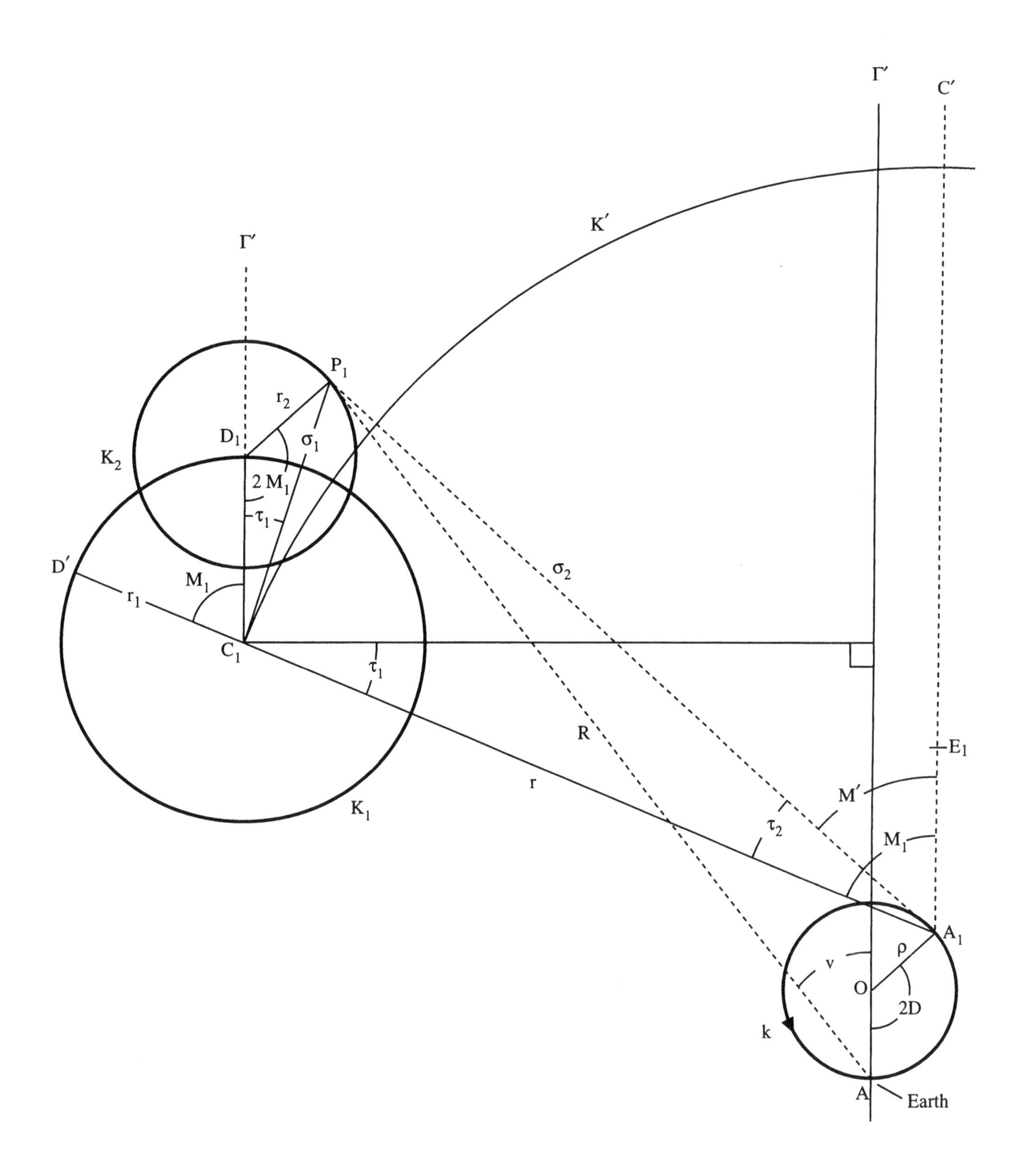

Figure VI.7 — Enlarged portion of Figure VI.6.

is given, from which τ_1. Further,

$$\sigma_1/r_2 = \sin 2M_1/\sin\tau_1, \tag{2}$$

from which σ_1, and

$$\angle D'C_1P_1 = M_1 + \tau_1. \tag{3}$$

Also, by inspection of $\Delta A_1C_1P_1$:

$$\sigma_2/\sigma_1 = \sin(M_1+\tau_1)/\sin\tau_2 \qquad \text{and} \tag{4}$$

$$\sigma_2\cos\tau_2 = r + \sigma_1\cos(M_1+\tau_1). \tag{5}$$

Solve equations (4) and (5) for τ_2 and σ_2. Also, from Figure VI.7:

$$\angle P_1A_1C' = M_1 - \tau_2\,. \tag{6}$$

Form $M' = M - \tau_2$, a better mean lunar longitude than M_1, since it includes correction for τ_2, which contains the evection, as well as for the prosthaphaeresis of the Moon due to the epicycles K_1 and K_2.

Now consider ℓ to be the Sun's true longitude counted from $A\Gamma'$, and D the true angular distance of the Moon from the Sun. Form $D = M' - \ell$. Now we can write $\angle AOA_1 = 2D$, since the motion in k is twice the solar rate. By taking projections perpendicular and parallel to $A\Gamma'$, we have from Figure VI.7 by inspection

$$R\sin v = \sigma_2\sin(M_1-\tau_2) - \rho\sin 2D \qquad \text{and} \tag{7}$$

$$R\cos v = \sigma_2\cos(M_1-\tau_2) + \rho - \rho\cos 2D\,. \tag{8}$$

Solving equations (7) and (8) for v and R and adding the "variation," $40.5'\ \sin 2D$, we have in principle completed Tycho's method of locating the Moon. Let the geocentric longitude of the lunar apogee be r', and the true geocentric longitude of the Moon at time τ be λ. Then

$$\lambda = r' + v + 40.5'\sin 2D\,, \tag{9}$$

together with a value for R, gives the longitude and distance at time τ. However, the time τ is in "mean solar" units derived from Tycho's slightly inaccurate "lunar Equation of Time."

Modern Development of Tycho's Method into a Series Involving the Mean Anomaly M_1 and Solar Phase Angle D

Inspection of Figure VI.7 shows that

$$R\sin v = r\sin M_1 - r_2\sin 2M_1 - \rho\sin 2D \qquad \text{and}$$

$$R\cos v = r\cos M_1 + r_1 - r_2\cos 2M_1 + \rho - \rho\cos 2D\,.$$

Eliminating $2M_1$ from these equations, we obtain

$$R\sin(M_1 - v) = (r_1 + r_2 + \rho)\sin M_1 + \rho\sin(2D - M_1) \qquad \text{and}$$

$$R\cos(M_1 - v) = r + (r_1 - r_2 + \rho)\cos M_1 - \rho\cos(2D - M_1)\,.$$

Form $\tan(M_1 - v)$, develop into powers of r_1/r, r_2/r, and ρ/r, neglect third and higher powers of these ratios, put $r = 1$, and solve for v (in radians):

$$v = M_1 - (r_1 + r_2 + \rho)\sin M_1 + \tfrac{1}{2}[(r_1 + \rho)^2 - r_2^2]\sin 2M_1 - \rho\sin(2D - M_1)$$
$$- \tfrac{1}{2}\rho^2 \sin(4D - 2M_1) + \rho(r_1 + \rho)\sin 2(D - M_1) - \rho r_2 \sin 2D\,.$$

We now substitute Tycho's values for r_1, r_2, and ρ, namely,

$r_1 = 0.058,\qquad r_2 = 0.029,\qquad$ and $\qquad \rho = 0.02174.$

We then obtain

$$v = M_1 - 6°14' \sin M_1 + 10' \sin 2M_1 - 1°15' \sin(2D - M_1) - 1' \sin(4D - 2M_1)$$
$$+ 6' \sin 2(D - M_1) - 2' \sin 2D.$$

Finally, apply equation (9) as in the previous section.

Change from Tycho's "Mean Solar Time" τ to Modern Mean Solar Time t

In equation (9) the unit of time for τ is a so-called mean solar day derived from observations of apparent solar transits but corrected by Tycho using his special Equation of Time table applicable to lunar work only. For purposes of comparing Tycho's theory to modern calculations, we shall change to modern mean solar time as follows.

The Equation of Time for the Sun is the correction applied to local apparent time (the time given by a sundial) to give local mean time. It has two main components, of which the first is the eccentricity or seasonal effect, expressing that part due to the Sun's nonuniform (elliptic) progression along the ecliptic. This component is

$$E_{\odot 1} = 8.2^{\mathrm{m}} \sin(v_\odot - 95°10'),$$

where the last constant is the longitude of the Sun's apogee at A.D. 1600. The second component, the obliquity effect, expresses the equivalent at various declinations of a constant increment for solar motion along the ecliptic as projected on the celestial equator: $E_{\odot 2} = 10.2^{\mathrm{m}} \sin 2v_\odot$. These terms together form the main portion of the Equation of Time, all other terms amounting to at most 1^{s}. A modern analysis of Tycho's (empirical) Equation of Time table for the Sun gives very closely $E_\odot = E_{\odot 1} + E_{\odot 2}$. For the Moon, however, he thought that the seasonal solar term should be omitted, and he constructed an empirical "lunar Equation of Time" table that modern analysis shows contains solely $E_{\odot 2}$. Now, if the results of deriving local mean solar times from apparent solar transits (as has been done ever since the time of Hipparchus) are to be comparable throughout the year, the *full* solar Equation of Time must be used, because to a first approximation the lunar orbit, freed from its perturbations, actually does move with the Earth's elliptic motion about the Sun. But in the Tychonic system the motion of the Moon is entirely independent of that of the Sun, both bodies having motions about the Earth, independently following their respective geocentric orbits and not influencing each other. Since Tycho's system is entirely without dynamics, there is no possible interaction between the motions of the Sun and the Moon.

Hence, Tycho's lunar Equation of Time table contained no seasonal (solar) term $E_{\odot 1}$ but only the term $E_{\odot 2}$.[1]

If now instead of the seasonally invalid "mean solar time" τ, we introduce the modern mean solar time t, the value of v must be augmented by the lunar motion in anomaly (i.e., in longitude) during the time interval given by the solar term $E_{\odot 1}$. The coefficient of this term in the lunar anomaly may be estimated as follows: The value of the coefficient in $E_{\odot 1}$ is 8.2^{m}. On average the Moon moves in celestial geocentric longitude approximately its own angular diameter per hour, which is $0.5'$ in 1 minute of time. Hence, in 8.2^{m} it moves $4.1'$. From a more accurate estimate we find that the additive term in v, consequent to using t in place of τ, is $4.5' \sin v_{\odot}$.

Analytical Comparison between Tycho's Expression
and the Modern Expression for the Moon's Longitude

The analytical expression for the Moon's true geocentric celestial longitude λ is given by Russell, Dugan, and Stewart (1926, pp. 287–88) as follows:

$$\begin{aligned}\lambda = L \;&+\; \underset{\text{[elliptic}}{6^\circ 17' \sin v} \;+\; \underset{\text{terms]}}{13' \sin 2v} \;+\; \underset{\text{[evection]}}{1^\circ 16' \sin(2\varphi - v)}\\ &+\; \underset{\text{[variation]}}{40' \sin 2\varphi} \;-\; \underset{\text{[annual equation]}}{11' \sin \ell_{\odot}} \;-\; \underset{\text{[parallactic inequality]}}{2' \sin \varphi} \;+\; \cdots,\end{aligned} \tag{10}$$

where

$L =$ the Moon's mean geocentric celestial longitude $= M_1 + 0.333^\circ$,

$v =$ the distance of the mean Moon from the mean lunar perigee,

$\varphi =$ the distance of the mean Moon from the Sun,

$\ell_{\odot} =$ the distance of the Sun from the (mean) solar perigee.

Due to the overlapping of a multitude of smaller terms, the perturbations are not clearly separated. Nevertheless the principal terms have received the indicated names.

In the above notation, Tycho's lunar theory would be expressed as follows:

$$\begin{aligned}\lambda = L \;&+\; 6^\circ 14' \sin v \;+\; 10' \sin 2v \;+\; 1^\circ 15' \sin(2\varphi - v)\\ &+\; 40' \sin 2\varphi \;-\; 5' \sin \ell_{\odot} \;-\; 1' \sin 2\varphi \;+\; \cdots.\end{aligned} \tag{11}$$

A comparison of equations (10) and (11) shows remarkable agreement for an orbit determined entirely from naked-eye observations. It illustrates Tycho's fine study of the variation and the moderate success he, or rather Longomontanus, had in accounting for the annual equation. It also shows that he made a start on the determination of the parallactic inequality! Kepler later amended the coefficient of the annual equation to be $11'$. With this improvement the Tychonic lunar theory became nearly as accurate as the observations on which it was based.

There are several other ways in which Tycho's results have a close similarity to modern values. As an illustration, consider his lunar Equation of Time table mentioned above. By analysis this table conforms closely to the expression

[1] Any increment at declination δ and parallactic angle q (the angle between the longitude and hour circle) was of course multiplied by $\cos q \sec \delta$ to find its equivalent effect in hour angle.

$$\odot - \alpha = 9^{m}56^{s} \sin 2\odot - 13^{s} \sin 4\odot + \cdots,$$

where $\odot$ and α denote the Sun's celestial longitude and right ascension, respectively. This should be identical to the obliquity terms in the modern expression for the solar Equation of Time, which according to Smart (1931) can be written

$$\odot - \alpha = 9^{m}52^{s} \sin 2\odot - 13^{s} \sin 4\odot + \cdots.$$

It may be of interest to recall that Ptolemy gave a value for lunar longitude equivalent to $\lambda = L + 6.3° \sin v + 1.3° \sin 2v$. The second term alone does not give elliptic motion, but rather a simple harmonic displacement from uniform circular motion. Its total amplitude, however, is well chosen. The third term is a slight improvement on Hipparchus's estimate of the size of the evection. We also note that Copernicus's lunar theory, though inferior to Tycho's, was an improvement on that of Ptolemy.

Numerical Example of Predicting the Moon's Place at Any Time in Tycho's System

To show a numerical example of Tycho's model, let us compute the celestial longitude of the Moon for an arbitrarily chosen time by the above equations, using for simplicity only a 10-inch slide rule and a four-place table of trigonometric functions. As a further simplification permitted by the smallness of the orbital inclination, we will compute the celestial longitude in the curtate orbit of the Moon; i.e., we will find the projection on the ecliptic of the Moon's celestial longitude.

Let us choose a position at about the first octant (measured from apogee), where large values occur for both the evection and the variation. Let the initial data be taken from the *Astronomical Almanac* for 1987.

Inspection shows that the best case in this almanac fulfilling these required conditions is the new moon on 1987 July 25.860 (Terrestrial Dynamic Time), and the closest apogee to this moment is at July 25.033. Thus, let the problem be "to find the Moon's celestial longitude at 1987 July 28.860 (at which time the Moon is about an octant past its apogee)."

The epoch is new moon plus 3 days. In the notation of the equations of the previous sections:

$$\mu = 0.230 \text{ rad/day} = 13.16°/\text{day},$$

$$\mu_1 = 0.2125 \text{ rad/day} = 12.19°/\text{day},$$

$$\tau = 3 \text{ days},$$

$$M = \mu\tau = 39.42°,$$

$$M_1 = \mu_1\tau = 36.59°,$$

$$2M_1 = 73.18°,$$

$$r_2 = \tfrac{1}{2} r_1 = 0.029 \text{ (given Tycho's value)},$$

$$\cos \tau_1 = 1 - \tfrac{1}{2} \cos 2M_1 = 1 - \tfrac{1}{2} \cos 73.18°,$$

from which

$$\tau_1 = 31.20°,$$

$$\sigma_1 = \tfrac{1}{2}\, r_1 \sin 2M_1 / \sin \tau_1 = 0.5 \sin 73.18^\circ / \sin 31.20^\circ = 0.922.$$

Form

$$M_1 + \tau_1 = 36.59^\circ + 31.20^\circ = 67.79^\circ.$$

To proceed, it is best to temporarily eliminate σ_2 from equations (4) and (5). We find

$$\tan \tau_2 = \frac{\sigma_1 \sin(M_1 + \tau_1)}{r + \sigma_1 \cos(M_1 + \tau_1)}.$$

An assumption must now be made about r, the radius of the Moon's deferent K. For the value of r we may use the average lunar σ_2 (since in the development of the above formula the center of the orbit was moved to A_1). For the sake of our illustration we will compute the error in celestial longitude for three cases, namely,

Case I: $r = 52.35$ Earth radii (Tycho's average value),

Case II: $r = 60.27$ Earth radii (modern average value), and

Case III: $r = 63.1$ Earth radii (instantaneous value at epoch).

For Case I, $\tau_2 = 0^\circ 46.2'$.

With equation (6) form

$$M' = M_1 - \tau_2 = 36^\circ 35.4' - 0^\circ 46.2' = 35^\circ 49'.2\,.$$

From equations (7) and (8) find

$$\tan v = \frac{\sigma_2 \sin(M_1 - \tau_2) - \rho \sin 2D}{\sigma_2 \cos(M_1 - \tau_2) + \rho - \rho \cos 2D}, \qquad \text{or} \qquad v = 34^\circ 42.5'.$$

Each value of v is affected by the variation, or $40.5' \sin 2D = 38.8'$. It also contains the effect of the advance of the line of apsides to the time of the epoch. This effect is found by adding $3 \times 0.111^\circ$ to the v computed for the stationary orbit illustrated in Figure VI.7. Thus we first form $v_1 = v +$ variation, then $v_2 = v_1 + 0.333^\circ$. Now let v_3 be the listed value of the celestial longitude of the Moon at the epoch July 28.860,[2] then the error in the computed longitude is $\ell' = v_2 - v_3$. The results for the three cases are shown in the following table:

Computed Errors of Predicted Curtate Lunar Celestial Longitudes, Using Tycho's Model

Case:	I	II	III
Assumed r	52.35	60.27	63.1
$\ell' = v_2 - v_3$	$-7.8'$	$+4.2'$	$+24'$

The table shows a dependence of the residuals, part of which may be due to our use of the Moon's curtate orbit. However, the effect of having omitted the correction for this projection is near a minimum at the chosen epoch. The difference between the instantaneous direction of the radius vector of the Moon and the line of apsides of its orbit is only 22° at this time, whereas the inclination of the orbit is at most 5°15′.

[2]Found by second-order interpolation on p. D14 of the 1987 *Astronomical Almanac*.

Wittich's Formula

In connection with the immense number of computations that were necessary to make use of his many accumulated observations, Tycho was always searching for the latest methods of computation. One of these was Wittich's formula, by which additions and subtractions of sines and cosines take the place of multiplication and division of numbers, thus greatly facilitating the work. Note that all of this was only a few years before the publication of the first logarithmic tables by Neper, Briggs, and Biirgi.

Wittich's formula is but one of the four well known addition formulas of plane trigonometry. It reads as follows:

Let $n_1 = \sin\alpha$ and $n_2 = \sin\beta$ be two numbers, and $n = n_1 \times n_2$, their product; then

$$n = \sin\alpha \times \sin\beta = \tfrac{1}{2}\Big[\cos(\alpha-\beta) - \cos(\alpha+\beta)\Big].$$

Example: To multiply $\sin\alpha = 0.85321$ by $\sin\beta = 0.12345$.

From a table find

$$n_1 = \sin\alpha = 0.85321, \qquad \alpha = 58°33.73',$$

$$n_2 = \sin\beta = 0.12345, \qquad \beta = 07°05.48',$$

$$\alpha+\beta = 65°39.21', \qquad \text{and} \qquad \alpha-\beta = 51°28.25'.$$

From the same table also find

$$\cos(\alpha+\beta) = 0.41226 \qquad \text{and} \qquad \cos(\alpha-\beta) = 0.62291,$$

from which $n = 0.10532$.

The Elliptic Terms

It is of interest to elucidate the meaning of "the elliptic terms" in the representation of the true anomaly λ of the Moon by a series of sines of multiples of its mean anomaly M_0.

The first three terms describe a pure elliptic motion and are

$$\lambda = M_0 + 2e\sin M_0 + \tfrac{5}{4}e^2\sin 2M_0\,,$$

where e is the eccentricity. In the case of the Moon, e varies continuously from about $\frac{1}{15}$ to $\frac{1}{22}$ in 1 anomalistic month. To illustrate the Moon's system, we take $e = \frac{1}{18}$ as an average value. Filling in numerical values, we have

$$2e = 0.111\,\text{rad} = 6°22' \qquad \text{and} \qquad \tfrac{5}{4}e^2 = 3.86\times10^{-3}\,\text{rad} = 13.3'\,.$$

Hence the orbit is represented (to three terms) by

$$\lambda = M_0 + 6°22'\sin M_0 + 13.3'\sin 2M_0 + \cdots\,.$$

Thus, we can fairly represent the motion in the unperturbed elliptic orbit of eccentricity e, namely,

$$\lambda = M_0 + 6°14'\sin M_0 + 10'\sin 2\;M_0 + \cdots$$

(cf. eqn. (11)), by the first three terms of equation (10), namely,

$$\lambda = M_0 + 6°17'\sin M_0 + 13'\sin 2M_0\,.$$

CHAPTER VII

Kepler (1571–1630)

Berry (1910) states that "there are few astronomers about whose merits such different opinions have been held as about Kepler." There is of course general agreement about the great importance of his three laws of planetary motion and of the *Rudolphine Tables,* which he computed on the basis of these. But it is said that most of his writings are encumbered with masses of wild speculations, irrelevant autobiographical remarks, mystical and occult fancies, astrology (he cast at least 1800 horoscopes, partly for a living), and weather prophecies. It should be noted, however, that part of his job during his first few years as regional mathematician at Graz was the annual preparation of an almanac, which in those days would perforce contain weather forecasts and astrological prognostications.

It has been said by an impatient reader that if Kepler had burned three-quarters of what he printed, we should in all probability have formed a higher opinion of his intellectual grasp and sobriety of judgment. This remark reveals a deplorable ahistorical viewpoint and reminds one of the narrow judgment of some of the French encyclopedists, who, intoxicated with perspectives gained during the Age of Reason, culminating in Newton's great achievements, exhibited the smallness of mind to accuse Kepler of being "a scatterbrained mystic." We shall see that, even if he had a deeply mystical temperament, he was far from scatterbrained. Others have called him a mathematical genius, the true father of non-Euclidean geometry, which is an equally mistaken characterization. Although he had great command of all useful ancient and medieval mathematics up to his day, he was not primarily a thinker who advances pure mathematics by devising new logical forms but an astronomer who brilliantly used mathematics as a tool to solve problems. The conjecture that he had ideas about the non-Euclidean nature of space is probably based on a misunderstanding of his "rectilinear orbits" of comets, as expressed in his *Mysterium Cosmographicum* and in another book on comets in 1619.

Berry (1910) gave a more accurate evaluation: "Kepler's name was destined to become immortal on account of the infinite patience with which he submitted his hypotheses to comparison with observation, the candor with which he admitted failure after failure, and the superhuman perseverance and ingenuity with which he renewed his attacks upon the riddles of nature."

In studying Kepler's works one finds, in addition to a clear description of his magnificent technical achievements, a wealth of materials permitting one to form a vivid and complete picture of a person with a gentle, though impulsive, fervently religious disposition. Although he was one of the few Protestants in the predominantly Catholic community of Graz in Styria (he was born in Weil der Stadt, Württemberg, in 1571), he never quite became an orthodox Protestant because he remained dissatisfied with one little detail of

Protestant dogma! His religion may best be described as a firm belief in a late version of Neoplatonism, if not straight Pythagoreanism. He says at one place: "Everything in nature is arranged according to measure and number, which can be elucidated only by dispassionate and laborious calculation." Also: "The circle which we draw with a compass is only an inexact copy of an idea which the mind carries as really existing in itself." Kepler vigorously argued against Aristotle's view that mathematical concepts cannot exist apart from sensible objects, and that such ideas are only gained by abstraction from these objects. Aristotle said that the mind is like a blank tablet on which nothing is written. Kepler believed that for the sake of the Christian religion he had to raise a protest against this. He was a follower of Plato, according to whom the human mind by itself learns all mathematical ideas and figures. Kepler quoted long passages from Proclus's commentary on Euclid to substantiate his idealistic viewpoint. All his work was prompted by a religious endeavor to reveal the Divine Mind through numerical relationships which he wrested from nature with superhuman exertion, limited only by periodic physical exhaustion. His whole life was a work of idealistic service aimed at revealing truth in nature. Research was to him a sacerdotal function. He had ideas so brilliant and profuse that many of them were unfounded, even fantastic, but he also had infinite patience for the sobering routine investigations necessary to disprove as often as to prove those ideas.

Kepler's productivity was so enormous that one could well write a series of books describing each of its several aspects, such as astronomy, astrology, optics, and mystical philosophy. In this chapter we shall describe only his main work (1601–5) contained in *Commentary on the Orbit of Mars*, also referred to as *Astronomia Nova* (1609), and summarized and extended in *Epitome of the Copernican Astronomy* (1618). In this work he revolutionized astronomical thought in proving conclusively from Brahe's observations that Mars moves with nonuniform velocity in an elliptical orbit of which one focus is in the real, physical Sun. It is not generally realized that this idea, of nonuniform elliptic motion – as opposed to perpetual uniform circular motion – gave rise to a cataclysmic change of viewpoint later matched only by Newton's universal gravity or Einstein's general theory of relativity. Hence, Alexandre Koyré characteristically and correctly refers to Kepler's work as an astronomical *revolution*, a revolution of thought. It does not seem a revolution to those of us brought up on the lucid consequences of Newton's elegant theory of gravitation, but it was a truly difficult change of viewpoint, one that even minds such as Tycho, Galileo, and Gassendi were not merely unwilling but *unable* to accept. To accept Kepler's views in 1610 was as difficult as to accept Einstein's views in 1916. In each instance very few men had the background to do so, and still fewer cared to do so.

Highlights of Kepler's Most Important Astronomical Books

Mysterium Cosmographicum (1596)

This volume, the first large book by Kepler, contains much that was mysterious and ill-founded even at the time of its publication. But as with the early products of many brilliant minds it contains in embryo many of the great ideas later worked out over a lifetime. The book is perhaps most popularly famous for its suggestion of a reason for the number of planets and for the actual spacing of the planetary orbits. Since in Euclidean geometry there can be only five regular solids, and since the Divine Mind creates only in perfect patterns, Kepler reasoned that only six planets could exist: Mercury, Venus, Earth, Mars, Jupiter, and Saturn. Kepler showed that if we imagine the five regular bodies, one inside the other, in such a way as to be centered on the Sun, thus forming a sequence of increasing sizes; and if we inscribe spherical concentric shells for each of these bodies in such a way that the corners of one lie on a shell touching the sides of the one exterior to it, and if the number of sides of successive bodies from the inside out is in the sequence 8, 20, 12, 4, and 6, then not only is the space around the Sun "full" of structure, but the planet orbits are of such size and so situated that they become great circles on the contemplated spheres. For instance, if the heliocentric orbit of Saturn were imagined to lie on the surface of a sphere and a cube inscribed in this sphere, the next inside sphere, the one touching the faces of the cube, would be just the size to accommodate Jupiter's orbit. Then if the corners of a tetrahedron were placed on Jupiter's sphere, the sphere just touching its faces would be one for which Mars's orbit would be a great circle, etc. The scheme stops of its own account at Mercury's orbit. According to Kepler the special love of Divine Providence for mankind, an idea that permeates his works, is evidenced by the placing of the Earth in the "best protected" position, between the dodecahedron and the icosahedron, the two regular bodies having the most faces.

Besides other mystical speculations the book stresses a sober fact (which must have been known to Copernicus): the angular diameters of the epicycles responsible for the retrograde motions of the planets are equal to the angular diameter of the Earth's orbit as seen from the distance of each of the planets. This brilliant proof of the possibility of the Copernican system ought to have been evident to every astronomer since Ptolemy, but Copernicus was the first to make proper use of it, and Kepler was the first to stress its immense importance. The book contains many other embryonic astronomical ideas, later elaborated in *Astronomia Nova.*

Copies of *Mysterium Cosmographicum*, beautifully printed and lavishly bound, were sent to both Galileo and Tycho immediately upon publication. Galileo probably read it and extracted what he could use without bothering to acknowledge the source, as was his wont. But its mystical style was probably repugnant to his aggressive and rational mind and may have inaugurated his disgust with Kepler's later works, the importance of which he never realized. He never referred to Kepler's main work, *Astronomia Nova*, nor did he bother to read it in detail. In fact it is questionable whether he believed in concentrated study of anybody's work. To him science progressed mainly by specious and arrogant debate between "famous" opponents. To the end of his days he continued to believe in the circular uniform motions of the celestial bodies and in the inapplicability of terrestrial dynamics to celestial bodies. In answering Galileo's "sidereal message" Kepler offered to continue the correspondence – a typical research procedure in those days – and to join him as a protagonist of the Copernican system, but Galileo never even acknowledged receipt of his letter. Later, however, in *The Assayer* (1623) Galileo tried to drag Kepler into a dispute over the physical nature of comets.

On the other hand Tycho graciously acknowledged receipt of the gift. He certainly read the book at once and immediately recognized its author as a fine astronomer, although a Copernican. Tycho needed a brilliant theoretical mind to help him interpret the significance of his many detailed observations. He later invited Kepler to become his assistant in Prague, which Kepler did in late 1599.

Astronomia Nova, Based on Celestial Physics with a Commentary on the Motion of Mars (1609)

This great volume, most often referred to simply as *Astronomia Nova*, is undoubtedly Kepler's main work, a monumental revision and extension of the astronomical knowledge of his day, of an importance so great that it is equaled, if at all, only by Ptolemy's *Almagest* and Copernicus's *Revolutionibus*. In this work Kepler rectified many standard procedures of his time, disproved the uniform circular motion of the celestial bodies, showed Mars's orbit to be an exact ellipse, and developed methods for finding its place in its orbit at any time. We shall devote most of the present chapter to a few highlights of the subjects treated in *Astronomia Nova* and to visualizing as far as possible Kepler's work as it was actually done, by illustrating its main features and processes with geometrical diagrams and evaluation of the accuracy obtained in his results.

Harmonice Mundi (1619)

Much of this large volume is repetition and further elaboration of many speculations contained in the *Mysterium Cosmographicum*, such as analogies between the proportions in the solar system and various musical intervals. Yet by a happy inspiration he thought of trying to relate the sizes of the orbits of the various planets with their times of revolution about the Sun. After a number of unsuccessful attempts, he discovered a simple and important relation commonly known as Kepler's Third Law: "The squares of the times of revolution of any two planets about the Sun are proportional to the cubes of their mean distances from the Sun."

The book contains a study of how the mean daily orbital angular motions of the planets vary with time or position in their orbits. For each planet the results are exhibited in a diagram of the musical notes corresponding in frequency to the number of seconds of arc of daily motion. This "music of the spheres" received fantastic, arbitrary, superstitious interpretations by Kepler. Yet, A. S. Eddington, in a speech dedicating a monument to Kepler at Graz in 1921, held that Kepler used this musical method of representation as an alternative to how we today would draw a graph of the relation between velocity and time. According to Dreyer (1905), Kepler himself says that he does not believe that there is any real "music of the spheres."

Astronomiae Copernicanae (In Parts: 1618, 1620, 1621)

This is a textbook on the Copernican system in the form of questions and answers, some ridiculously simple and others technically advanced. This is a tedious style because no individual reader will fit the greatly varying qualifications assumed for the questioner. Fortunately the answers are all given with a high degree of precision and elaboration. They also incorporate Kepler's latest improvements in the numerical values of astronomical quantities.

The *Epitome of Copernican Astronomy* contains the first clear statement (although without adequate proof) that Kepler's First and Second Laws, established in *Astronomia Nova* for the case of Mars, also apply to the other planets, to the motions of Jupiter's satellites, and probably to the Moon's motion about the Earth, although for the Moon the hypothesis of simple elliptic motion was obscured by enormous complications due to the annual equation (which Kepler correctly estimated as 11′) and other perturbations. Being the nearest to a complete and unadulterated exposition of the Copernican system, in 1633 it was placed on the Roman Catholic Church's Index of Prohibited Books – and not removed until 1822!

Tabulae Rudolphinae (1627)

The greatness of Kepler's "New Astronomy" was nowhere better illustrated than in his construction of a monumental work permitting relatively easy computation of the positions of the planets and the occurrence of eclipses. As early as 1601 Tycho, in an audience with Rudolph II, emperor of Bohemia, obtained permission to construct a new set of planetary tables and to use the title *Rudolphine Tables* for this great volume. This work was to replace the *Prutenic Tables* (1551) of Rheinhold, computed on the Copernican theory but showing by then errors in the positions of the planets of up to 5°. It was of course originally understood that the tables would be constructed on the basis of Tycho's theory of the solar system.

After Tycho's death in the autumn of 1601 Kepler, now appointed Imperial Mathematician in his place, inherited this obligation along with sole access to most of Tycho's extensive observational journals. Although he had this obligation constantly in mind and periodically worked on it, it was not until 1627 that the tables were published. By then, Kepler's other important work had been done and Tycho's theory had been proven entirely inadequate in comparison with the theory of elliptic motion, something that Tycho, in spite of his haughtiness, would probably have been the first to admit. The high accuracy of the computed planet positions, constructed on the hypothesis of pure elliptic motion, represented a triumph for Kepler's theory. With these tables Kepler reached his supreme goal. As Berry (1910) comments, "at one stroke he had blown away all the ancient cobwebs of about 240 eccentrics, epicycles, and equants that described motions upon motions about mathematical points having no physical existence, but serving merely as a mental convenience for calculation." In their place he had substituted pure elliptical motions about a real, physical body, the Sun. The tables were remarkably free from errors and represented the best in astronomy for about a century.

Ever since the completion of *Astronomia Nova* in 1609, Kepler, with his medieval mania for harmony in nature, must have entertained complete confidence in the elliptic shape of the planet orbits. However, instead of going directly to the task of working out the details of the outer planet orbits, as required for the *Rudolphine Tables*, he spent a year (1602) on pure optics, and a much longer time on speculations involving magnetic and so-called motrix forces and purporting to prove theoretically that the shape of an orbit must be elliptical. Only in about 1618 did he clearly formulate his Third Law, which strongly suggested a similarity among all the planetary orbits. It seems reasonable that this should have inspired him to work more diligently on the outer planets. Perhaps he was anxious first to finish his *Epitome* establishing his version of the Copernican system, knowing that the rest was mere routine.

Shorter Books and Pamphlets

In a number of shorter books, pamphlets, and letters Kepler treats a multitude of subjects, showing himself to be exceedingly well informed on the latest astronomical, optical, and physical advances.

In a small pamphlet often referred to as *Stereometria Doliorum*, published about the time of his first marriage (1597), he developed a mathematical method for computing the volumes of wine barrels with curved sides to any desired degree of accuracy. This method shows his high ability in pure mathematics and is said to be almost identical to the kinds of constructions which later led to the development of calculus by Leibniz and Newton.

With *Optica* (1606) he became practically the founder of geometrical optics as we now know it. He considered the effects obtainable with lenses and mirrors of various mathematical shapes, and "almost discovered" the law of refraction in his studies of the rainbow. He was always poor at manual work, and thus the first actual "Keplerian" telescope[1] was probably constructed by Christoph Scheiner about 1623. He invented the term *focus* for the convergence point of rays reflected from a conical surface of revolution.

In *Conversations with the Sidereal Messenger* (1610), an answer to Galileo's *Sidereus Nuncius* (1610), he states his views on a multitude of subjects ranging from refraction and irradiation to mystical cosmogony, the latter being occasioned by Galileo's discovery of precisely four Jovian satellites.

A Brief View of Kepler's Accomplishments

A Common Popular (Under)statement of Kepler's Work

In astronomy texts one often finds statements similar to the following: "Using Tycho's exact measurements of the positions of Mars, Kepler found its distance from the Sun in various parts of the orbit. It turned out that the orbit is an ellipse with the Sun at one focus."

While this statement is true as far as it goes, one is left with the impression that Kepler merely used a cut-and-dried process, repeated perhaps a time or two, and then arrived at his final result. On the contrary, we shall see that his work was a hundred times more laborious, was well planned, and involved a prodigious amount of computation. This chapter is an attempt to bring into manageable perspective not only what Kepler did but also the magnitude of his activity, and so to permit a fair evaluation of his astronomical greatness.

A More Detailed List of Kepler's Contributions

1. He gave reasons for adopting the Copernican system from the sizes of the retrograde arcs of the planets.
2. He showed how to find the sidereal period of a planet directly by observing the times of node passages.
3. He showed how to find the distance of a planet from the Sun from pairs of measured elongations separated by a whole multiple of synodic periods.

[1]The Keplerian telescope was an improvement on the Galilean telescope, primarily in that it gave a larger field of view.

4. He showed that Mars's orbit is a plane curve, is an ellipse, and that its radius vector covers equal areas in equal time intervals.
5. He showed that these laws hold true for the other planets as well.
6. He discovered that for any two planets the squares of the periodic times are to each other as the cubes of the mean distances.
7. He showed that this harmonic relation also holds true between any two of Jupiter's satellites.

Semipopular Statement of Kepler's Work on Mars's Orbit[2]

Kepler's material consisted of nearly 20 years of Tycho's observations of Mars's geocentric celestial latitude and longitude, including 10 oppositions from c. 1580 to 1600 and 2 later oppositions observed by himself.

From Tycho's ledgers he selected six observations near zero geocentric celestial latitude and from these derived

a. the sidereal period (for checking a quantity already known), and
b. the position of the line of nodes, $\Omega = 46\frac{1}{3}^\circ$.

Tycho assigned Kepler the problem of improving an earlier orbit for Mars that had been based on Tycho's theory. Kepler found a fairly good representation of the geocentric celestial longitudes (good to $2'$) but radically wrong values for the celestial latitudes. This problem with the latitudes was no different from what Ptolemy had found, and it had driven all astronomers since Ptolemy to assume either variable orbital inclinations or epicycles oscillating in both size and tilt.

Kepler now obtained Tycho's permission to work out the orbit on Copernicus's theory. In both Copernicus's and Tycho's theories the opposition times were computed for the moment when the *mean* geocentric longitudes of the Sun and Mars differed by 180°. Kepler objected to this and persuaded both Tycho and Longomontanus to abandon the corresponding practice in their work on the Moon's orbit. He argued that the correct definition of opposition was the moment when the *true* geocentric longitude of the Sun and Mars differed by 180°. This amounted to as much as 5° difference in the case of Mars if the orbits of the Earth and Mars were "centered" on the current Sun position rather than the mean Sun position. Thus it was really Kepler, rather than Copernicus, who put the Sun at the focus. In Copernicus's system the center of the Earth's orbit was the mean position of the Sun, not its actual position.

Kepler then used three methods of finding the inclination (Fig. VII.1):

a. Directly, by latitudes of Mars when at the position M_1, which is at right angles to the line of nodes. At such times T_1, the distance Earth-Mars R equals the distance Sun-Mars and $i_1 = i$.
b. Directly, by latitudes at quadratures at such times T_2 when Sun and Earth are in the line of nodes and $i_2 = i \cos d$.

[2]This condensation is partly gleaned from J. L. E. Dreyer's *History of the Planetary Systems* (1905). It summarizes about five years of Kepler's work, starting in 1600 and detailed in *Astronomia Nova* (1609).

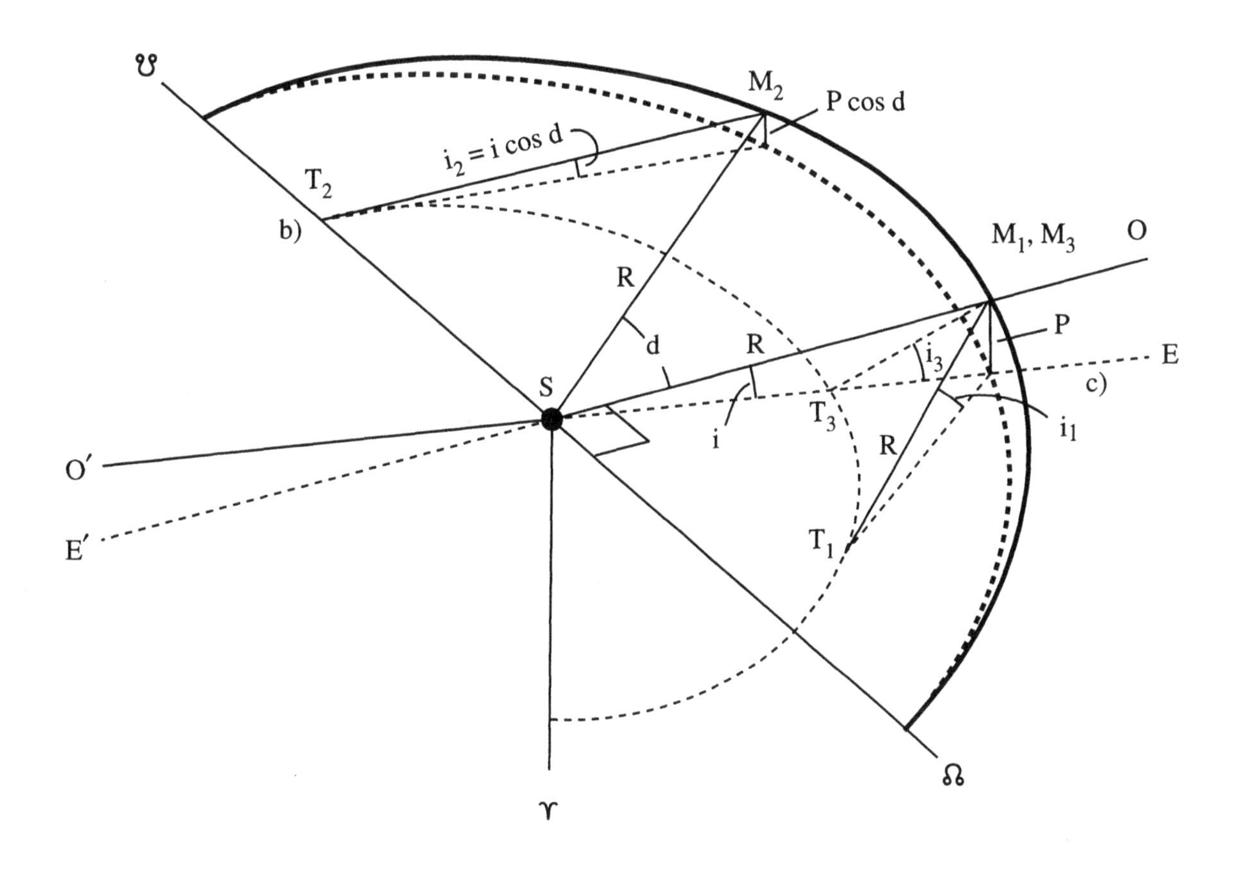

Figure VII.1 — Kepler's three ways of determining Mars's orbital inclination, the longitude of the node, and the time of node passage: (a) $i = i_1$; (b) $i = i_2/\cos d$; (c) $i = i_3(R-1)/R$.

c. By Copernicus's method of latitudes at opposition M_3, where $i_3 = iR/(R-1)$.

All three methods yielded $i = 1°50'$. The fact that the three agreed proved that the plane of the planet orbit passed through the Sun (the first part of Kepler's First Law).

To determine the line of apsides (i.e., the longitude of perihelion $\tilde{\omega}$), the eccentricity e, and the mean anomaly for a given time t, Kepler selected four of Tycho's oppositions for which each observation gave the time t and the geocentric longitude and latitude λ and β. From preliminary work he had found all of these quantities using Tycho's theory and approximate values of ω and e. With these he laid off the approximate line of apsides and the equant point, at first setting $CE = CS$, where C = center of orbit, S = the Sun, and E = the *punctum equans*. The observations directly gave the geocentric longitudes, and then applying the heliocentric longitudes of the Earth, he derived the heliocentric longitudes of Mars at each of the four oppositions. Assuming that the positions of the planet lay on a circle, center at C, the angles around E added up, not to 360°, but rather to 365°, which disproved Tycho's circular orbit of Mars.

The Vicarious Hypothesis

Kepler now let go of the Ptolemaic assumption that C bisects SE, but still assumed that the orbit is some kind of a circle; this was his "vicarious" hypothesis. By laborious computations with slightly different circles, he found that for the four oppositions the angles at E added up to 360° only if he made $EC = 0.07232$ and $SC = 0.11332$ (instead of $EC = SC = 0.09284$, as was found on the usual equant scheme). Then he computed the remaining eight oppositions and found that all 12 heliocentric longitudes were well represented (i.e., to within $2'12''$, and usually to $1'$). But when he computed the latitudes of those oppositions at greatest northern and southern heliocentric latitudes, he found an $8'$ error in the latitudes. (This could not be due to an error in the inclination i, because the maximum heliocentric latitudes at opposition were obtained from the geocentric opposition latitudes, which depended even more on Earth's position than on Mars's orbital position.) The error was as bad as Ptolemy's, but while Ptolemy was justified in dismissing the subject, having at his disposal an observational accuracy of only about $10'$, Kepler, who was certain that Tycho's observations were good to $1'$ or $2'$, decided that an error lay in his theory.

Kepler now computed the longitudes for the other observations (those not at opposition) and found large errors for these. The complete failure of the vicarious hypothesis, which had cost such a prodigious amount of computation, was nevertheless a step forward, for it proved that either no circle whatsoever can represent the orbit of Mars or, if the orbit is a circle, there exists no fixed point from which the planet's motion will appear uniform. Furthermore, assuming that it was a circle, either there is no *punctum equans* at all or, assuming that there is a *punctum equans*, this point itself would have to oscillate back and forth on the line of apsides. (This is what Ptolemy tried for the Moon when he introduced his prosneusis.)

Next, using Tycho's theory and observations of Mars at the same point of its orbit (Kepler's Method), he proved that the eccentricity of the Earth's orbit was 0.01800, or 1/56 (later $1/60 = 0.0167$), half of Tycho's value. (This new value of the eccentricity was later confirmed observationally by measurement of the annual variations in the semidiameter of the Sun made possible by the development of the telescope.)

Now Kepler began to speculate on the law of the speed of the Earth in its orbit and revived his 1596 view (derived from Aristotle) that all the planets are moved by circularly "projectile" forces, emanating from the Sun and inversely proportional to their distances from it. Thus he said that the time to cover any arc is longer the farther from the Sun is this arc. The time necessary to traverse each element of a segment is directly proportional to its distance from the Sun. The total time for each segment is therefore directly proportional to the sum of all the radius vectors to the segment. These radii "fill in" the area of the segment passed over by a varying radius vector to the planet moving on its arc. Hence Kepler's Second Law, the law of areas. (There exist at least two fallacies in this proof: (1) Speed varies inversely as focal perpendiculars, not inversely as focal radii; and (2) the sum of an infinite number of adjacent line segments never constitutes an area!)

Kepler now substituted the law of areas for the equant motion in both the Earth's and Mars's orbits and recomputed the vicarious hypothesis, still using eccentric circles. This, as before, gave excellent agreement in the longitudes.[3] But he still found the latitudes to be in error by $8'$.

The First Oval

Kepler now let go of any preconceived ideas of the shape of the orbit, and using his trigonometric method of finding the heliocentric distance of any planet from two observations separated in time by one sidereal period, he computed the distances from observation and trigonometry (i.e., directly and independently of any theory of Mars's orbit) for three places having true anomalies of $10°$, $31°$, and $104°$. He also found the distances from his vicarious hypothesis, using the Second Law for the motion on the eccentric. The directly determined distances fitted almost exactly at the apsides but were too short at the chosen anomalies; in fact, "circle minus observed" had excesses of 0.3%, 0.8%, and 0.9%. This proved that the real orbit was an oval of some sort, inside the circle, touching it at the apsides, somewhat broader toward aphelion, more pointed near perihelion.

To better fit this oval, Kepler now gave up the vicarious hypothesis, which, as we have seen, was centered on C, such that $SC = 0.11332$. Instead he proposed an eccentric centered on B, such that $SB = EB = \frac{1}{2}(SC + CE) = \frac{1}{2}(0.11332 + 0.07232) = 0.09282$. On this eccentric he added a retrograde epicycle. The planet itself moved uniformly on the epicycle, and the center of the epicycle, not the planet, moved on the eccentric in accordance with the law of areas about the Sun. The net result of this experimentation, along with many adjustments of the numerical values, was to reduce the errors in the latitudes only from $8'$ to $7'$!

The "Auxiliary Ellipse"

Kepler now for the first time introduced an "auxiliary ellipse," which was an auxiliary elliptical deferent. This was done not to replace the oval, which was still taken as the real orbit, but to facilitate computation of areas under the oval and to replace the circular deferent. Using the law of equal areas of the center of the epicyclic motion on the ellipse, he now could speed up the planet at its mean distance (above what equal-areas motion along

[3]We now know why this was so. By Ward's Principle (see Fig. VII.14), the equant is good enough for small eccentricities. The difference was but a few arc seconds in the case of the Earth.

the circle would yield) while keeping its speed much the same at the apsides. This ellipse had its center at the same place as the deferent, and its $(a-b)/a$ was 0.00858. Figuring now the areas from the ellipse, and the positions from the oval or its equivalent deferent and epicycle, the error in the latitude still had a maximum of 7′! The maximum errors were now in the octants and with opposite sign to those in the preceding arrangement. This showed that (a) the true orbit was intermediate between the enveloping circle and the auxiliary ellipse, (b) the line of apsides was correct, and (c) the line of apsides passed through the real Sun, not through the mean position of the Sun as had always been previously maintained. In addition, all the distances were too small, on average by 0.66%.

The "Orbital Ellipse"

The maximum difference between the enveloping circle and the oval fitted to the observations was 0.00432 (on a scale of 1.000). (Kepler noted that this was very close to one-half of the 0.00858 above.) The equation of center for an ellipse of eccentricity 0.0928, then called "the optical equation," is 5°18′. Kepler noted that if he multiplied all distances on the major eccentric circle perpendicular to the line of apsides by $\cos 5°18' = 0.9957$, and abandoned the epicycle, the eccentric circle itself would turn into an exact ellipse of eccentricity 0.0928 and major axis coincident with the line of apsides.[4] By this process the eccentric circle became the elliptic orbit itself, and perfect agreement ensued in both longitude and latitude for all parts of the orbit. This proved the second part of Kepler's First Law: the orbit of Mars is an ellipse of eccentricity 0.0928. (The first part, that the orbit is in a plane passing through the Sun, was discussed earlier.)

The preceding semipopular statement of Kepler's work is correct as far as it goes. In spite of its conciseness it gives only an inkling of the enormity of his activity, especially during 1604–5, when most of the basic work on Mars's motion was done. To the nonprofessional reader it may seem hard to follow. Undoubtedly a number of diagrams could be added to facilitate understanding of the details. Feeling this need I was induced to try to rewrite the account in slightly more detail. While starting on this task a reprinting (1963) of Robert Small's (1804) volume appeared. It contained more than 80 diagrams, 28 of which refer to Kepler's work. Small's volume is undoubtedly a superb account of Kepler's large technical tomes, but it seemed to me in some places unnecessarily detailed for my more limited purpose. The accounts of some of Kepler's struggles seemed not absolutely necessary. I therefore decided to discuss Kepler's investigations from my own standpoint, using several new diagrams.

Kepler's Work on the Lunar Theory

With regard to the orbit of the Moon, Kepler found the introduction of elliptic motion very troublesome, owing of course to the Moon's constantly varying eccentricity. He made trials, but to no avail. He discovered the annual equation independently of Tycho and determined its correct value, 11′, whereas Tycho had estimated it as only 4.5′. In a letter to Herwart von Hohenburg he attempted two alternative explanations:

[4]The way he hit upon this was as follows: dividing the distances perpendicular to the line of apsides by $\sec 5°18' = 1.00429$, so near 1.00432, gave correct distances all around. This transformed the circle into an ellipse.

1. Like Tycho, he first thought that the mean solar days, figured by a lunar equation of time rather than the solar one, were a little shorter in winter, so that the Moon in winter seemed to pass over equal arcs in a longer time than in summer. Unlike Tycho, he thought that maybe the Earth actually rotated a bit quicker in winter than in summer, the speed depending on its distance from the Sun. This idea was, however, soon abandoned.
2. The Moon might have its motion about the Earth retarded by a force emanating from the Sun, which would be the greatest in winter when the Earth is nearest the Sun.[5] As early as 1600 Kepler accepted this nonuniform motion of the Moon in its orbit. No such theory had ever been advanced. This principle became of supreme importance in Kepler's later studies of planetary motion.

Kepler's Solar Theory

Kepler's refinement of solar theory was inspired by his attempts to improve his orbit of Mars. Having failed to obtain satisfactory results with theories such as his vicarious hypothesis (1601), which started from the assumption of uniform circular motion, his attention next turned toward the details of the defects of the representation of the retrograde arcs of the planets caused by the orbital motion of the Earth. He realized (as had Tycho) that insufficient accuracy in the Earth's (or the Sun's) motion would cause a troublesome effect in the observed motion of Mars in all parts of the orbit, not only at the stations and retrogressions. Tycho's 20 years of excellent solar observations produced a table of its motion accurate to $10''$–$20''$. Kepler thus knew the length of the year to an accuracy of about 1 second of time, and the eccentricity and orientation of the Earth's orbit (Tycho's solar orbit) with a correspondingly high accuracy. This would suffice for a good representation of the Sun's yearly motion, but the size of the Earth's orbit, which would directly affect the apparent planetary motions, was known with much less accuracy.

Kepler had long suspected that the Earth's orbit required an equant in the same manner as each of the other planets. This suspicion was strongly confirmed by his correspondence with Tycho, who, as early as 1598, wrote to Kepler that the orbit ascribed to the Earth by Copernicus did not always appear to be of the same size but sensibly varied its size relative to those of the three superior planets. Tycho also pointed out that in some particular situations, the angles which the semidiameter of the Earth's orbit subtended at the position of Mars differed by no less than $1°45'$! About 1602 this was verified by Kepler, who by then, after his appointment as Imperial Mathematician, had full access to all of Tycho's crucial observations. He showed from two observations of Mars (in 1585 and 1591), separated by three sidereal periods of Mars, that it was impossible that the Earth's equant and the center of its orbit could coincide. As we shall see presently, he showed that Tycho's total eccentricity of 0.03584 was very good, and that the center of the Earth's orbit was so situated as to require a bisection of the eccentricity so that the eccentricity of the equant was 0.01837, almost equal to 0.01792, or half of Tycho's orbital eccentricity. He thus introduced an equant point into the Earth's orbit at a position on the line of apsides so that the center of the (circular) orbit exactly bisected the distance between the equant point and the Sun.

Thus, the Earth's (or the Sun's) motion was uniform, not with regard to the center of its orbit, as assumed by Hipparchus, Copernicus, and Tycho, but with regard to a *punctum*

[5]Note that this is the modern explanation.

equans, as in the Ptolemaic theory of the planets. Observation of the Sun's motion alone (without the use of a telescope) could never have revealed the insufficiency of this simple eccentric uniform circular motion involving an equant because, due to the small eccentricity of the Earth's orbit, the maximum difference of longitude between (a) elliptic motion and (b) uniform circular motion with a properly situated equant is only $14''$. If the question were to distinguish between true elliptic motion and a uniform circular motion with no equant but with Tycho's eccentricity, the difference would be $56''$. It would thus have been barely possible for Tycho to disprove the simple eccentric solar orbit of Hipparchus from his table of solar longitudes. Was this the cause of Kepler having "long suspected" that an equant was necessary? No, the idea was probably rather a consequence of the viewpoint presented in *Mysterium Cosmographicum* that the Earth was a planet and therefore, like the other planets, required an equant point.

Kepler's Determination of the Equant Point of the Earth's Orbit

We will now show in more detail how Kepler perceived that the discrepancy of $1°45'$ in the angle subtended by the semidiameter of the Earth's orbit as seen from a point in Mars's orbit was due, not to a variation of the diameter or of the eccentricity of the Earth's orbit, but to the existence of an equant point situated on the Earth's line of apsides. This equant was on the opposite side of the center from the Sun, at a distance from the center equal to that of the Sun.

In Figure VII.2, let A be the Sun, B the center of the Earth's orbit, C the equant point (whose position is to be determined), DE the line of apsides of the Earth's orbit, and F the position of Mars, on the line perpendicular to the line of apsides of the Earth's orbit and near quadrature as seen from the Earth. Then the Earth would be at K or at L, near greatest elongation as seen from Mars, and the conditions would be most favorable for studying the inequality of the eastern and western areocentric elongations of the Earth resulting from the eccentric position of the equant point C.

As a preliminary consideration suppose the equant point coincides with the center of the circular orbit at B, and that at the beginning of the motion Mars was at F', and the Earth at K', east of BF' by $\frac{1}{2} \times 43° = 21\frac{1}{2}°$. After one Martian sidereal period ($686.98^{\rm d} \approx 687^{\rm d}$) Mars will again be at F', while the Earth will have performed two full revolutions ($730\frac{1}{2}^{\rm d}$) less Mars's motion, i.e., $730\frac{1}{2}^{\rm d} - 687^{\rm d} = 43\frac{1}{2}^{\rm d}$, or $43°$. The Earth will therefore be at L', $21\frac{1}{2}°$ west of BF'. After a second Martian sidereal period, starting with the Earth at K'', it will be at L'', such that $K''BF' = L''BF' = 43°$. After three Martian periods we have $K'''BF' = L'''BF' = 64\frac{1}{2}°$, etc. In all these cases the elongations from B, measured at F', would be equal in pairs, and $F'BK' = F'BL'$; $F'BK'' = F'BL''$; $F'BK''' = F'BL'''$. In particular, $\angle L'''BF' = \angle K'''BF'$, but Tycho had found the first angle greater than the second by $1°40'$.

Kepler now assumed that the equant was not at B but at some point of DE, such as C, and proceeded to determine its position as follows: Searching through Tycho's ledgers of observations he found no pairs separated in time by exactly a whole multiple of Martian sidereal periods and at the same time with Mars exactly at a quadrature. But these conditions were approximately satisfied by observations taken near the positions K and L in the Earth's orbit, with Mars at F. The first observation, near K, was on 30 May 1585, and the second, near L, was made 20 January 1591, about three Martian sidereal periods later.

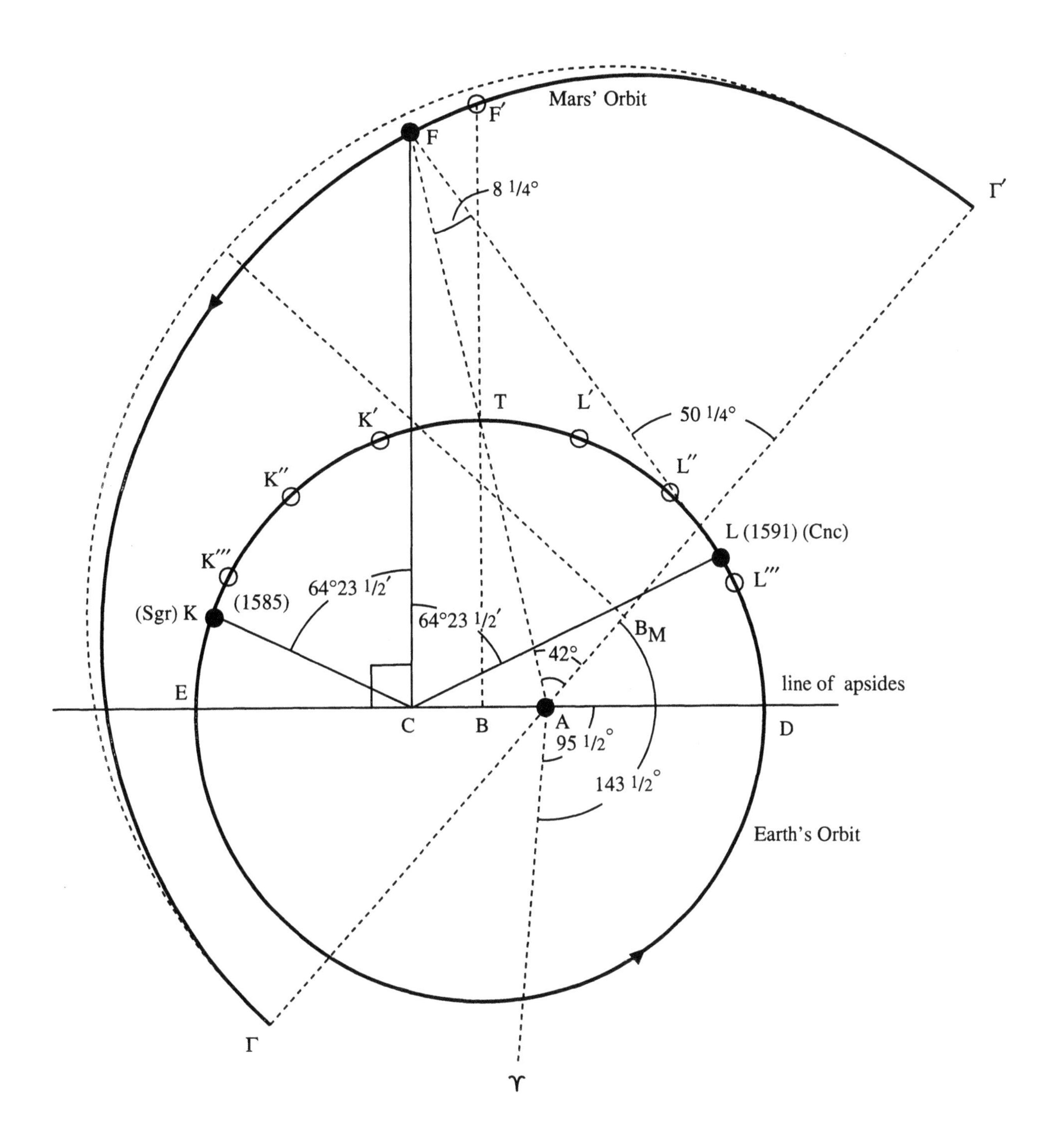

Figure VII.2 — Determination of equant point of Earth's orbit by two observations of Mars at the same heliocentric longitude (not to scale).

From Tycho's orbit of Mars (good to about 2′ in longitude), and using Tycho's position of the Martian apogee, $143\frac{1}{2}^{\circ}$, Kepler could find the mean anomaly of Mars from the heliocentric Martian places given by Tycho. Also, from Magini's planetary tables, giving daily and hourly angular rates of motion in all parts of the Martian orbit, he could derive fictitious places of the planet for moderate interpolation intervals with a high degree of accuracy. From Tycho's own observational table of the solar motion (good to perhaps 10″, certainly to 20″) he could interpolate very good positions of the Earth in its orbit at any time. After laborious trials he derived the fictitious situation shown in Figure VII.2, in which Mars was at the same point F in its orbit at the moments when in 1585 the Earth was at K, and in 1591 at L (corrected of course for six years of advance of the line of apsides of the orbits of both Mars and Earth). This situation realized an ideal configuration in which FC, the line connecting Mars with the equant of the Earth's orbit, was perpendicular to DE, the line of apsides of the Earth's orbit, and at the same time the angles of commutation $\angle KCF$ and $\angle LCF$ were both equal to 64°23′30″. (To effect this adjustment the observation of 1585 near K had to be carried forward 12 days, and the one of 1591 near L backward 2 days.)

The angles of parallax $\angle LFC$ and $\angle KFC$, although not of maximum value, differed by 1°15′, showing immediately why (as Tycho had found) the supposed radii LB and KB of the Earth's orbit seemed to vary in size, since if one erroneously took the center of the orbit to be at C, one would find $CL > CK$.

Kepler now proceeded to compute the distance BC, the amount the equant point lay beyond the orbital center away from the position of the Sun at A. An outline of his geometry is as follows:

In Figure VII.3, join BK, BL, KL; from L drop a perpendicular LP to KC extended. Draw from B and C perpendiculars BM and CN to KL; and from N draw the line NO parallel to the Earth's orbital line of apsides DE.

a. By observation and adjustment of the geocentric longitudes $\lambda_{geoc.}$ of Mars, the "angles of parallax" $\angle KFC$ and $\angle LFC$ of the Earth as seen from Mars could be found as follows (see Fig. VII.2):

$$\angle KFC = 95.5^{\circ} + 90^{\circ} - \lambda_{geoc.}\ (KF) = 36^{\circ}51'\,, \qquad \text{and}$$

$$\angle LFC = \lambda_{geoc.}\ (LF) - 185.5^{\circ} = 38^{\circ}06'\,. \quad [\text{Check: } \angle LFC - \angle KFC = 1^{\circ}15']$$

b. By adjustment from the known rates of motion, $\angle KCF = 64^{\circ}23'30''$.

c. Calling $FC = 1.00000$, find $KC = 0.61148$ by the sine formula.

d. Similarly find $LC = 0.63186$.

e. In ΔKCL find $\angle KCL = 2 \times 64^{\circ}23\frac{1}{2}' = 128^{\circ}47'19''$, by the abovementioned exact adjustment from times and rates of motion (Fig. VII.3).

f. In ΔLPC, $\angle PCL = 180^{\circ} - \angle KCL = 51^{\circ}12'41''$, and $\angle PLC = 90^{\circ} - \angle PCL = 38^{\circ}47'19''$. Find $PC = CL \cos \angle PCL = 0.39583$, and $PL = CL \sin \angle PCL = 0.49251$.

g. By addition find $PK = PC + CK = 1.00731$.

h. In ΔLPK find $\angle LKP = \tan^{-1} \frac{PL}{PK} = 26^{\circ}03'21''$. Find $KL = KP \sec \angle LKP = 1.12126$, so that $LM = \frac{1}{2}KL = 0.56063$.

i. $\angle CLK = 180^{\circ} - (\angle KCL + \angle LKP) = 25^{\circ}09'21''$.

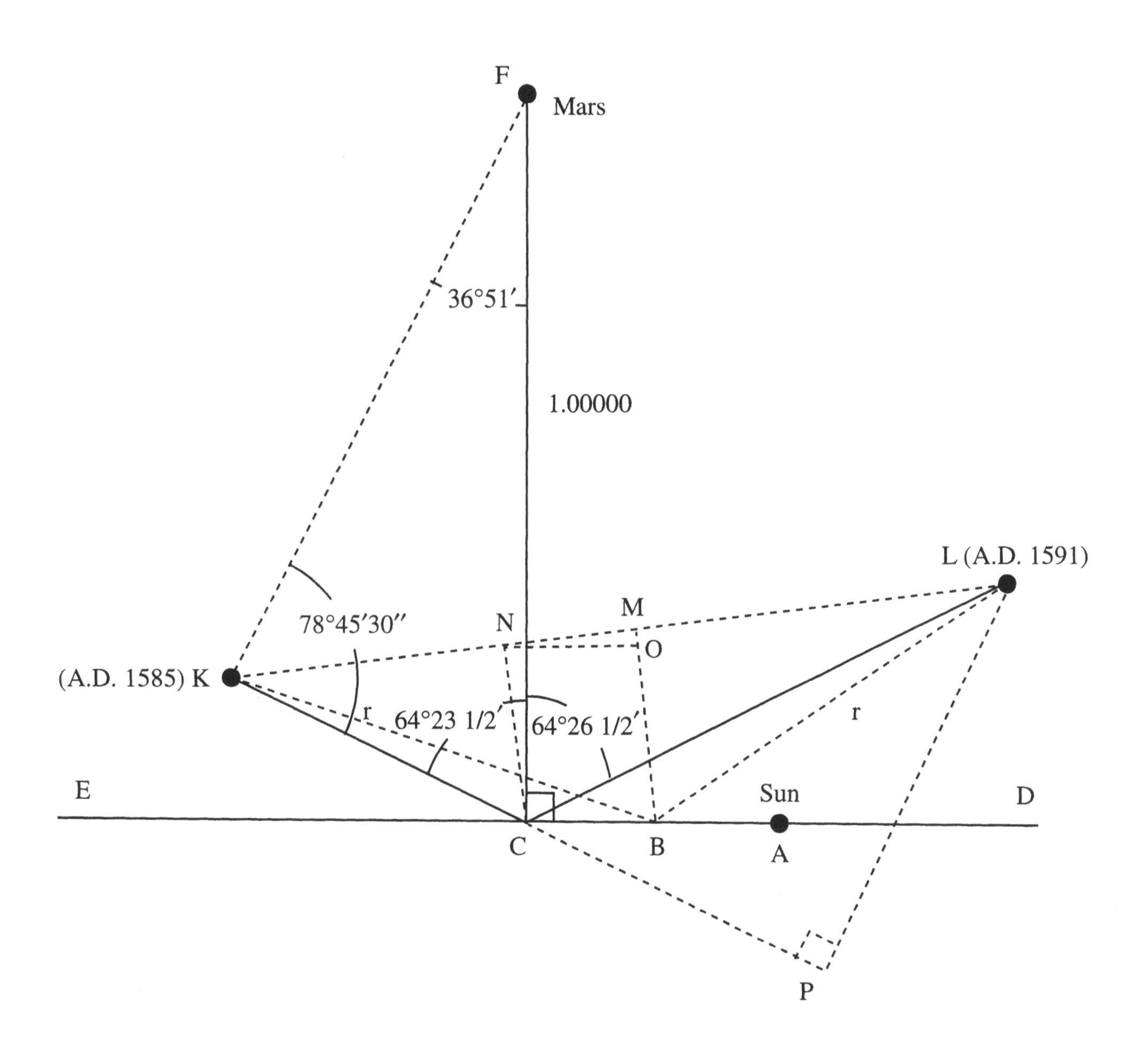

Figure VII.3 — Kepler's computation of the position of the equant point of Earth's orbit (not to scale).

j. $CN = CL \sin \angle CLK = 0.26859$; $LN = CL \cos \angle CLK = 0.57193$. By subtraction $MN = LN - LM = 0.01130$.

k. $\angle KCE = 90^\circ - 64^\circ 23'30'' = 25^\circ 36'30''$,

$\angle LKC = 180^\circ - (128^\circ 47'19'' + 25^\circ 09'21'') = 180^\circ - 153^\circ 56'40'' = 26^\circ 03'20''$,

$\angle ONM = \angle LKC - \angle KCE = 8^\circ 25'21''$,

$MO = MN \tan \angle ONM = 0.00167$,

$NO = MN \sec \angle ONM = 0.01143 = BC$ (with FC taken as unity),

$BM = CN + MO = 0.27025$,

$KB = \sqrt{KM^2 + BM^2} = 0.62237$, and finally,

$BC = 0.01837$ (with KB taken as unity).

Some Further Determinations of the Equant Position in the Earth's Orbit

Realizing the crucial importance of the solar theory for the further progress of his planetary theory, Kepler used four methods to test the position of the equant in the Earth's orbit:

1. He developed a method wherein any number of observations separated by intervals of 687 days could be used. From four such well-situated observations with Mars at a particular point of its orbit, Kepler checked that the equant point E was so situated that the orbital center C was, as he had always maintained, very nearly halfway between it and the Sun S. Furthermore, computations of the distances of the Earth from the equant point (which he later found to be the second focus of the Earth's ellipse) proved conclusively that the Earth's orbital speed was accelerated when approaching the Sun and retarded when receding.

2. Using the same four observations, but applying greater refinement in their preliminary reduction to the moments when Mars was at the same point of its orbit, he now solved for the distance SC and the distances of the Earth from the Sun. The idea was that if SC should also, like CE, turn out to be one–half of Tycho's eccentricity, the evidence would be complete for the bisection of the Sun-equant distance by the orbital center. He found this to be true within the high accuracy of his computations, and he computed the four distances of the Earth from the Sun with a relative accuracy of better than 0.1%.

3. Assuming that the four positions of the Earth at the times considered all lay on the same circle (the Earth's orbit), he realized that in the quadrilateral drawn between these points, the two angles subtended by any side at the points opposite this side should be exactly half the angle subtended by the chosen side at the center of the orbit. This construction depended upon one heliocentric longitude of Mars and four of the Earth. He computed the peripheral angles in the quadrilateral assuming values for these five positions using "the method of false position." This required repeated trials while varying slightly the position of Mars obtained from the vicarious hypothesis, and consequently the positions of the Earth. After satisfactorily close agreement was achieved between the central angles and double the relevant peripheral angles, he computed the distance

of the center of the Earth's orbit from the Sun to be 0.1653. Since his experience showed that although a closer estimate could be made, the value would never exceed 0.1800, the proof was thus already satisfactory.

4. He applied "Kepler's Method" of triangulation for one distance of Mars from the Sun, assuming the positions of the Earth to lie on a circular orbit eccentric by 0.1800 with an equant point of eccentricity 0.3600.

Using the same four observations and one more reduced in the same manner, Kepler found that three independent determinations agreed within three units in the fifth decimal place. This proved that an eccentricity of 0.1800 gave excellent results, and therefore that the Earth's orbit required this value, and not Tycho's value of 0.3600.

Kepler's Preliminary Work on the Orbit of Mars

As indicated above in the brief statement of Kepler's work in Prague, he began in 1600 to make fundamental and enormously detailed corrections of much of Tycho's reduction work. For instance, he proved that if, as in Tycho's theory, the orbit planes were to pass through the *mean* position of the Sun, the error in Mars's longitude near opposition would amount to $1°3'32''$. That Tycho needed the help of an energetic younger man, a theoretical astronomer, is confirmed by the fact that Tycho, the otherwise proud and superior autocrat, was willing to agree with Kepler's improvements.

First, Kepler used Tycho's extensive and accurate observations and found no diurnal parallax of Mars, whereas Tycho erroneously had used $4'48''$.

Next he determined the inclination of the Martian orbit to be $1°50'$.

He also determined the longitude of the node and checked the sidereal period of the orbit by his method of timing the planet at two similar node passages.

He next found that the calculated longitudes of Mars for the oppositions of August 1598 and August 1608, even using the fairly recent *Prutenic Tables*, were too small by 4° and 5°. Even larger errors were found from all other tables. He correctly ascribed these disagreements to the practice of calculating oppositions to the *mean* position of the Sun, as Copernicus and all others before him had done, rather than to the *true* place of the Sun.

By comparing Tycho's accurate observations with data from Ptolemy, he correctly determined a value of $13''$ for the average annual advance of the line of apsides of the Martian orbit.

Tycho had observed changes in the geocentric latitude of Mars all the way from $+4°33'$ to $-6°26'$ and, following Ptolemy, had accounted for this by introducing an epicycle of varying tilt. But instead of helping the situation the calculated latitudes and distances were unavoidably deranged. Unlike Copernicus, Tycho considered the orbital inclinations of the planets constant, but his failure to recognize that the planets were observed from the Earth, which moved in a plane (that of the ecliptic) slightly inclined to those of the planetary orbits, caused great confusion in understanding the latitudes.

Kepler's Vicarious Hypothesis

As shown in the *Mysterium Cosmographicum* (1596) Kepler early maintained that the equant concept ought to operate universally. In introducing this concept Ptolemy had assumed a "bisection of the eccentricity," i.e., placement of the equant point on the line of apsides, and as far beyond the center of the eccentric as that orbital center was from the Earth's center. But Kepler never felt comfortable with the notion that the motions of a body, be it the Moon or a planet, should be determined by its relation to an *empty* point in space. He was always seeking a physical cause for the motion of the celestial bodies.[6]

Upon joining Tycho in Prague in 1600 Kepler performed many abortive tests on Mars's orbit using all three systems: Ptolemaic, Copernican, and Tychonic. They all assumed circular uniform motions, an orbital eccentricity, and a method for finding the equant – and all were proved unsuitable. He then decided to abandon the third assumption and instead to determine directly from observations the necessary position of the center of Mars's orbit and its equant, assuming only that it lies on the line of apsides for which $\Gamma' = 148°48'55''$ (Tycho, 1587) and that the eccentricity of the equant is 0.18 (half of Tycho's eccentricity of orbit). Upon checking Tycho's worksheets Kepler found to his great satisfaction that Tycho had also at one time departed from Ptolemy's authority and computed an orbit on his own system good to $2'$ in the opposition longitudes. For this orbit, if translated to eccentric theory, the distance between the Earth and the equant point was divided by the center of the eccentric in a ratio different from unity. Even Copernicus, who, except for the distinguishing heliocentric feature of his system, had greatly deferred to Ptolemy's authority, had also substituted in his orbit of Mars a ratio different from unity.

Kepler called this theory, requiring no assumption of bisection of the equantal eccentricity by the orbital eccentricity, the "vicarious hypothesis." It was constructed from Tycho's longitudes at four oppositions (1587, 1591, 1593, and 1595) by a cumbersome method of approximation, there being no possible direct method. All the initial quantities were borrowed from Tycho's orbit of Mars, which Kepler had recomputed in great detail. The work fell into three steps:

1. to make the sum of the angles of an inscribed quadrilateral equal 360°,
2. to make the equant point fall on the line of apsides, and
3. to compute the resultant position of the center of the orbit.

Kepler continued these adjustments for four years, slightly varying some of the initial values at each new trial. He performed the full set of computations, each time to an accuracy of $1''$, at least 72 times before he was satisfied with the result – and this was (just) before the invention of logarithms! The result was the best circular orbit possible. It represented to an accuracy of $2'$ all the observed 12 opposition longitudes, 10 by Tycho and 2 by Kepler.

A simple outline of Kepler's scheme of operations is as follows. In Figure VII.4, let S be the Sun and D, E, F, and G four positions of Mars at four well-distributed oppositions in its "eccentric circular" orbit, of center C. The Earth is simultaneously at the positions d, e, f, and g, projected on the plane of the Martian orbit, and has the same heliocentric longitudes as D, E, F, and G. Let IH be the line of apsides and Υ the vernal equinox (for 1587, say).

Let C be the center of the final Martian orbit (whose position is to be determined), and A be the equant point of this orbit.

[6]As pointed out by A. S. Eddington, Kepler can be considered the first theoretical astrophysicist or, better, physical astronomer.

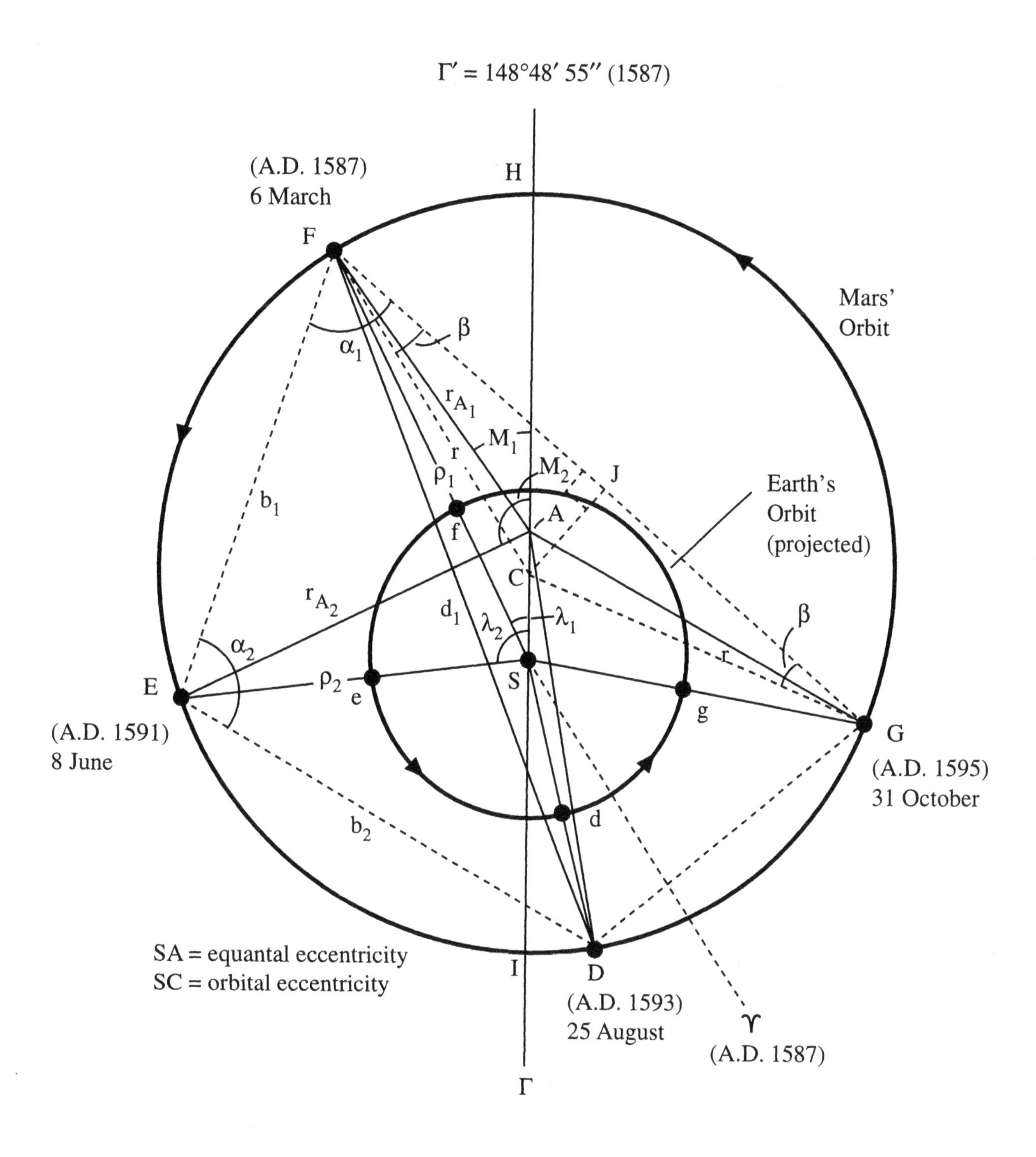

Figure VII.4 — Kepler's vicarious hypothesis.

Let M_1, M_2, etc. be the mean anomalies (counted from the aphelion H of Mars's orbit, as was customary until the nineteenth century). Let λ_1, λ_2, etc. be the heliocentric longitudes of Mars.

Let C_o, r_o (not shown) be the center and radius of Tycho's orbit of Mars. Let r be the radius of the final Martian orbit centered at C.

Let ρ_1, ρ_2, etc. be the radius vectors to the given Mars positions (from the Sun).

Let $r\mathrm{A}_1$, $r\mathrm{A}_2$, etc. be the distances of Mars from the equant point of its final orbit. Let $e = SC$ and $e_\mathrm{A} = SA$ be the orbital and equantal eccentricities of Mars.

Let α_1, α_2, etc. be the angles of the quadrilateral $DEFG$, as shown.

Let b_1, b_2, etc. be the sides of the quadrilateral $DEFG$, as shown.

Let Γ' be Tycho's heliocentric longitude of the aphelion of Mars's orbit, or $148°48'55''$ (1587).

Let ΔM_1, ΔM_2, etc. $= M_2 - M_1$, $M_3 - M_2$, etc.; and L_1, L_2, etc.. $= \lambda_2 - \lambda_1$, $\lambda_3 - \lambda_2$, etc.

From the observations near opposition the exact moments of opposition in longitude can be obtained by interpolation, supplying the solar longitudes from Tycho's accurate table of the Sun's yearly motion. The heliocentric opposition longitudes of Mars thus obtained are then further reduced for precession to a common epoch (1587). From the known time intervals between the oppositions and the mean daily motion of Mars ($n = \frac{2\pi}{P}$, where P is the sidereal period), the mean anomalies M_1, M_2, etc. are simply found. The initial values r_o, C_o, e_o, and M_o for Kepler's calculations were taken from Tycho's previous solution for the orbit.

The problem can now be stated as follows:
Given: ΔM_1, ΔM_2, etc.; L_1, L_2, etc.; and assuming $e_A = 1.00000$.
Find (by trial): $\alpha_1 + \alpha_3 = \alpha_2 + \alpha_4 = 180°$, and $\angle FSC = \angle FSH$.
Compute SC and r (in terms of $SA = 1$).
Procedure:

1. Given in ΔASF: $AS = 1.00000$; $\angle SAF = 180° - M_o$; $\angle ASF = \lambda_1 - \Gamma'$.
   Find: r_{A_1}, ρ_1; similarly for the other subscripts.
2. Given in ΔEAF: $\angle EAF = M_2 - M_1$ $(= M_2 - M_0)$; r_{A_1}; r_{A_2}.
   Find: b_1, $\angle AFE$, $\angle AEF$; similarly for the other sides and angles of the quadrilateral $DEFG$.
3. Given in quadrilateral $DEFG$: $\angle EFG = \angle EFA + \angle AFG$, and similarly for the other angles of the quadrilateral.
   Find and check: $\alpha_1 + \alpha_3 = \alpha_2 + \alpha_4 = 180°$. If this check is not close enough, try a slight adjustment of Γ_o and repeat the work until satisfactory. Success ensures that the positions of Mars lie on a perfect circle. Only then go on to determine the center C so that it lies on the line of apsides HI.
4. Given in isosceles ΔFCG: $FJ = \frac{1}{2}b_4$; $\angle FCG = \text{arc } GHF = 2 \times \angle FEG$. Lay off FC by making $\beta = 90° - \frac{1}{2}\angle FCG$, thus locating C. Check by laying off $JC = p = \frac{1}{2}b_4 \tan \beta$. Compute r (in terms of $AS = 1$).

5. Given in ΔCSF: $\angle SFC = \angle SFG - \angle CFG$; r; ρ_1.

   Find: CS; $\angle FSC$ ($= \angle FSH = \lambda_1$, for a check). A perfect check occurs when C lies on the line of apsides SH. The final location of C is found on this line using the computed value of SC. Similar determinations can be made for any of the other three corners of the quadrilateral.

The final results adopted by Kepler in 1604, after more than 70 trials, were
$\Gamma' = 148°48'55''$ (1587); $AC = 0.07232$; $CS = 0.11332$.
Eccentricity of equant $= e_A = AS = 0.18564$ ($r = 1.00000$ and $a = 1.00000$).

A rough comparison with the modern value of the position of the line of apsides is as follows:

Duncan's (1926) *Astronomy* lists Γ' (1935) $= 155°$ to the nearest degree. During 348 years (1587–1935) the precession of the equinoxes amounts to about 5° and the advance of perihelion of Mars to about 1°. Applying the sum of these two effects to the modern value of Γ', one arrives at 149°, Tycho's value to the nearest degree.

Estimate of the Accuracy of the Vicarious Hypothesis in Longitude

In the vicarious hypothesis Kepler had arrived at what we now know to be the best possible representation of Mars's motion in longitude by a single eccentric circle with an equant point arbitrarily placed at the most favorable position on the line of apsides. He found that it represented the 12 opposition longitudes exceedingly well, the largest residual (which may have been due to an observational error) being 2′12″. This occurred near minimum distance when the sighting was the most difficult because of increased brightness, with consequent increased spurious optical diameter.[7]

Other longitudes (not at opposition) were, however, very poorly represented, the discrepancies amounting to several tenths of a degree! They were mainly due to erroneous distances caused by the large value of $e_A - e_o$ (see below).[8] Mixed up with these differences were the much smaller dynamical residuals, which only became evident with the next step, when the bisection of eccentricity hypothesis was adopted.

At this stage it should perhaps be noted that the use of opposition longitudes by all early astronomers was favored not so much for any ease in observing them as for the fact that, in deriving heliocentric longitudes from observed geocentric ones, the distances of bodies (Sun, Earth, planet) from each other matter but little; in fact, they are absolutely irrelevant if the opposition happens exactly at the node of the orbit. In other words, opposition longitudes are practically independent of any kind of orbit assigned to the planet or to the Earth. On the other hand, when deriving heliocentric longitudes from geocentric ones *not* at opposition, distances, orbital inclination, and longitude of the node all enter the reduction, with values for distances being the most crucial when inclinations are small.

A mathematical study of the possible accuracy of various equant theories has been undertaken by Herz (1897, p. 87). Equations are given for a general equant theory, and by inserting certain values in these equations, the accuracy of the vicarious and other hypotheses can be exhibited. The equations for the true anomaly v_o (measured at the Sun eastward from aphelion) and radius vector r_o (measured from the Sun) in a general equant

[7]The real increase in diameter of a few arc seconds could hardly have been of consequence at a time when a first-magnitude star was believed to exhibit a diameter of 2′ or 3′.

[8]As we shall see later, perturbations by the other planets were by chance insignificant.

theory were developed into series in terms of multiples of the mean anomaly and powers of the eccentricity, and compared with the corresponding quantities v and r in pure elliptic motion about the Sun. The symbols illustrated in Figure VII.5 are as follows. Let Γ, Γ' be the vertices, C the center, S and F the occupied and empty foci of an ellipse of semimajor axis $C\Gamma' = a$ and eccentricity e. Let C_o be the orbital center of a circular orbit having a general equant at A. Let $C_oA = C_oF = SC = e =$ the eccentricity of the ellipse.

$SC_o = e_o =$ the orbital eccentricity of the general equant hypothesis, measured from the Sun,

$AC_o = e_o' =$ the orbital eccentricity of the general equant hypothesis measured from the equant (or the empty focus).

At any time, $M_0 =$ the mean anomaly $= (2\pi/P)(t - t_0)$, where

$P =$ sidereal period of the motion,

$t =$ the epoch considered,

$t_0 =$ the initial epoch ($M_0 = 0$, aphelion passage).

(By the above definitions we have in all theories $SA = e_A = 2e = e_o + e_o'$.)

To an accuracy defined by neglecting the third and higher powers of the eccentricity, the general equant equation is

$$v_o = M_0 - (e_o + e_o') \sin M_0 + \tfrac{1}{2}e_o(e_o + e_o') \sin 2M_0 + \ldots \qquad (M_0 \text{ from } \Gamma'), \tag{1}$$

$$r_o = 1 + \tfrac{1}{4}(e_o^2 + 2e_oe_o') + e_o \cos M_0 - \tfrac{1}{4}(e_o^2 + 2e_oe_o') \cos 2M_0 + \ldots. \tag{2}$$

Similarly, for an elliptic motion, we have

$$v = M_0 - 2e \sin M_0 + \tfrac{5}{4}e^2 \sin 2M_0 + \ldots, \tag{3}$$

$$r = 1 + e \cos M_0 + e^2 \sin^2 M_0 + \ldots. \tag{4}$$

To test the accuracy of any equant theory we must compare its values of v_o and r_o with the v and r values given by pure elliptic motion. Combining equations (1) and (3), making use of the identity $e_o + e_o' = 2e$, and simplifying, one finds

$$v_o - v = e\left(\tfrac{5}{4}e - e_o\right) \sin 2M_0 + \ldots \approx \tfrac{1}{4}e^2 \sin 2M\,. \tag{5}$$

Similarly,

$$r_o - r = (e_o - e) \cos M_0 + \tfrac{1}{2}(2e^2 - e_o') \sin^2 M_0 + \ldots \approx e^2 \sin^2 M\,. \tag{6}$$

For a perfect first-order agreement of the longitudes ($\approx$ the true anomalies) we should require the coefficient $\frac{5}{4}e - e_o = 0$, or $e_o = \frac{5}{4}e$. Kepler, by his 72 computations of the vicarious hypothesis and much additional "curve fitting," found empirically that the best agreement of the longitudes required $e_o = 0.11332$ and $e = 0.09282$. The closeness of his value of $\frac{5}{4}e = 0.11258$ to his value for e_o explains the near-perfect agreement in longitude he attained with his vicarious hypothesis. In fact, according to equation (5) his error due to the first and most important single term was only $e(\frac{5}{4}e - e_o) = 0.09828 \times 0.00014 = 0.0000687$ rad $\approx 14''$. However, according to Herz the neglected terms may occasionally

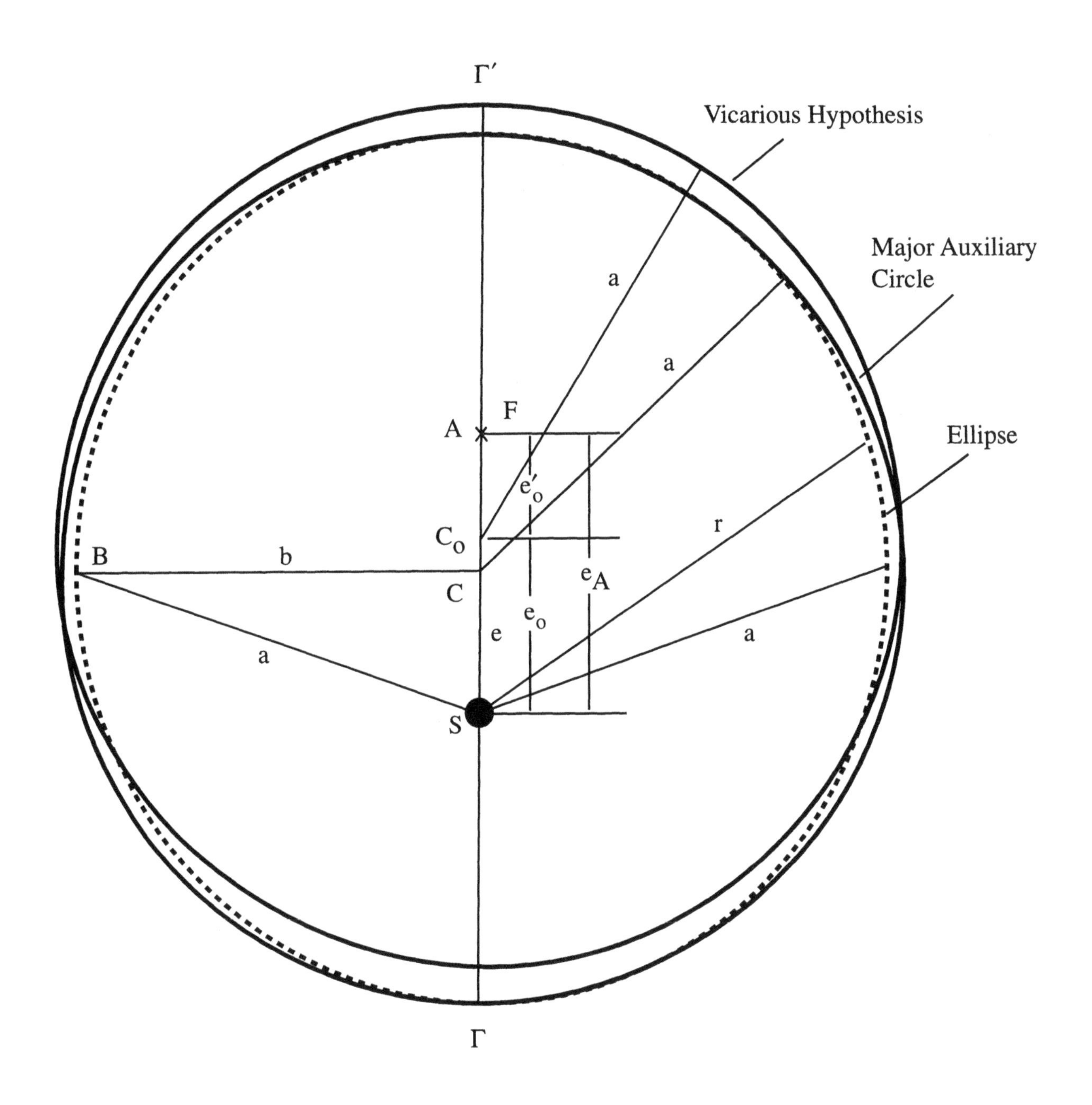

Figure VII.5 — The vicarious hypothesis and ellipse compared, following Herz (1897).

sum up to $\frac{1}{4}e^3$, or 42″. Thus it is safe to estimate that Kepler's vicarious hypothesis gave reliable opposition longitudes (in fact, all longitudes) to better than 1′ accuracy.

For the distances Herz estimated that the errors in equation (6) may increase to about 2.5%. As we shall soon see, this is the major cause of the very poor representation by the vicarious hypothesis of the celestial latitudes and of the nonopposition celestial geocentric longitudes.

The following three adaptations of the general equant theory are relevant to our future discussion:

a. The "vicarious hypothesis" proper is obtained from the general equant theory by letting $e_o = \frac{5}{4}e$. As we saw, it is the best possible approximation of equantal circular orbital motion to elliptic angular motion.

b. The bisection of the equantal eccentricity by the orbital eccentricity, or the "bisection of equant" case, is found by setting $e_A = 2e$. This is the case commonly said to be introduced by Ptolemy in both his lunar and his planet orbits. It is essentially Ward's Principle, described later. It is a definite improvement on uniform circular motion, though inferior to the result of Kepler's searching analysis.

c. The case of simple eccentric uniform circular motion (Hipparchus's theory) is obtained by letting $e_o = 2e$. The equant point coincides with the center of the orbit.

Kepler's First Refutation of the Vicarious Hypothesis (from the Latitudes)

When Kepler made general use of his vicarious hypothesis he found immediately that discrepancies between calculated and observed positions were far too great to be explained by observational errors. The errors in the celestial latitudes near the apsides were nearly 9′, and those in the nonopposition celestial longitudes near the octants of anomaly were about 8′. Based upon his firm belief that Tycho's observations were always individually accurate to 2′, he had resolved upon joining Tycho in Prague to find an orbit for Mars which would represent the observed longitudes and latitudes with an accuracy better than 2′ and the distances to better than 0.1%. Since neither of these conditions was fulfilled, the vicarious hypothesis, which looked so excellent for the opposition longitudes, had to be abandoned except as an aid at opposition. As we shall see presently, its failure was mainly due to erroneous distances and an erroneous eccentricity.

Estimate of the Accuracy of the Vicarious Hypothesis in Latitude

Having just shown that the vicarious hypothesis gave such excellent values for opposition longitudes, it now remains to show the reason for its complete failure to supply good celestial latitudes.

Adapting equation (6) to the vicarious hypothesis, i.e., letting $e_o = \frac{5}{4}e$, and therefore $e'_o = \frac{3}{5}e_o = \frac{3}{5} \cdot \frac{5}{4}e = \frac{3}{4}e$, we write

$$r_o - r = (\tfrac{5}{4} - 1)e \cos M_0 - \tfrac{1}{2}(2 - \tfrac{9}{16})e^2 \sin^2 M_0 + \ldots = \tfrac{1}{4}e \cos M_0 - \tfrac{23}{32}e^2 \sin^2 M_0 + \ldots \quad (7)$$

For small eccentricities this expression has maxima and minima in the neighborhood of the line of apsides (where M_0 is a multiple of 90°) of absolute value about $\frac{1}{4}e$. On the scale of

Mars's orbit (with $a = 1.52$), $e = 0.141$, so that the error in r is about $\Delta r = \frac{0.141}{4} = 0.035$, with r being too short by this amount at perihelion, and too long at aphelion (Fig. VII.5). This entails errors in the celestial latitudes, especially at those oppositions where the line of sight from Earth to Mars is steepest with the plane of the ecliptic. Thus the computed latitudes of the planet will be too far south at perihelion (if the latitude is south) and the distances too small. They will also be too far south at aphelion, the latitude now being north, and the distances too large. The approximate magnitudes of these errors may be estimated as follows:

In Figure VII.6 let

HH' = the ecliptic plane,

MM' = Mars's orbital plane,

☊ = the ascending node of Mars's orbit,

S = the Sun,

E = the Earth (whose orbital eccentricity is here neglected),

A = Mars (at aphelion),

O = Mars's position as determined with the erroneous distance SO,

$SE = R = 1.00$, the astronomical unit,

$SA = r = 1.66$, aphelion distance of Mars,

$EA = \rho = 0.66$, geocentric distance of Mars at opposition,

$AO = \Delta r$ = the assumed error in distance,

$\angle ESA = i = 1.85°$ = heliocentric latitude (approximate orbital inclination) of Mars,

$\angle H'EA$ = geocentric latitude of Mars (observed and corrected for instrumental errors and refraction),

$\angle H'EO = \beta$ = geocentric latitude of Mars computed from i and SO,

$\angle AEO = \Delta\beta$ = error in geocentric latitude caused by Δr,

$OP = x$ = projection of Δr on a perpendicular to the line of sight EA, extended,

$\angle SAE = \angle OAP = \alpha$ = "angle of commutation" of the Earth with the Sun as seen from Mars.

With these specifications we see by inspection of the figure that

$$\beta + \Delta\beta = \frac{1.66}{0.66} \times 1.85° = 4.63°,$$

$$\alpha = \beta + \Delta\beta - i = 4.63° - 1.85° = 2.78°,$$

$\Delta r = 0.035$ A.U. (given in the vicarious hypothesis),

$$x = \Delta r \sin\alpha = 0.035 \times 0.048 = 0.00168 \text{ A.U.},$$

$$\Delta\beta = 0.00168/0.66 = 0.00254 \text{ rad} = 9'.$$

For a position at Mars's perihelion ($r = 1.37$) the error is of course larger. Using again $\Delta r = 0.035$ as before, but setting $r = 1.37$, we find $\Delta\beta = 16'$.

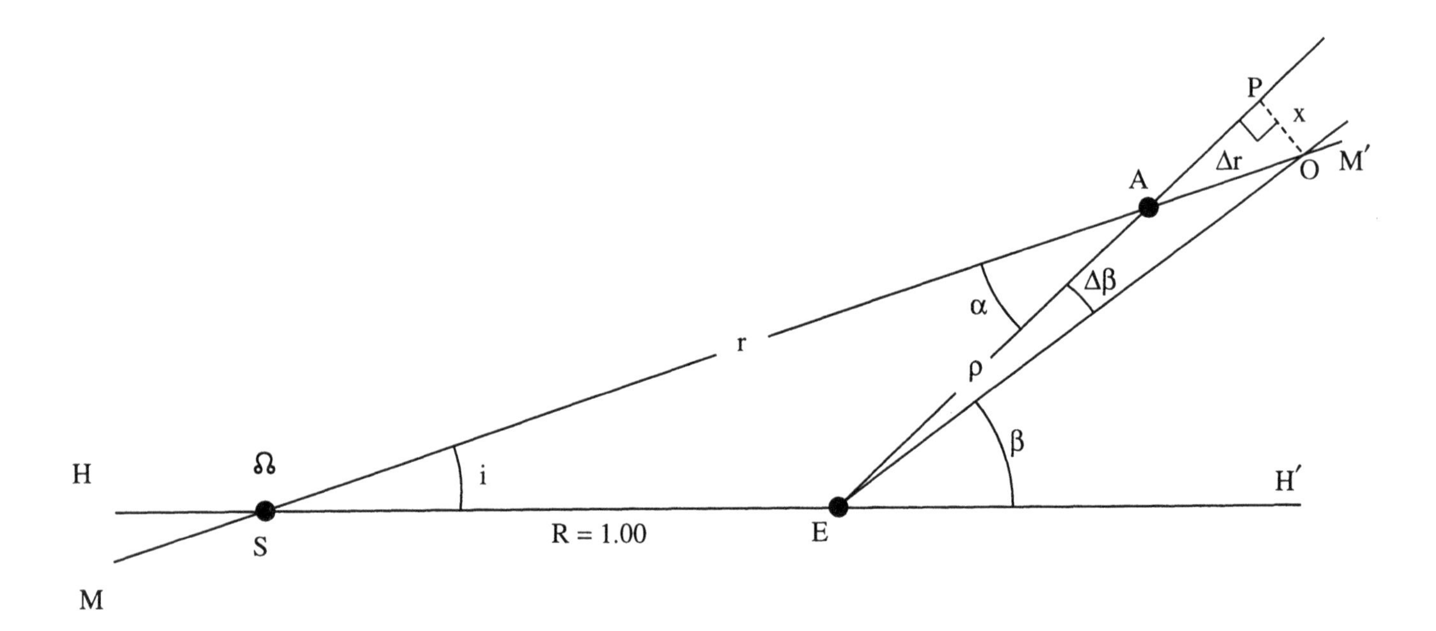

Figure VII.6 — Error in latitude (of Mars) on the vicarious hypothesis.

As already noted these estimates are based entirely upon the error in the distance caused by the eccentric position of the vicarious circular orbit with respect to the apsides of the Martian orbit. Due to this cause (shifting of the circle along the major axis of the ellipse by 0.035) the maximum values of the error occur at the apsides, and zero values occur nearly at right angles to these, i.e., at quadrature.

The difference of shape between the elliptic orbit and the circle also causes an error in distance. This effect is much smaller in latitude, being zero at the apsides and having maxima situated nearly at right angles to the major axis. In the case of Mars, by mere chance the closeness of the direction of the line of nodes to that perpendicular to the line of apsides means the distance error reduces almost to zero.[9]

In summary, the vicarious hypothesis, although well nigh perfect for computing all heliocentric and geocentric opposition longitudes, was nevertheless completely unreliable for computing opposition latitudes, as well as all latitudes except the zero values at exact node passages.

Kepler's Estimate of Mars's Orbital Eccentricity (from the Latitudes)

Kepler knew of the reluctance with which earlier astronomers had abandoned the principle of the "bisection of the eccentricity," even though it had never worked for them. Modifications had been tried, but with no real success. Ptolemy and Copernicus had tackled the problem with no satisfactory results, and then Tycho had tried it but gave up and instead turned to work on the Moon's orbit. Now Kepler came to a similar critical situation. He began to consider that a theory with "bisection of the eccentricity" should be worked out in full detail. He therefore started to investigate the eccentricity of Mars's orbit from the celestial latitudes alone, i.e., by a method entirely independent of the way in which he had derived it from the position of the circular orbit on the vicarious hypothesis. To illustrate his method consider the following simplifying conditions:

In Figures VII.7(a) and (b):

a. Let the orbits of the Earth ⊕ and Mars ♂ be circles of radii 1.00 and 1.52, centered on the Sun ⊙.

b. Let Mars's orbit be in a plane through the Sun's center, of constant inclination i to the ecliptic.

c. Let the line of nodes of Mars's orbit be constant in space.

Let two observations be made of the geocentric celestial latitude of Mars at opposition, one 90° before an ascending node passage, the other 90° after a similar passage. These geocentric latitudes should be numerically equal but of opposite sign ($\beta_1 = -\beta_2$ in Fig. VII.7(a)).

In Figure VII.7(b) let Mars's orbit be centered on a point C on a line in its orbital plane at right angles to the line of nodes, a distance ae from the Sun. (For the sake of definiteness let this center be north of the plane of the ecliptic.) The orbit now has a line of apsides with aphelion at heliocentric nodal coordinates $\ell = 90°$, $\beta = i$. By the geometry of the problem, the geocentric latitudes now should be unequal, the one $\beta_3 < \beta_1$ at $\ell = 90°$, and

[9]This distance error at the quadratures will, however, be of enormous importance at a later stage, when Kepler does away with all circular, eccentric orbits and minutely investigates the true shape of the orbit.

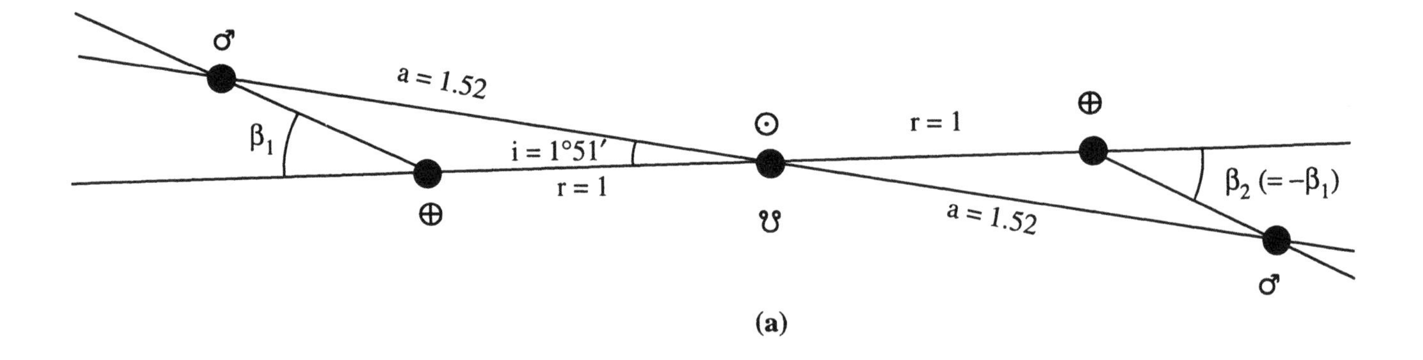

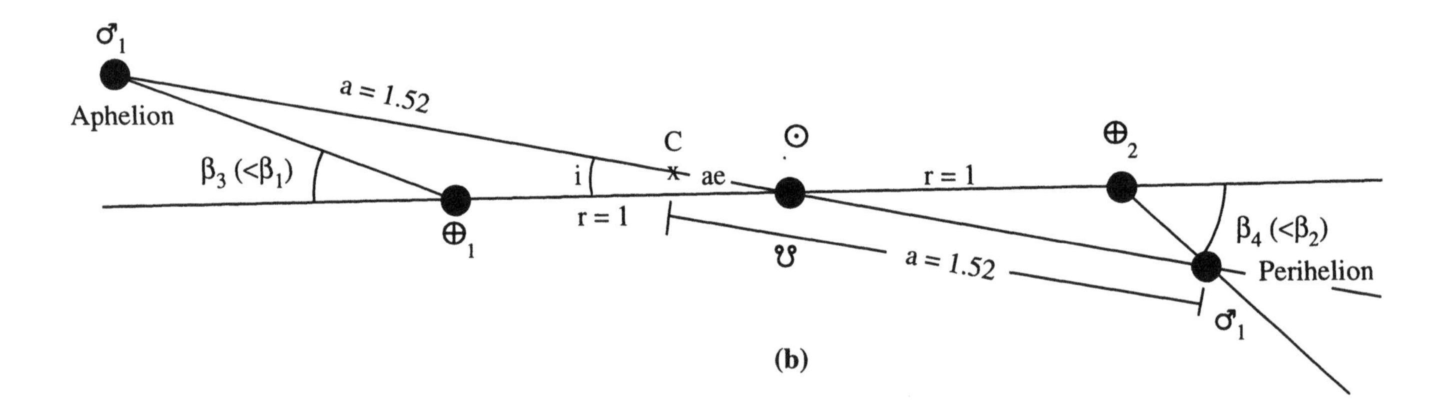

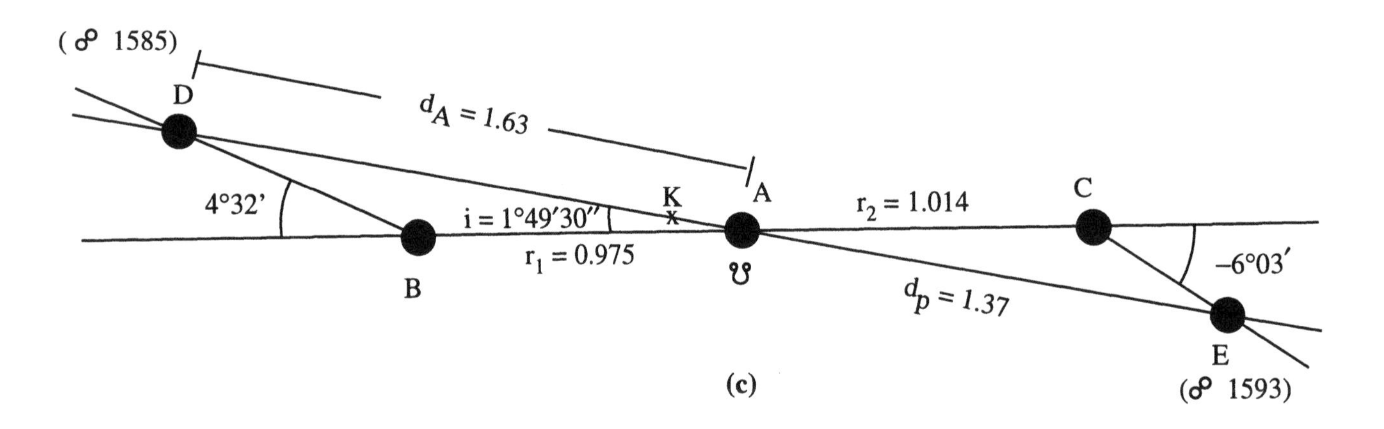

Figure VII.7 — Kepler's determination of Mars's orbital eccentricity from celestial latitudes. (a) Oppositions 90° before and after an ascending node passage. (b) Kepler's analysis of the general situation. (c) Kepler's analysis for the oppositions of 1585 and 1593.

the opposite one $|\beta_4| > |\beta_2|$, and negative. (The values of β_3 and β_4 can be computed from the oblique plane triangles $\odot \oplus_1 ♂_1$ and $\odot \oplus_2 ♂_2$, in which $\odot \oplus = 1$, $\angle \oplus \odot ♂ = i$, $\odot ♂_1 = 1.52 + ae$, and $\odot ♂_2 = 1.52 - ae$.) One finds of course that $|\beta_4| > |\beta_3|$. Conversely, in $\Delta \odot \oplus_1 ♂_1$ one may assume $\angle \odot \oplus_1 ♂_1 = 180°$ minus the measured geocentric latitude, $\odot \oplus_1 = 1$, and $\angle \oplus_1 \odot ♂_1 = i$. Solve for $\odot ♂_1$ and find (neglecting the inclination i) $e = \frac{1}{2}[(a + ae) - (a - ae)]$.[10]

From this kind of simple latitude analysis yet rigorously allowing for the projections, Kepler soon proved that the eccentricity of the vicarious hypothesis could not be correct. The conditions of this method are of course never precisely fulfilled, but the method nevertheless gave excellent results for Kepler.

For example, he used two oppositions well observed by Tycho (Fig. VII.7(c)): for 30 January 1585, $\beta = 4°32'$ ($\ell = 7°$ before aphelion and 94° past ascending node), and for 25 August 1593, $\beta = -6°3'$ ($\ell = 16°$ after perihelion and 85° past descending node). From Tycho's orbit of the Earth (the best then available) he found $AB = 0.975$ and $AC = 1.014$. By solving the relevant planet triangles he found $AD = 1.63$ and $AE = 1.37$. Let K be the midpoint of ED, thus the center of Mars's orbit. He found $KA = e = 0.0800$.

From several such pairs of observations he found that the eccentricity of Mars's orbit was certainly between 0.0800 and 0.09943. These results would be but slightly altered if he were to adopt half of Tycho's value (as he thought correct) for the Earth's orbital eccentricity, i.e., 0.018 instead of 0.036. Thus the value of the orbital eccentricity derived from the vicarious hypothesis, 0.11332, was in flagrant contradiction with the results of the latitude studies. The vicarious hypothesis had to be rejected as a good representation of Mars's orbit. As we now know, the vicarious hypothesis was a mere fortunate interpolation device for the heliocentric longitudes, which worked because of the fortuitously close coincidence between 0.11332 and $\frac{5}{4} \times 0.09282 = 0.11138$. In the vicarious hypothesis the eccentricity of the equant was 0.18564. Half of this, 0.09282, is so close to the average of 0.0800 and 0.09943, or 0.08972, that bisecting the eccentricity led to some success.

Kepler's Second Refutation of the Vicarious Hypothesis (from the Longitudes)

As noted above, when Kepler used the vicarious hypothesis to represent observations not at opposition, he immediately found errors in the celestial longitudes of upward of several tenths of a degree!

There are two main causes of the differences in observed and calculated nonopposition longitudes obtained from the vicarious hypothesis:

1. The eccentric position of the vicarious (circular) orbit (center C_0 in Fig. VII.5) with respect to the major auxiliary circle of the elliptic orbit of Mars (center C).
2. The deviation of the elliptic orbit from its major auxiliary circle.

[10]Since for Mars's orbit the line of apsides is not exactly at right angles to the line of nodes, the heliocentric latitudes for Mars at the apsides would be only approximately i and $-i$. In the actual orbit, since $☊ = 46°$ and $\Gamma = 149°$ (1587), the line of apsides made an angle of 103° with the line of nodes. This alters slightly the values to be used in the following illustrations. Calling the true angle between the major axis and its projection on the plane of the ecliptic i_1, we have closely, since i is small, $i_1 = i \sin(\Gamma - ☊) = i \sin 103° = 0.9741$. The resultant errors generated in using i for i_1 are less than 3%, which is unimportant for illustrative purposes.

For the first case, the eccentric position of the vicarious orbit necessitates corrections of a purely geometrical nature of as much as 1° and can be exactly computed from the assumed constants and the position of the Earth. This is a part of the reduction process characteristic of using the vicarious hypothesis out of opposition. This error is zero at the quadratures and reaches maximum values at the apsides. It will be zero at opposition or conjunction and reach a maximum at quadrature if Mars is then on the line of apsides.

In Figure VII.8, M_1 and M_2 are positions of Mars at aphelion and perihelion, while T_1 and T_2, are positions of the Earth with Mars near quadrature. V_1 and V_2 are the intersections of the line of apsides with the vicarious orbit. Inspection will confirm the algebraic signs of the residuals (observed minus computed) noted in the figure, and the arrows indicate the direction in longitude of the corresponding $\Delta\lambda$, in the sense of the error to be applied to a longitude ST_2V_2 computed by the vicarious hypothesis to correct it to the true value ST_2M_2.

Consider the triangle ST_2M_2 in which $\angle ST_2M_2 = 90°$, $ST_2 \approx 1.00$, and $SM_2 = 1.44 = 1.528 - 0.093$. Let $\alpha = \angle SM_2T_2$, the commutation of the Earth with the Sun from Mars. Then by definition $\alpha = \sin^{-1}0.6944 = 44°$. In Figure VII.8, we have given $V_2M_2 = C_oB = 0.035 \approx V_2F_2$ (since $\alpha \approx 45°$). Also $V_2P_2 \approx 0.7V_2F = 0.0245$; and $V_2T_2 \approx ST_2 \approx 1$. Hence $\angle V_2T_2P_2 = \Delta\lambda_{geoc.} = 0.0245$ rad $= 84'$ and $\Delta\lambda_{helioc.} = \frac{0.035}{1.44}$ rad $= 84' = \angle F_2SV_2$.

Thus the maximum difference of heliocentric longitudes caused by using the erroneous perihelion distance given by the vicarious hypothesis can amount to over 1°! It is obvious that as long as the vicarious circle is considered an orbit, and not merely an equant, by varying a one can greatly change the numerical values of the differences, although the residuals of distance will normally have the same algebraic sign for two consecutive quadrants.

In the case of these longitudes Kepler now reversed the process and computed the orbital eccentricity necessary to represent the observation without sensible error. His process is illustrated schematically in Figure VII.9, in which T_1 and T_2 are positions of the Earth in its orbit, here considered circular and centered on the Sun S. M_1 and M_2 are the positions of Mars at aphelion and at perihelion, observed from T_1 and T_2 respectively.[11] The positions of T_1 and T_2 are known from the times of observation and the constants of the Earth's orbit. Given

$\angle M_1ST_1 = 30°00'28''$ (from the time of observation and the known position, from Tycho's orbit, of the line of apsides), and

$\angle M_1T_1S = 123°26'39''$ (from Tycho's observation of the elongation of Mars).

We find

$\angle T_1M_1S = 180° - (123°26'39'' + 30°0'28'') = 26°32'53''$, and

$T_1S = 0.99302$ (from Tycho's orbit of the Earth).

By the sine formula,

$$SM_1 = 0.99302 \frac{\sin 123°26'39''}{\sin 26°32'53''} = 1.65680.$$

[11]The observed positions were not exactly at these points but 0°12′ and 3°11′ away from M_1 and M_2. Corrections were applied to reduce the computed distances to the exact positions on the line of apsides.

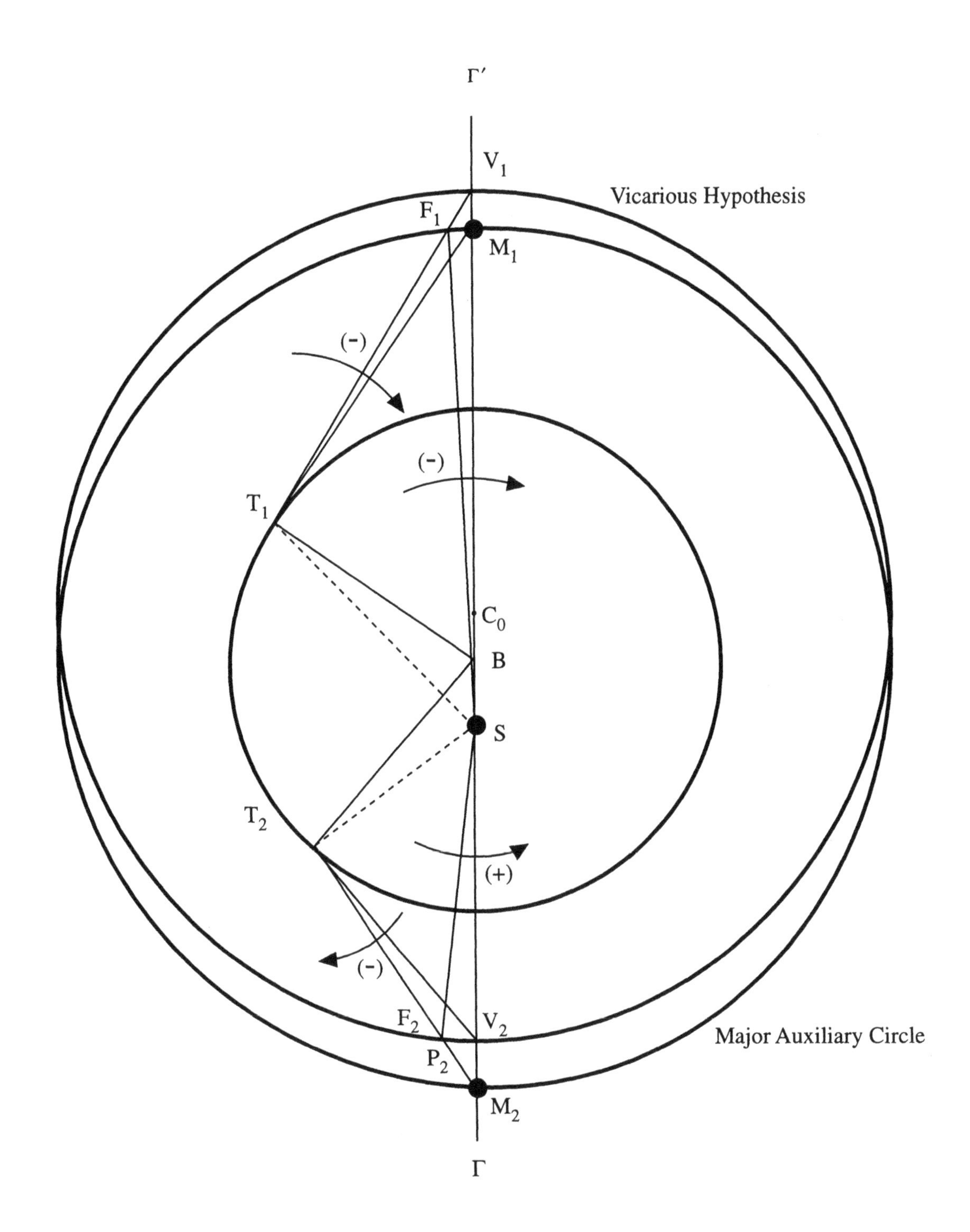

Figure VII.8 — Geometrical corrections to the computed longitudes from the vicarious hypothesis.

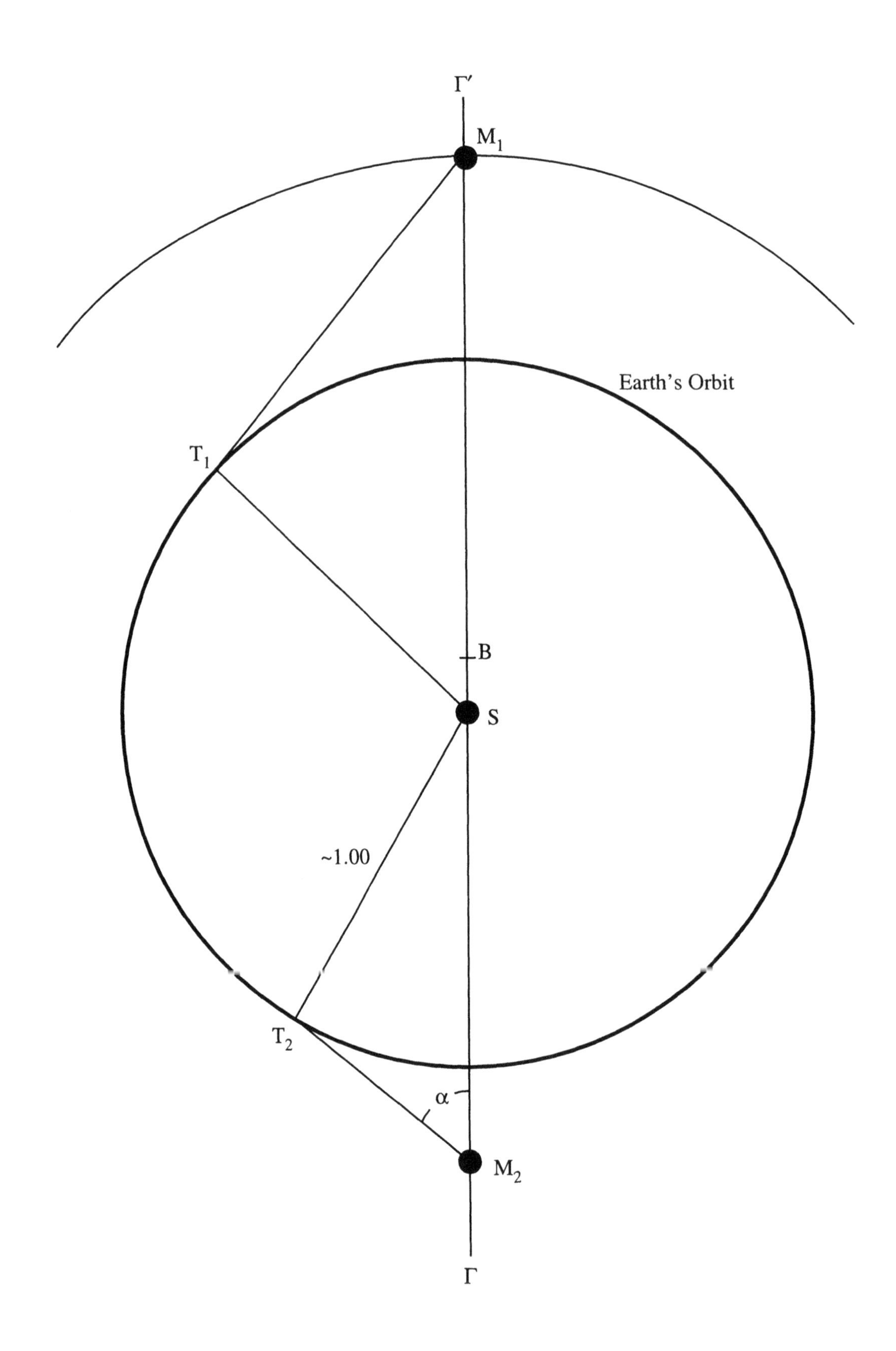

Figure VII.9 — Kepler's second refutation of the vicarious hypothesis.

Similarly, from ΔT_2SM_2 we find $SM_2 = 1.40080$.

Corrections of small fractions of a percent were applied by Kepler for the respective heliocentric latitudes, $1°48'$ and $1°36'$.

Finally, Kepler found

$$M_1M_2 = 3.05899, \qquad a = M_1B = \tfrac{1}{2}M_1M_2 = 1.52950, \qquad \text{and}$$
$$e = SB = SM_1 - M_1B = 0.12812.$$

This value of e refers to $a = 1.52$, etc. If referred to a value $a = 1.00000$, Kepler found that values of the orbital eccentricity ranging from 0.08377 to 0.10106 were required for the best results. Again, as in the case of the latitudes, the average 0.09242 of these values, found by this entirely independent method, was so close to 0.09282 that it seemed to indicate a true orbital eccentricity bisecting the equantal eccentricity of 0.18564. Thus he arrived at the same conclusion as from a study of the latitudes. After additional pages of trial computations Kepler informs us:

> When the longitudes are computed for observations not at opposition, the needed eccentricity differs markedly from that of the vicarious hypothesis. Also, (1) if the orbit is a circle, then there can be no fixed point within the circle about which the planet moves with uniform angular and linear velocity, or (2) if the orbit is a circle and we retain a point about which the planet has uniform angular and linear velocity, this point oscillates back and forth on the line of apsides, thus producing the varying eccentricities needed, or (3) the orbit is not a circle.

The Bisection of Eccentricity Hypothesis

Immediately upon discarding the vicarious hypothesis and trying out the alternative possibility of temporarily introducing a circular orbit with eccentricity 0.09282 (the bisection theory), Kepler at once ran into large errors even at opposition longitudes (which the vicarious hypothesis had fitted to perfection). The greatest errors of this kind were at the opposition longitude of 1582, which differed by $7'41''$ from that given by the vicarious hypothesis, and by $9'15''$ from Tycho's observed (and corrected) value. This residual is often referred to as "Kepler's famous eight arcminutes." This opposition occurred near the first octant of the orbit, 41° from aphelion. We shall see later that work on observations near the octants soon became of paramount importance for further progress on the shape of the orbit.

Kepler's Improvement of the Earth's Orbit by Bisecting Its Eccentricity

Kepler's solar theory involved the introduction of an equant with bisection of the eccentricity for the Earth's orbit. It was undertaken partway through the grueling work on Mars after a seemingly complete failure of circular orbits for Mars. We shall see, however, that the solar theory of Kepler, though still based upon circular uniform motion of the Earth with an equant point, was an excellent representation of the Earth's annual motion because of the small eccentricity of the Earth's orbital ellipse. It made superfluous any further worry about the error introduced into the planetary motions by insufficiently known corrections arising from the Earth's motion or position. Kepler was later fully able to attain the accuracy necessary to prove the elliptical nature of the Martian orbit even while he retained a circular orbit for the Earth. Nor was it necessary for him to substitute an ellipse for the Earth's orbit in order to reach his goal for the orbits of other planets. Eventually he did substitute an ellipse, partly from his belief in "the unity of nature" (an important consideration in all his work) and partly as an aid to improved accuracy.

Bisection of the Eccentricity for Mars's Orbit

The great importance of Kepler's side step to develop an accurate solar theory was not only the production of a set of excellent elements for the Earth's orbit but also the five or more proofs that in this orbit the equantal eccentricity is bisected by the orbital one. The result gave values for the Earth's orbit of $a = 1.00000$ and $e = 0.01837$, nearly half of 0.03584, the value used by Tycho for a simple eccentric with no equant.

Having now proved that an equant is necessary and that the eccentricity is bisected for the Earth's orbit, Kepler extended these principles to the orbit of Mars because he took it to be a physical necessity for all the planet orbits. He also (for the first time) computed the Martian anomalies by his law of areas (as applied to the eccentric circle on the bisection of eccentricity hypothesis, and in its first, slightly approximate form, namely the sum of distances from the Sun computed for each degree of anomaly). Thus he left the orbit circular. Based on observations mainly near the apsides, this process gave excellent elements for Mars's orbit: $a = 1.5264$, $e = 0.09264$, $\Gamma' = 148°39'46''$. The longitudes near the apsides and quadrants agreed beautifully with those of the vicarious hypothesis (the best thus far obtained). For example, there was only a $24''$ difference in the true anomalies computed by the two theories at mean anomaly 90°.

Suspicion of a Law: Estimate of the Accuracy of the Bisection of Eccentricity Hypothesis

But at the first octant (eccentric anomaly = 45°, reckoned from Γ') the computed true anomaly was $8'21''$ in advance of the vicarious hypothesis, and at the third octant (135°) it was $8'01''$ behind. Mars in the new orbit appeared to be moving too rapidly near the apsides and too slowly near the mean distances. Instead of considering this an indication of a lack of orbital circularity, Kepler at first suspected his new law of areas method (first version) for computing anomalies. In reality this approximate application of the areal velocity method was in error, not so much because of its approximate form, but because it was applied to an eccentric circle and not to an elliptical orbit. Kepler next showed (by careful calculation for every degree of a semicircle) that his approximate areal method of adding computed distances to represent areas gave the (known) area to within 7 parts in 18,000, and that this discrepancy could never affect the computed longitudes (around the quadrants) by more than $4\frac{1}{2}'$. Moreover, the computed planet positions would be ahead of the observed ones near the quadrants and not at the octants. The $8'$ error in longitude had to be due to something else.

We now know what caused this famous $8'$ residual in longitude. If in the general equant equation (1) we put $e_o = e'_o = e$, we obtain the case for bisection of the eccentricity, namely

$$v_o = M_0 - 2e \sin M_0 + e^2 \sin 2M_0 + \cdots \qquad (M_0 \text{ from } \Gamma'). \tag{8}$$

If from this we subtract equation (3):

$$v = M_0 - 2e \sin M_0 + \tfrac{5}{4}e^2 \sin 2M_0 + \cdots,$$

valid for pure elliptic motion, we obtain

$$v_o - v = -\tfrac{1}{4}e^2 \sin 2M_0, \tag{9}$$

the amount by which equant motion is advanced on pure elliptic motion. Inspection shows that maximum values of $v_o - v$ occur at the octants. Taking $e = 0.093$, we have $(v_o - v)_{max} = \frac{1}{4}e^2 = 0.00218\,\text{rad} \approx 8'$. Hence the $\pm 8'$ residuals are due to erroneously considering Mars to follow a concentric circular orbit of radius a, rather than an elliptical orbit of $e = 0.093$ and semimajor axis a.

Kepler's genius was characterized by two dissimilar tendencies: first, precipitous adoption of some happily inspired new principle, of which we shall see examples later, and secondly, tenaciousness with which he kept to his old ideas, giving them up only after he had laboriously established several incontrovertible proofs of their utter impossibility. He now set to work to see if it might not be possible to rescue the idea of uniform circular planetary motion by preserving the double epicycles of Copernicus and Tycho. This scheme represented the path of Mars as an oval running beyond the eccentric circle at the quadratures, presumably slowing down the motion at the quadratures. Instead of using the combination of circular motions that these systems required, he used his areal principle for the motion of the epicyclic center along the circular deferent of the planet orbit. These computations resulted in a complete fiasco, making the planet move much faster at the quadratures than anything so far tried, whereas observations indicated that it ran slower there.

Direct Determination of the Distances of Mars from the Sun by Tycho's Observations (1602)

The above experience made Kepler abandon the epicycle method for a while. He had previously abandoned the equant method for his areal method. He now acted on a third conviction of his genius, namely, that truth emerges only from the experience gained in correcting results obtained by the committing of errors. As a result of all this apparent failure, he began to suspect, for the first time seriously, that the orbit of Mars might not be an exact circle. He immediately set out to prove this from the observations themselves, independently of any theory of motion. To do this he selected three of the best observations and computed their relative distances from the Sun, assuming known the line of apsides and the eccentric position of an enveloping circle determined from eight other observations. Since the line of apsides and major axis were the best determined quantities, he adjusted the position of the circle by trial and error until the central angles (at B in Fig. VII.10) were exactly twice the peripheral angles to the same arcs, i.e., to the differences separating the observations. Thus, by adjustment, he made $\angle FEG = \frac{1}{2}\angle FBG$, $\angle FGE = \frac{1}{2}\angle FBE$, and $\angle GFE = \frac{1}{2}\angle GBE$. He next computed the three radius vectors of Mars from the observations by a triangulation method and compared them to the corresponding distances of points on the circle intersected by the radius vectors. He found that in every case the radius vector to the point on the circle exceeded that to Mars, the more so the greater the true anomaly. For the three values of the true anomaly (measured from the aphelion) of 10°, 38°, and 104°, he found that the computed distances from the Sun by a circular orbit with bisection of the eccentricity exceeded the distances computed from the observations by 0.00350, 0.00783, and 0.00789, respectively (on a scale of $a = 1.52$). From these few values Kepler ventured the following conclusion: "The orbit is a curve which coincides with the circle at the apsides, then retires more and more within it until, at 90° and 270° of eccentric anomaly, it comes to its greatest deviation from the circle. The orbit is broadest at aphelion, more pointed at perihelion." He called it an oval.

Three of Kepler's Efforts to Retain an Epicycle and a Deferent: Kepler's "Ovoid" Orbit

Instead of assuming a noncircular orbit of Mars, Kepler entered upon a new spate of attempts based on uniform circular motions. Many fruitless attempts were made, and even Small (1804) considers it worthless to enter into all of these details. Yet, to show the genesis of Kepler's ideas and better to appreciate the enormity of his efforts, we will briefly mention three of his attempts to reproduce an oval by uniform circular motions.

1. In the first attempt Kepler precipitously assumed that the planet moved uniformly on a retrograde epicycle, the center of which shifted eastward nonuniformly on a circular deferent in such a way as to produce the optical equation of the planet. (The *optical equation* is the angle at the planet subtended by lines of sight from the planet to the center of its orbit and to the Sun.)
2. In the second attempt Kepler tried to construct the shape of the oval by motions on an eccentric (to which the two motions, one on the epicycle, the other on the deferent, had been converted). The retrograde motion on the epicycle, due to a "planetary virtue," was uniform, like the rotation of the Earth, but the motion on the deferent, due to a "solar virtue," was nonuniform and governed by the law of areas under the deferent – measured with apex in the Sun. It led to geometrical constructions from which no exact predictions could be made because it involved – quite apart from the inaccuracy of the primitive law of areas – the insuperable problem of dividing an angle in any ratio.
3. The third attempt, often referred to as Kepler's "ovoid" orbit, or simply Kepler's oval, had a closer resemblance to elliptic motion than any previously attempted construction. It was a curious hodgepodge of two conflicting theories. First, the vicarious hypothesis was used only for the longitudes (it had, as we know, an equant 0.07 beyond the center of its circular orbit at $e = 0.11$). Second, for the distances Kepler used a circular orbit centered on the line of apsides at 0.093 from the Sun with an equant twice as far out. Yet the motion in this combination orbit was governed, not by the bisection of eccentricity principle, but by a complicated areal principle applied to the still unknown orbit.

First Construction of Kepler's Ovoid Orbit

Kepler used this combination of two orbits to yield points on the observational oval, and it became clear to him that the resulting curve was not an exact ellipse. Yet the resulting oval was almost indistinguishable from an ellipse, here called the "auxiliary ellipse," having center and axes the same as the oval.

Kepler's method of construction of his ovoid orbit was as follows. In Figure VII.11, let

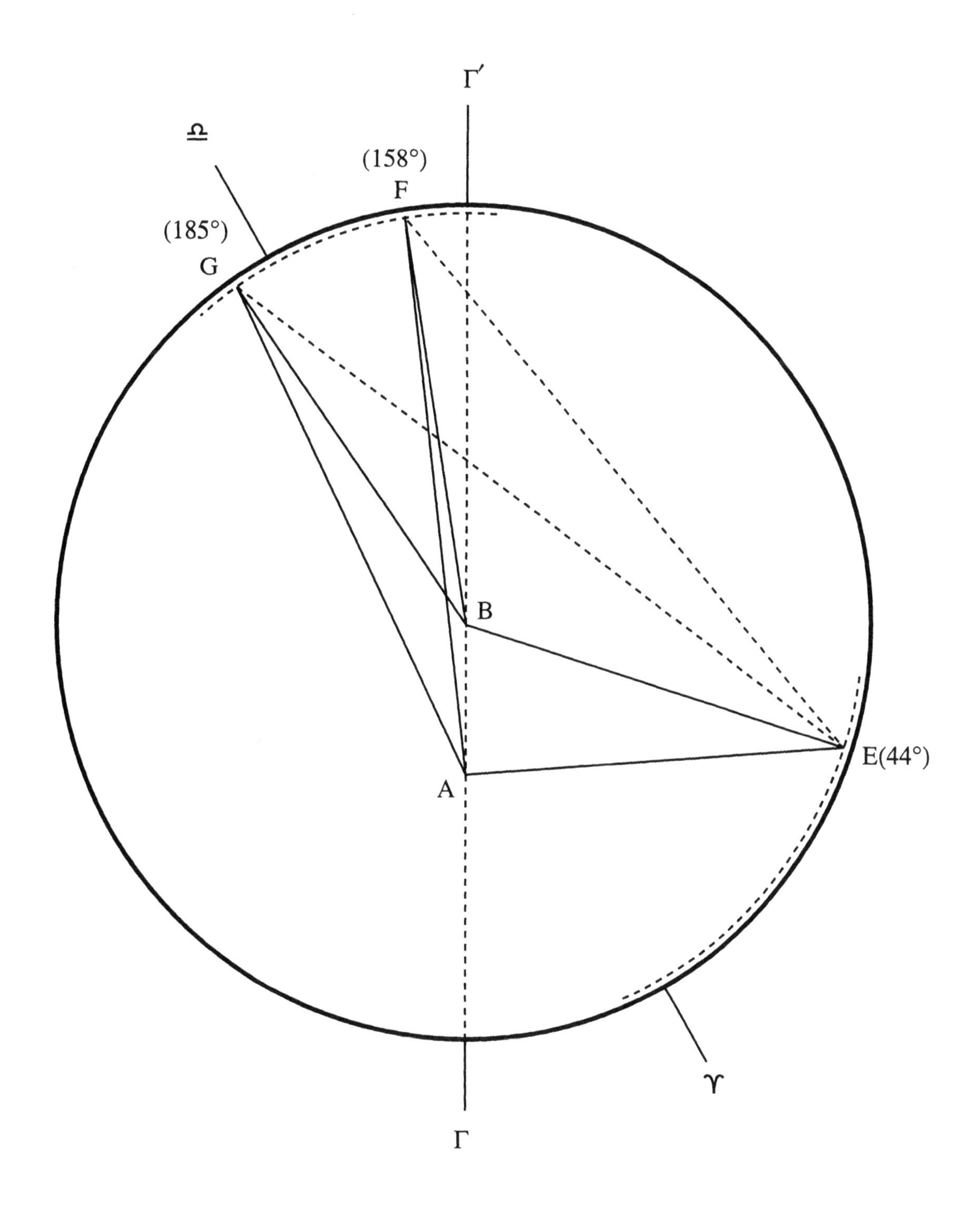

Figure VII.10 — Kepler's first observational orbit of Mars.

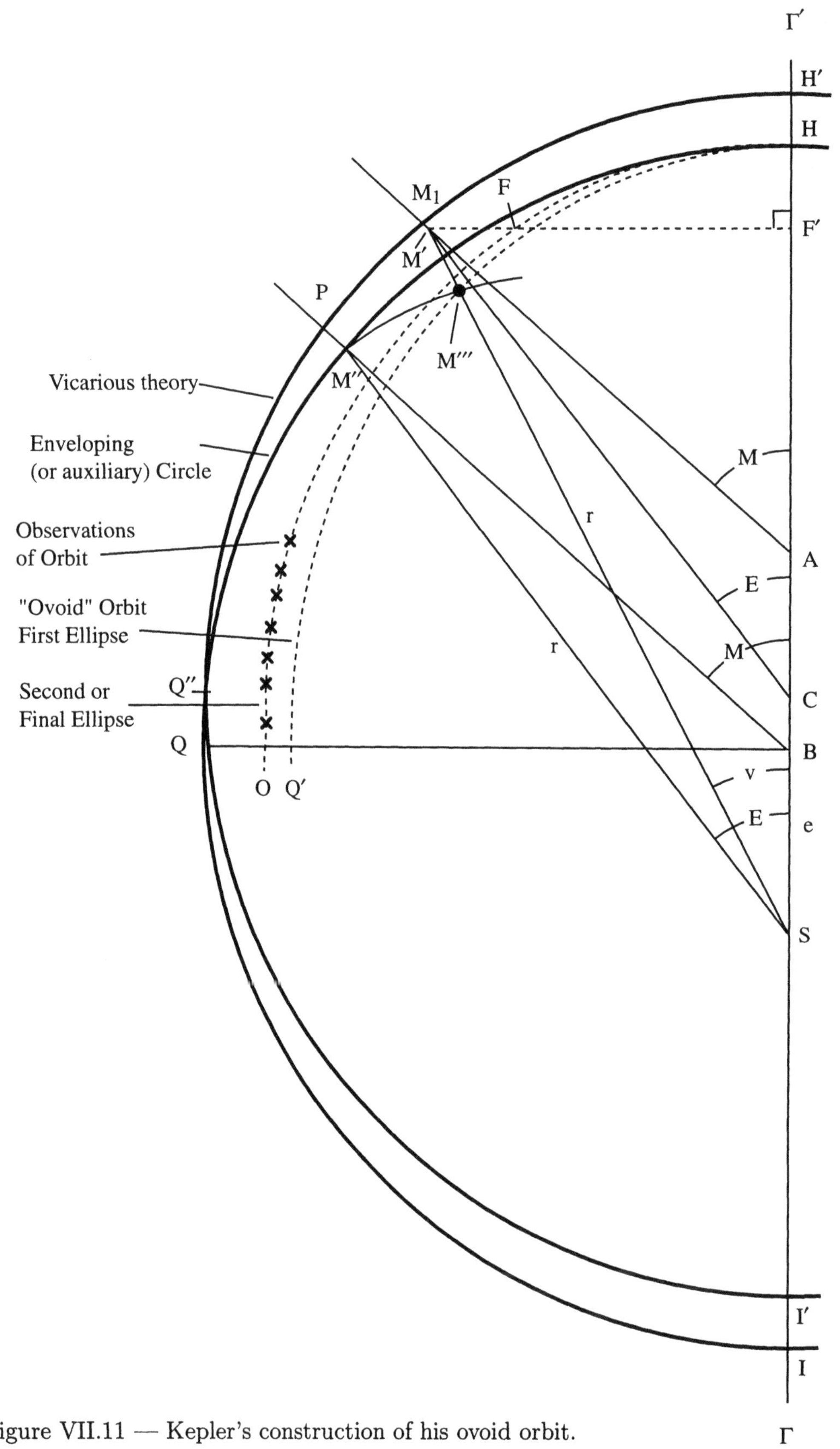

Figure VII.11 — Kepler's construction of his ovoid orbit.

IH = the line of apsides,

I = the perihelion,

H = the aphelion,

S = the Sun,

B = the center of the ovoid orbit = center of the "bisection of eccentricity" orbit,

HOI = the elliptical orbit (still unknown),

$HQ'I$ = the "auxiliary ellipse" or "first ellipse" ≈ the "ovoid" orbit,

xxx = observations of the orbit,

C = the center of the vicarious orbit,

A = the equant both for the vicarious hypothesis and for the "modified bisection of eccentricity" hypothesis (i.e., with his primitive areal principle).

This yields the following constants:

$BH = BI = BM'' = BQ = 1.00000$,

$SB = BA = 0.09282$,

$SC = 0.11332$; and therefore

$BC = QQ'' = HH' = II' = 0.02050$,

$CA = 0.07232$.

Also let

HQI = the major auxiliary circle of the observed orbit,

$HQ'I$ = the auxiliary ellipse = the ovoid orbit,

$H'Q''I'$ = the vicarious orbit.

Then

HQI = the enveloping circle of the above four curves plus HOI.

We shall further find

$QQ' = 0.00858$ = sagitta of ovoid and of auxiliary ellipse,

$QO = 0.00429$ = sagitta of orbital ellipse (still unknown).

Also,

$QO = OQ'$.

Process:

1. Draw AM_1 so that $\angle HAM_1 = M$, the mean anomaly.
   Draw BP so that $\angle HBP = M$.
2. Draw $CM' = 1$, thus locating M' by intersection with AM_1 (not with $H'Q''I'$).

3. Draw SM', the heliocentric *direction* of Mars (giving $HSM' = v$, the true anomaly).
4. On BP, drawn parallel to AM_1, take mean distance $BM'' = 1$.
5. Draw $SM'' = r$, the heliocentric *distance* of Mars.
6. Draw $SM''' = r$. Then M''' is the heliocentric *position* of Mars.

Geometrical Estimate of the Sagitta of Kepler's Ovoid Orbit

This construction leads to a complicated mathematical equation for the ovoid from which the sagitta $M''O$ in Figure VII.12, or the maximum width of the lune $HM''IOH$ separating the auxiliary (enveloping) circle from the curve itself, could be found. It is, however, unnecessary to find this equation. An approximation to the width of the sagitta of the ovoid can be made as follows: Construct a point on the ovoid for $M = 90°$ by following the process outlined above for a general case. This results in Figure VII.12, in which for a small eccentricity e:

a. the distance $M'M'''$ may be considered an approximation to the sagitta $M''O$, and
b. the small circular arcs $M'M''$, $M'M'''$ may be regarded as straight lines.

Taking $\angle M'M'''M''$ to be a plane right angle, we have approximately $\angle M'M''M''' = \angle BM''S = \frac{SB}{BM''} = \frac{e}{1} = e$. Thus $M'M''' \approx e \sin e \approx e^2$. On a scale of the semimajor axis $BH = a = 1$, $OM'' \approx M'M''' \approx (0.09282)^2 = 0.00862$. Thus the sagitta of maximum width in the lune of the ovoid is ≈ 0.00862.

Replacement of the Ovoid by an Epicycle and Deferent

Kepler supplied a concoction of two physical causes for the motion of a planet along its oval. First, it was moving uniformly in the retrograde direction on an epicycle of radius e. This part of the motion was due to a "planetary virtue," akin to the Earth's rotation or to its driving action on the Moon. Second, the center of the epicycle was moving on a circular deferent centered on the Sun. This was a nonuniform motion caused by a nonuniform "solar virtue," which acted less strongly the farther the distance from the Sun. The motion of the epicycle on the circular deferent (radius = 1.00000, centered on the Sun) was governed not by an equant, but in such a way that the motion of the radius vector from the Sun to the planet caused a uniform increase of the area under the ovoid (*not* the deferent).

By difficult constructions and laborious calculations, adding increment to increment, Kepler could approximate, from degree to degree of true anomaly, the areas under the oval and the longitudes of the planet. In these values there were, in all cases, errors of 4′ to 5′ at the quadrants. He thought that these were due to a defective computation of the areas under the oval.[12] He soon noticed, however, that an ellipse tangent to the oval at the ends of its axes was almost indistinguishable from the oval itself all along its perimeter. This so-called first ellipse was a mere mathematical fiction used to figure areas more easily and was never adopted as an actual orbit, which at this stage was the ovoid. But before passing to this improvement of the area technique, let us see the basic construction whereby the oval was

[12]He realized that he was involved in a "bootstrap" procedure because the computation of the areas depended upon knowing the total length of the oval, which was unknown at the start, but had to be nevertheless initially estimated.

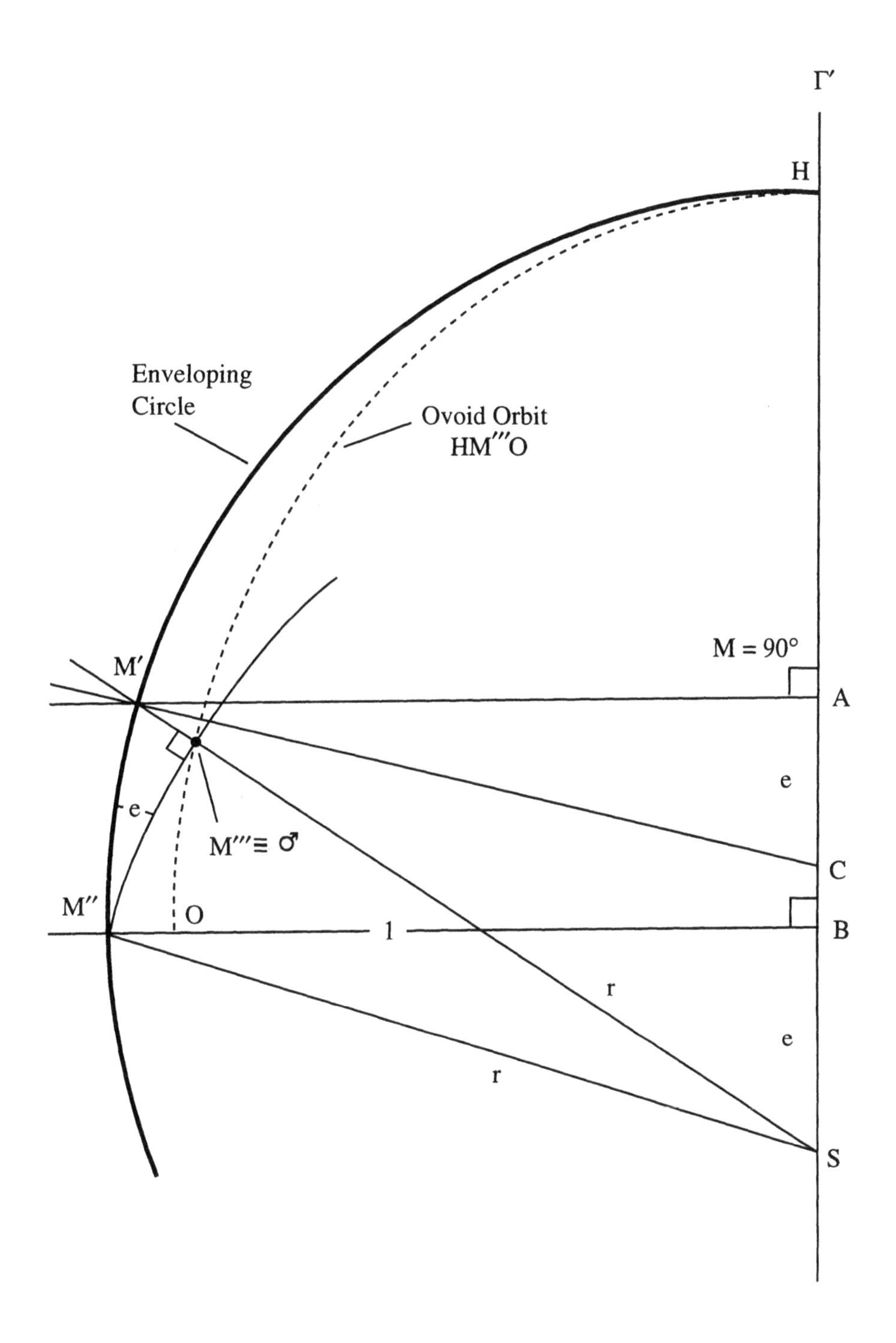

Figure VII.12 — Estimating the sagitta $M''O$ of Kepler's ovoid orbit.

replaced by a retrograde epicycle on a circular eccentric deferent. The basic construction is suggested in Figure VII.13, in which

AB = the line of apsides,

A = the aphelion of the ovoid orbit,

B = the perihelion of the ovoid orbit,

C = the center of the ovoid orbit and of its reference (enveloping) circle,

S = the Sun, center of circular deferent $A_1Q_2B_1$,

E = the equant of the bisection of eccentricity theory (not used in constructing the ovoid, which takes its angles from the vicarious hypothesis),

$e = SC = CE = 0.09282$ = the "eccentricity" of the ovoid ($= QP = Q_1P_1 = Q_2P_2 = Q_3P_3 = A_1A = B_1B$),

R = a point on the reference circle of the ovoid,

Q, Q_1, Q_2, Q_3 = various positions of the epicyclic center on the circular eccentric deferent A_1QB_1,

APB = the ovoid,

ARB = the reference circle of the ovoid,

$\delta = \angle SP_2C$ = the "optical equation" (in this position also the "equation of the center"),

$CA = CB = CR = SA_1 = SQ_2 = SB_1 = SQ_3 = 1.00000$,

$SP = r$ = the radius vector to the Sun.

The point P moves uniformly on the circular epicycle, center at Q and radius e, in a retrograde direction starting at A, with QP in the position A_1A. The angle α, shown in the figure, increases uniformly in the retrograde direction. If n = the mean daily motion, t = the epoch, and t_0 the moment when $\alpha = 0°$, then $\alpha = n(t - t_0)$.

The angles ϕ and ψ increase nonuniformly such that the area bounded by SP, SA, and arc AP of the ovoid increases uniformly, i.e., such that $\psi = \int_0^\psi r^2 d\phi = c\alpha$, where c is a constant. This fixes the length of r at each step of, say, 1°. After one-quarter of the period the planet is nearly at P_1 and the center of the epicycle is at Q_1. Also, because of the uniform areal velocity, approximate equality of areas ASP_1 and BSP attains, each being about one-quarter that of the oval. At this moment $\angle SCQ_1$ is nearly a right angle because $Q_1P_1 = CE = e$. Therefore $EP_1 \perp SA$, nearly. This can also be illustrated by the following consideration: Regarding for a moment the curve $APP_1P_2P_3B$, which is an exact ellipse of small eccentricity e, with foci S and E, we can apply Ward's Principle, illustrated in Figure VII.14, namely: In an ellipse of small eccentricity, uniform areal motion about one focus is very nearly equivalent to uniform angular motion about the other focus. With this approximation permitted, we see by inspection of Figure VII.13 that the planet starting at A reaches P_1 so that $\angle AEP_1 = 90°$ in one-quarter period, and since $Q_1P_1 = e$, then at the same time $\angle ACQ_1 \approx 90°$ and $\alpha_1 \approx 90°$. Since at this point we have that the rates of increase of α and δ are very nearly the same, when the planet is at P_2 a short time later we then have $\angle ASQ_2 \approx 90°$. Hence $\alpha_2 \approx 90° + \delta$ and $\angle Q_2P_2S \approx 90°$. Since $SQ_2 = 1$ and $Q_2P_2 = e$ in ΔQ_2P_2S, we have $SP_2 = \sqrt{1 - e^2}$. Now, in ΔSCP_2 where $SC = e$, we have

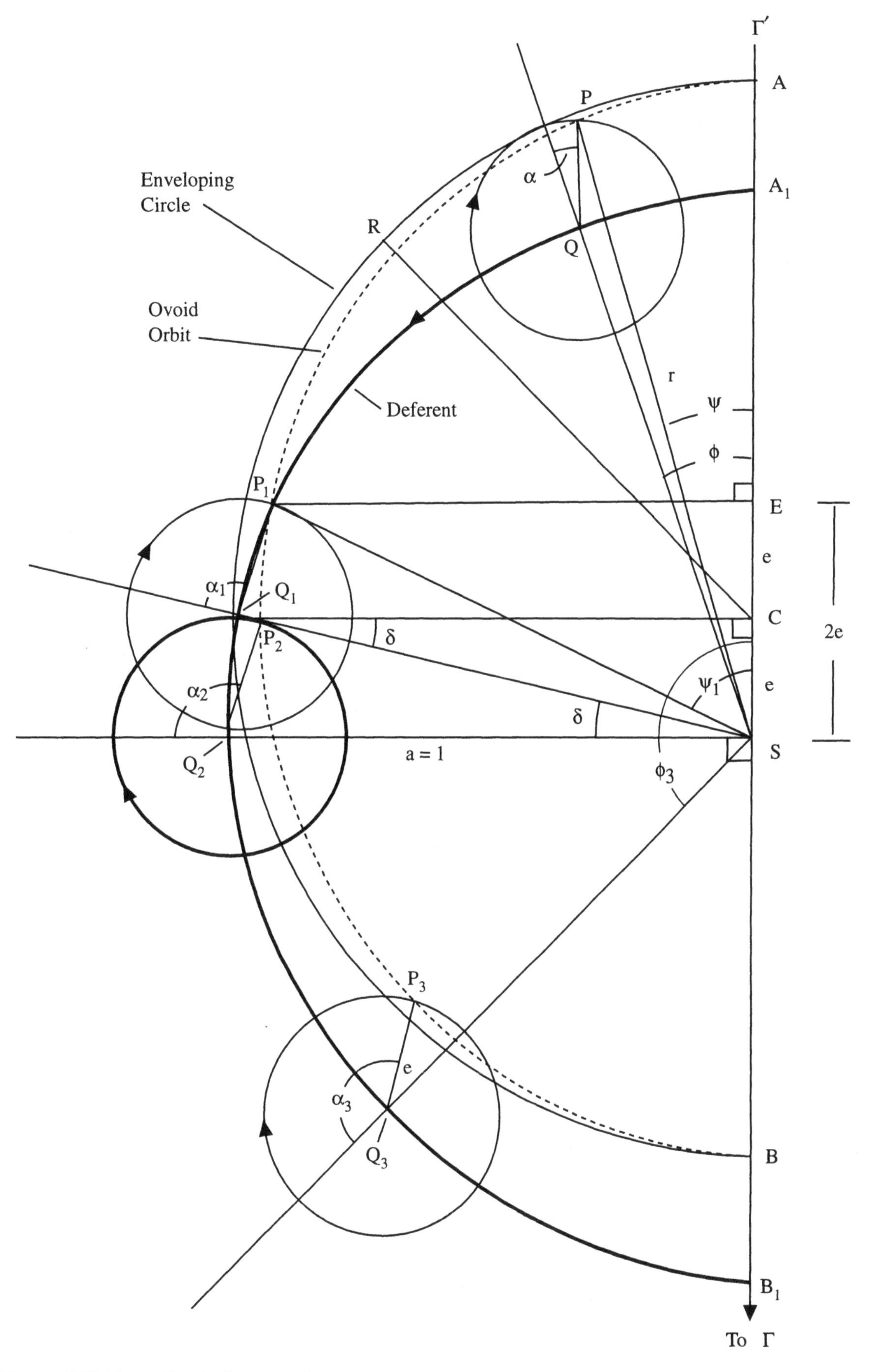

Figure VII.13 — Kepler's epicyclic representation of his ovoid orbit.

$CP_2 = \sqrt{1 - e^2 - e^2} = \sqrt{1 - 2e^2}$ = the short semiaxis of the oval, on a scale where the longer axis $CA = 1.00$.

Sagitta of the Ovoid Construction

We may now find the sagitta of the ovoid as given by this construction, namely $CA - CP_2$ or $1 - \sqrt{1 - 2e^2}$. This is approximately $1 - (1 - \frac{1}{2} \times 2e^2) = e^2$, by the binomial theorem. Thus the sagitta of the lune produced by the epicyclic representation of the ovoid is $(0.09282)^2 = 0.00862$, the same as was found to be the value for the ovoid itself from its geometrical construction.

Why Kepler Considered the Ovoid Theory to Be a Physical Theory

One reason for Kepler's attempt to describe, as in Figure VII.13, the ovoid motion by a retrograde uniform circular motion in an epicycle, which in turn is moving directly with a nonuniform circular motion on an eccentrically situated circular deferent, was that he thought of both these motions as "physical." As noted above, the uniform epicyclic motion was due to "planetary virtue," an innate tendency of a body to have uniform circular motion, like the rotation of the Earth, which was a permanent whirlpool in the ether within which the planet was embedded. Later, after reading Gilbert's *De Magnete* (1600), and in order to avoid the repugnant concept of motion influenced by an empty point in space, he considered forces analogous to magnetic forces acting on the planets. The deferential motion was due to "solar virtue," a force sweeping all the planets eastward about the Sun with a force (and therefore, on the Aristotelian viewpoint, with a velocity) which varied inversely as the solar distance. The planet thus had two forces acting on it, one tangential to the epicycle and the other to the deferent. This concept required a composition of velocities (a novel idea in Kepler's time) and therefore (in the Aristotelian view) of the "projectile forces" along the circles. Kepler was the first to consider a combination of forces pushing a planet along its orbit, which was a distinctly physical concept for his day. This is in striking contrast to his contemporary Galileo, who never extended his brilliant terrestrial concepts of inertia and acceleration to the celestial realm, but who held that the natural motion of the Moon and other celestial bodies was circular, while bodies in the terrestrial realm had a tendency to straight-line motion (vertically for solid bodies and horizontally for liquids). We can perhaps now appreciate why A. S. Eddington asserted that Kepler was the first astrophysicist in history. He was always seeking physical causes, although of course always in the context of his pre-Newtonian world.

Kepler's First, or Auxiliary, Ellipse: Its Eccentricity and Properties

It has already been pointed out that, as an approximation for finding areas under his ovoid, Kepler used an exact ellipse having the same axes as his geometrically constructed ovoid. According to Dreyer (1905) this is the first ellipse ever introduced into astronomy. This ellipse must, however, not be taken strictly as an orbit, even a preliminary one, for the ovoid construction itself, not the points of the ellipse, was used to compute the positions of the planet for comparison with observations. But it was useful in the following way. By an approximation process called "the assumption of a false position" Kepler was able to find

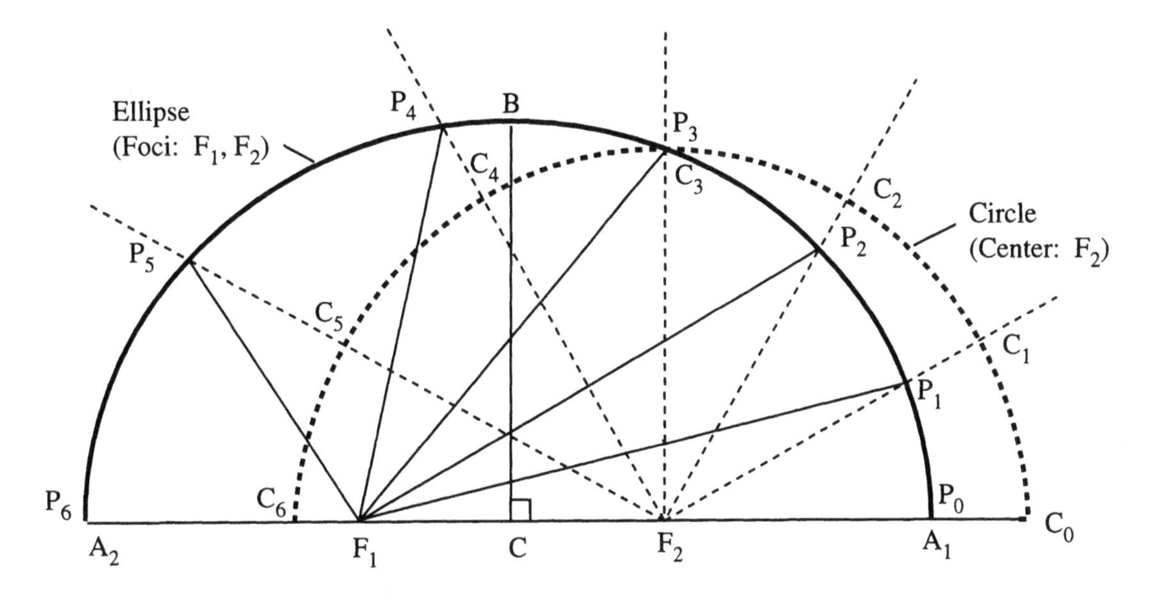

Given: Uniform areal motion about F_1. Areas $P_0F_1P_1 = P_1F_1P_2 = ... = P_5F_1P_6 \approx \frac{1}{6}\ (A_1BA_2)$

Find: Nearly uniform angular motion about F_2.

$\angle C_0F_2C_1 \approx \angle C_1F_2C_2 \approx \cdots \approx \angle C_5F_2C_6 \approx 30°$

Figure VII.14 — Ward's Principle.

areas under this ellipse more easily than the corresponding ones under his oval. The major axis was the same as in the vicarious hypothesis, and the eccentricity was 0.131, as we shall show presently. In an ellipse the sagitta, or maximum width of the lune between the curve and its major auxiliary circle, is the difference between the major and minor semiaxes. To find which ellipse is tangent to the oval at its apsides we equate the sagittas of the two curves. For the enveloping ellipse $a_1 - b_1 = a_1 - \sqrt{a_1^2 - a_1^2 e_1^2}$. Letting $a_1 = 1$, we find the sagitta $a_1 - b_1 = 1 - \sqrt{1 - e_1^2} = 1 - (1 - \frac{1}{2}\, e_1^2) = \frac{1}{2} e_1^2$. Setting this equal to the sagitta of the ovoid, we find $\frac{1}{2} e_1^2 = e^2$, or $e_1 = \sqrt{2e}$. Here $e_1 = 1.4142 \times 0.09282 = 0.131$, on a scale of $a_1 = a = 1$. This means that the distance between focus and center of the enveloping ellipse is 0.131 and that its equation of center ϕ' is $7°51'$. Such an ellipse practically duplicates the ovoid. To find areas is relatively simple, since all areas under the ellipse are b_1/a_1 times the corresponding areas under the auxiliary circle (of radius a_1).

Kepler's first, or auxiliary, ellipse is shown in Figure VII.15. For future reference the figure also shows the second, or final, orbital ellipse and the orbit derived from observations, indistinguishable from it. In particular,

HI = the line of apsides,

H = the aphelion,

I = the perihelion,

S = the Sun,

C = the center of the circular orbit on the vicarious hypothesis,

B = the center on the bisection of eccentricity theory
= the center of the ovoid
= the center of the auxiliary ellipse
= the center of the final ellipse
= the center of the major auxiliary circles of the last four curves,

A = the equant of both the vicarious and bisection of eccentricity hypotheses,

F_1', F_2' = the foci of the auxiliary ellipse,

F_1, F_2 = the foci of the final ellipse,

a = the semimajor axis of all ellipses and orbits,

$HB = BI = BQ = SO = F_2'Q' = a = 1$,

ϕ = the equation of center of the orbital ellipse = $5°18'$,

ϕ' = the equation of center of the auxiliary ellipse = $7°51'$,

QO = the sagitta of the orbital ellipse = 0.00429,

QQ' = the sagitta of the auxiliary ellipse = 0.00858,

HQ'I = the ovoid orbit and its auxiliary ellipse,

HOI = the final ellipse and observational orbit (by triangulation),

HQI = the reference circle, or the major auxiliary circle for all the curves except the vicarious hypothesis,

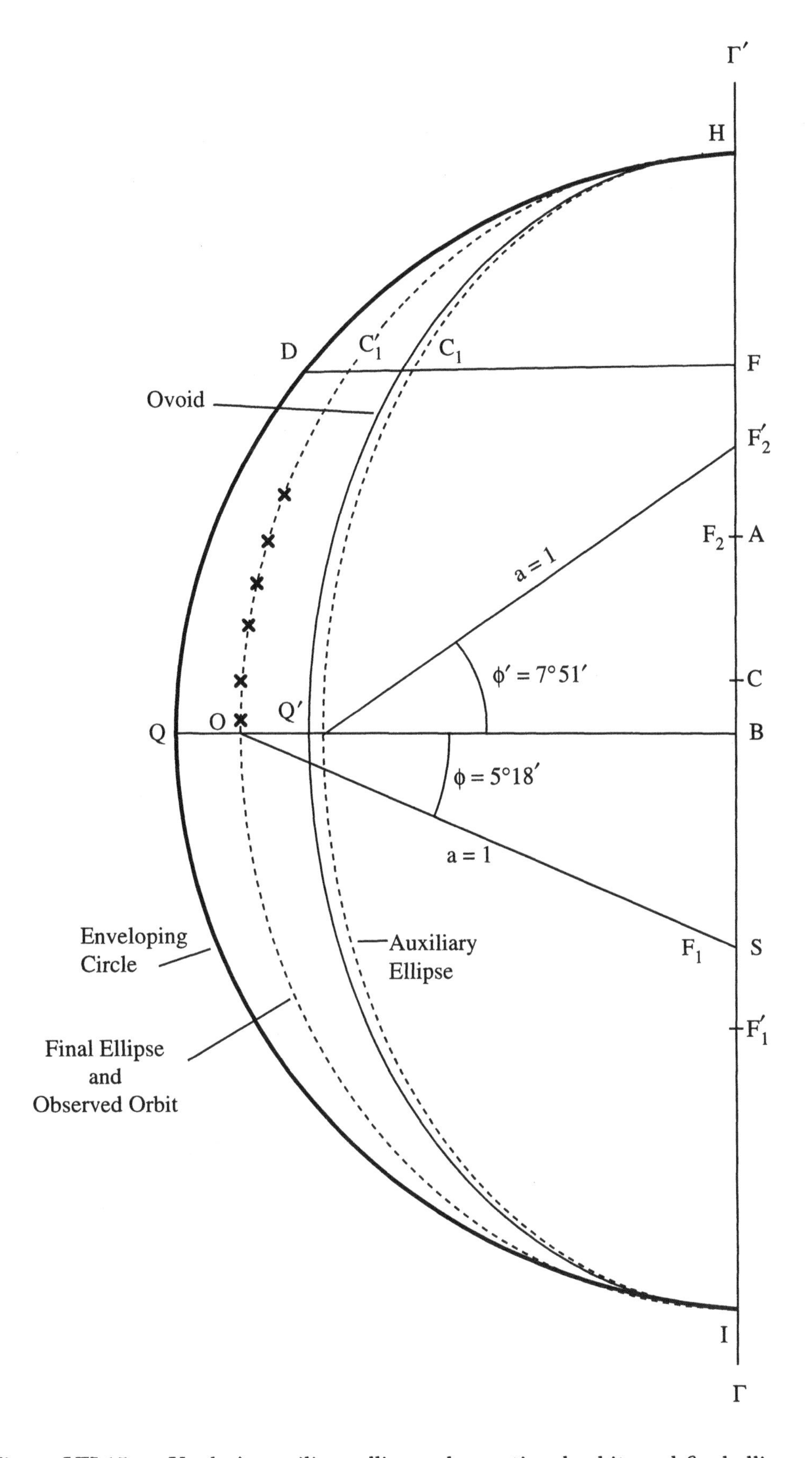

Figure VII.15 — Kepler's auxiliary ellipse, observational orbit, and final ellipse.

XXX = observations of the orbit.

Figure VII.15 also serves to summarize some of the highlights in Kepler's work leading to his final orbit:

a. Construction of the ovoid and determination of its sagitta $QQ' = 0.00858$.

b. Introduction of an auxiliary (nonorbital) ellipse of the same sagitta, i.e., $e = 0.131$, $\phi' = 7°51' = \sin^{-1}e_1$.

c. Easier determination of areas under the ellipse, since $HC_1'F \approx HDF \times (b/a)$, etc.

d. Construction of the observational orbit, measuring its sagitta $QO = 0.00429$, i.e., $e = 0.09263$, $\phi = 5°18' = \sin^{-1}e$. The true anomaly at the quadrants is $(90° - \phi)$.

e. sec $\phi = 1.00429$, so that, since $a - b = \sec\phi - 1 = 0.00429$, the observational orbit fits the ellipse of $e = 0.9282$ at the apsides and quadrants.

f. Computation of the *distantiae diametrales* for all the observations showed that the whole orbit is the ellipse in (e) (see Fig. VII.18).

Some of Kepler's Further Attempts with Ovals or Combination Orbits

Kepler next expended an almost superhuman amount of energy (often nearly causing his death by exhaustion) trying orbits having a circular epicycle of radius 0.00429 with retrograde uniform motion and a deferent, either circular eccentric or central elliptic or oval, with uniform areal motion.

First, he tested his oval theory by observations. By using the approximation process called the method of "false position" he computed areas under the auxiliary ellipse. For example, taking an observation at mean anomaly 95°18′28″ he found (by approximation) that the true anomaly was 84°39′42″, only 2′20″ different from the value given by the vicarious hypothesis and therefore considered satisfactory. (No longer did the planet move too fast at the quadratures.) This also fixed (by observation) the breadth of the lunula between the major auxiliary circle of the oval and the ellipse as 0.00858 and therefore the eccentricity of the ellipse as 0.131, as shown above.

But when Kepler tried the same process for an observation at mean anomaly 48°45′12″, the difference of true anomaly from that given by the vicarious hypothesis was 6′24″ – entirely too much for his standards.

The detailed results are shown in his calculations of true anomalies in seven ways for an observation at each of three positions: first, second, and third octant (counting from aphelion). The results are shown in Table VII.1.

The theories referenced in the first column are the following:

(1) The vicarious hypothesis (taken as standard in the discussion below).

(2) A circular orbit of radius a, eccentric by $e = 0.09282$, with direct (nonuniform) linear and angular motions, but uniform areal velocity of the radius vector to the Sun.

(3) A circular epicycle of radius e moving with uniform retrograde motion, on a circular deferent, eccentric by e, with uniform direct areal motion of the radius vector to the planet under the "auxiliary" ellipse (practically the ovoid motion) (Fig. VII.13).

Table VII.1. Tests of Various Martian Orbits at Crucial Anomalies

	First Octant	Second Octant	Third Octant
Observations of true anomalies	48°45′	95°18½′	138°45′
Theoretical Predictions			
(1) Vicarious hypothesis	41°20½′	84°42′	131°07½′
(2) Circle, constant areal velocity of radius vector	41°29′	84°42½′	130°59½′
(3) Ovoid, constant areal velocity using the auxiliary ellipse	41°14′	84°40′	131°14′
(4) Mean of circle and ovoid	41°21½′	84°41′	131°07′
For comparison, Kepler added:			
(5) Hipparchus	41°40′	84°40′	130°41′
(6) Bisection of eccentricity	40°46′	84°38′	131°45′
(7) Ptolemy	41°15′	84°41′	131°15½′

(4) The mean of motions (2) and (3).

(5) Direct uniform circular motion in a simple eccentric with no equant (Hipparchus's method).

(6) Direct circular motion in an eccentric governed by uniform circular motion in an equant twice as eccentric (bisection of eccentricity method).

(7) Same as (6), but using Ptolemy's method of computation.

Kepler now proceeded to make a comparison between the values in the above table, drawing conclusions leading to further work. First of all, fine agreement at the quadrants is possible in almost any orbit by the proper adjustment of its constants. It shows merely a good start. Unless there is agreement also at the octants with the vicarious hypothesis, which Kepler knew from experience gave excellent heliocentric longitudes (Herz [1897] has shown mathematically why this is so), the orbit cannot be true. The mean (4) of (2) and (3) very closely approaches the observations (1), even at the octants. This *could* have suggested to Kepler what he often said in his letters, that he now *believed* that the orbit is an ellipse, though not the "auxiliary ellipse" used in (3). Yet, Small (1804) points out that a minute change of eccentricity makes (7) as good as (3). However, with such a change (7) gives much too slow a motion at the apsides, and too quick at the mean distance. This is why Kepler temporarily adopted (3).

A comparison of (2) and (3) with (1) bears out the statement that the residuals in the octants, from the circle and from the oval (which is practically the auxiliary ellipse) are about equal and of opposite signs, i.e., $\sim \pm 8'$, and that therefore the real orbit must be about halfway between the auxiliary ellipse and its major auxiliary circle.

Kepler's Further Experiments with Circular Uniform Motions

If Kepler had now just taken a second ellipse halfway between his auxiliary ellipse and its major auxiliary circle, his goal would have been accomplished. Although he says in a letter to Fabricius, "I *believe* the orbit is an ellipse," and in one to Moestlin, "The *only* [smooth] curve intermediate between a circle and an ellipse is another ellipse," he still did not attempt such a second ellipse! It seems that he could not yet give up the deeply rooted notion of uniform circular motions. He wanted to run down every last possibility and adopt an ellipse only if forced by computation to abandon all simple combinations of uniform circular motions. There were still several possibilities of errors in his calculations, and he made many more attempts. First, Kepler was uncertain whether his area estimates along the auxiliary ellipse approximated areas under the oval orbit with sufficient accuracy. He thus returned to the very difficult computation of the sums of the distances in the first oval itself, which, as we know, was no simple geometrical figure. He had to use the laborious approximation method of finding the distances from the Sun for each degree of mean anomaly, then adding the results to find the corresponding true anomalies. With his irrepressible tendency to make quick guesses he adopted a simplifying, seemingly fairly plausible, but nonrigorous and entirely unproved working principle: he would pass from sums of solar distances of points on the circumference of the enveloping circle to the corresponding sum of distances of points on the oval. Recall that he took the linear speed of a planet at any point along its orbit as inversely proportional to the distance of that point from the Sun. It follows that the time taken to cover unit distance along the orbit at any point was proportional to the distance of that point from the Sun. He therefore assumed that the time taken to cover any interval of the orbit was proportional to the sum of the distances of all the points of the orbit from the Sun. He realized that no rigorous method existed to find the sum of all the distances from a given point (the Sun) of the infinite number of points along a finite stretch of a continuous curve, but following Archimedes an arbitrarily close approximation could be made by increasing the finite number of distances. As an approximation to the sum of the infinite number of distances making up the area, he substituted where possible the area between the given curve and two radius vectors from the Sun to the orbital points.[13]

As a beginning Kepler computed the solar distances for each degree of mean anomaly in one-half of the enveloping circle. Starting at aphelion and taking the radius of the circle as unity, the sum for the 180 divisions in a semicircle was 180.3778, slightly more (as expected) than 180. On the other hand, for an oval of small adopted eccentricity, whose total length and mathematical shape were unknown, but whose process of construction was definite, the corresponding result of adding 180 laboriously computed solar distances of points along the oval was 179.2375. Kepler now intuitively made the (erroneous) assumption that the area under the oval (from aphelion to any point) was to the corresponding area under the semicircle as 179.2375/180.3778. On this assumption he computed the area under three other ovals for each degree of true anomaly and thus produced new calculated positions. The trial ovals were

1. eccentricity of 0.09165 and "length" of the semi-oval equal to 180.0000,

[13]Contrary to Dreyer's (1905) criticism of this concept, Kepler was well aware that "even an infinite number of geometrical lines laid side by side does not constitute an area." However, he believed in and referred to what he called "physical lines" of slender triangles, or even conical strings, in this case slender tapering cones having their vertices in the Sun.

2. eccentricity of 0.09165 and "length" equal to 179.2375, and
3. eccentricity of 0.09230 and "length" equal to 179.2375.

The third oval gave the best representation of the observations. The residuals in the sense "orbit 3 minus vicarious hypothesis" are given in Table VII.2. These residuals show that the planet was moving too fast at the mean distances and too slow at the apsides. Yet this was the best representation obtained, and for a while Kepler flattered himself with the hope that it might be improved if ever the exact area under the oval could be calculated. Four years before, when he first joined Tycho in Prague and took over the Mars problem from Longomontanus, he had adopted the goal of representing Tycho's observations with errors of not over 2′ in longitude or latitude, nor over 0.001 A.U. in distance. As he still had not achieved this, he set out immediately to improve his work.

Table VII.2. Results of Trial Oval Orbits

Mean Anomaly	True Anomaly (Vicarious Hypothesis)	Residual
45°	37°04.9′	−2.5′
90°	79°27.7′	−0.9′
135°	126°52.0′	+4.4′

Even if the above representation had been close to the theory on which it was based, it would not have satisfied Kepler in the long run, because his oval theory was based on a frustrating circular reasoning: the length of the oval up to the point occupied by the planet at a given time could not be obtained without determining the position of all the points occupied at successive intervals of time, but these points in turn could not be computed without knowing the exact length of the oval up to any particular position.

Kepler now made at least seven more laborious orbit computations, using an epicycle of retrograde uniform circular motion moving directly on a circular deferent centered on the Sun, and according to a principle of uniform areal velocity. Some results of these attempts are shown in Table VII.3. The orbits were as follows:

Orbit 1. This orbit uses a radius of 1.00000 for the deferent and 0.09264 for the epicycle. The true anomalies of the epicyclic center on the deferent were reduced in the ratio 360.75562/360., a correction factor resulting from the summation of the 360 distances. The planet still moved too fast at the mean distances and too slow at the apsides. An increase of the eccentricity would have resulted in increased residuals.

Orbit 2. This orbit is a peculiar oval having an eccentricity of 0.09165 for the orbital center and an equant at eccentricity of 0.18564 as in the vicarious hypothesis. To compute distances Kepler introduced a new concept, the "distancial anomaly," which was the average of the mean and true anomalies. He computed for each of the 360 degrees of distancial anomaly a corresponding distance as a ratio to the radius of the enveloping circle. The sum of the computed distances should be exactly 360. but was 359.24252. The distances used for defining the orbit were then adjusted so that their sum was 360. The residuals computed with this orbit were entirely too large.

Orbit 3. An orbit with the same eccentricity and distancial anomalies as in orbit 2. The distances were derived by making each one a *third proportional* between the corresponding

Table VII.3. Results of Further Trial Orbits

Orbit	Mean Anomaly	True Anomaly	Residual (Orbit minus Vicarious Hypothesis)
1	45°		−8.2′
	90		−0.4
	120		+9.6
	150		+8.8
2	48.7°		−7.8′
	95.3		+5.0
	138.7		+16.3
3		45°	−13.3′
		90	−5.3
		135	+1.8
4	48.6°		+13.9′
	95.2		+8.1
	138.8		−5.4
7		41°	+5.0′
		81	−3.5
		91	−5.4
		131	−11.4

distance in orbit 2 and the semidiameter a of the orbit. In such a third proportional, if d_2 is any distance in orbit 2, and d the corresponding distance derived from it, $d = d_2^2/a$. Residuals here were also very poor!

Orbit 4. An orbit with true anomalies calculated in the usual way (i.e., with the vicarious hypothesis except that the eccentricity was as in orbit 2), corrected for an earlier error in their computation.[14] But residuals were still no better!

Orbit 5. An orbit using distances that were third proportionals to those in orbit 4, computed for whole degrees of eccentric anomaly by the constants of orbit 2. Residuals still worse!

Orbit 6. An attempt to combine a table of whole degrees of true anomalies computed from the original vicarious hypothesis, reduced by the proportion 357.70014/360 to the required true anomalies, with distances obtained from the ovoid theory. Completely unsuccessful!

Orbit 7. In this orbit, distances are third proportionals to those of orbit 6 and the corresponding semidiameters of the ovoid, and anomalies are true anomalies from the vicarious hypothesis, corrected by 356.92403/360. Thus, if d_2 is any distance in orbit 2, S the semidiameter of the ovoid at the anomaly of d_2, and d the corresponding distance used in orbit 6, then $d = S^2/d_2$. Increasing the eccentricities of the orbital center and the equant

[14]This error had arisen when Kepler used unequal divisions on the arc of the eccentric (i.e., on the orbit itself), caused by his choice of equal divisions (whole degrees) of distancial anomaly.

diminished the errors, but the planet still moved too slowly at the apsides and too rapidly in the intermediate parts.

Not only were all these trials frustrating, but at first glance they seem (to us) aimless and arbitrarily chosen, perhaps even ill-conceived. Yet historical research shows that not one was undertaken without a compelling logical reason revealed by a previous result. There was also Kepler' s resolve to prove the impossibility of any attempt which did not turn out to be a true solution of the problem. "Truth emerges from the correction of errors," he said. One may even understand why, to a critic who does not realize this aspect of his character and the logic of his choices, he might appear scatterbrained. Yet for his time the enormity of his efforts seems reasonable, as they all had the object of eventually justifying a combination of uniform circular motions. One must remember that for over a thousand years this principle was ingrown into the consciousness of every astronomer. To abandon it for elliptic, or in fact any nonuniform, motion at the beginning of the seventeenth century was as difficult a change of concept as was the acceptance of Einstein's theory of relativity three hundred years later.

Kepler's Construction of an Empirical Orbit of Mars Directly from Tycho's Observations

From only three distances little can be known empirically about the actual shape of an orbit. Yet this was all Kepler had when he embarked upon the harrowing computations of his many ovoid orbits and undertook the herculean effort of constructing and testing these. Incredible as it seems, it was only in 1604 that he turned to making full use of his greatest asset – unlimited access to all of Tycho's accurate observations. He now set about to construct an empirical orbit from the observations themselves. For this purpose he developed two simple methods of computing distances, one from two observations in nearly the same heliocentric longitude, the other from two observations at convenient geocentric elongations from the Sun nearly symmetrically spaced about the line of apsides of Mars's orbit. Figure VII.16 illustrates the first method, which involved triangulating to Mars from two positions of the Earth, with Mars at exactly the same position in its orbit, i.e., at an interval of one Martian sidereal period. Since in general such a situation can be expected to occur only rarely, the motions of the two bodies sometimes had to be carried forward or backward for a few hours, or even days, to qualify as a basis for this method. This could be effected by the use of Giovanni Magini's tables for Mars's daily motion and Tycho's accurate solar table. Using this method, Kepler found from 24 observations the distances of Mars from the Sun in six directions in space, independently of any assumed orbit for Mars (Table VII.4).

Kepler's second method, that depending upon the observation at any two convenient elongations, is illustrated in Figure VII.17. The best results are obtained from observations before and after opposition when the angular distances of the planet east and west of the line of apsides are about equal. In Figure VII.17, let

S = the Sun,

SA = the line of apsides of Mars's orbit,

M_1, M_2 = positions of Mars at the times of observation,

T_1, T_2 = positions of Earth at the times of observation,

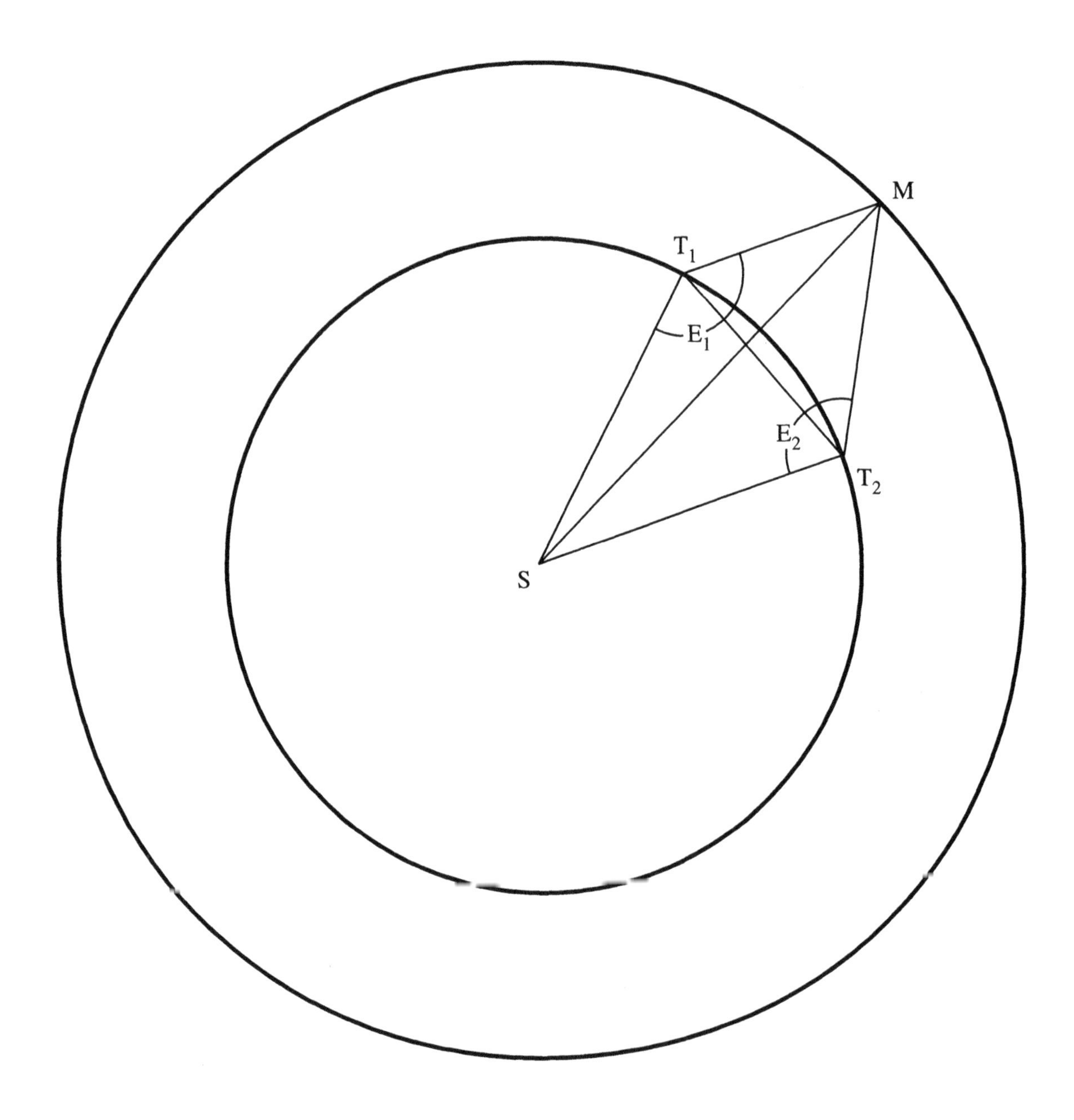

Figure VII.16 — Kepler's first method of finding distance, from two observations of geocentric elongations E with Mars at the same point in its orbit.

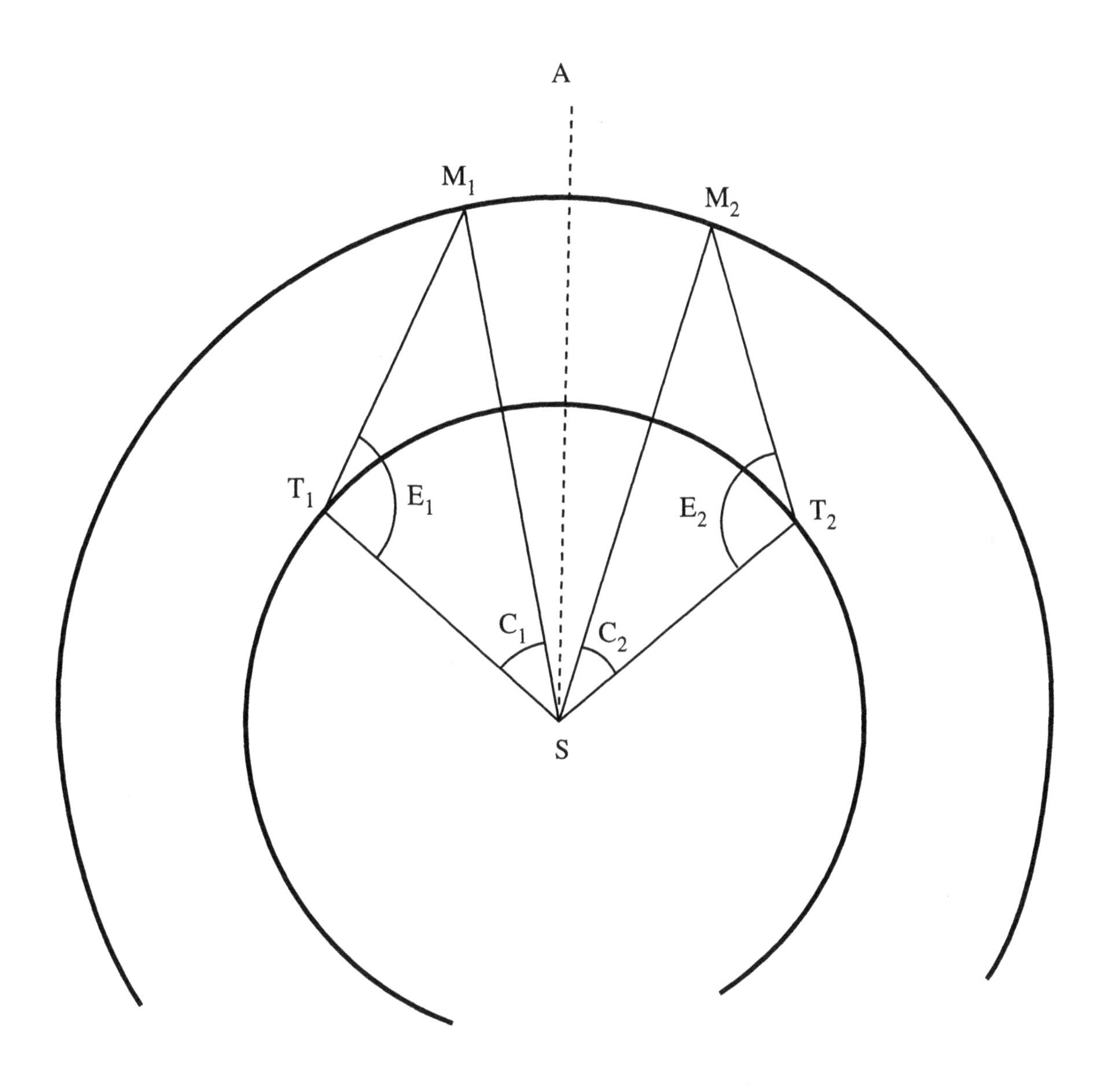

Figure VII.17 — Kepler's second method of finding distance (to Mars), from any two observations of geocentric elongations.

Table VII.4. Six Opposition Distances of Mars from the Sun
Computed by Kepler's First Method

Point	Distance of Mars from Sun		Col. 2 minus Col. 3	Mean Anomaly
	From observation[a]	From ovoid orbit[b]		
1	1.66212 A.U.	1.66179 A.U.	+0.00033 A.U.	11°37′
2	1.63064	1.62895	+0.00169	43 23.5
3	1.58158	1.57530	+0.00528	70 55
4	1.54394	1.53697	+0.00697	87 09.4
5	1.47676	1.46928	+0.00748	113 00.0
6	1.38977	1.38991	−0.00014	162 00.0

[a]Average of four observations about equally distant from the aphelion, using Kepler's first method. The individual points fell about equally far inside the major auxiliary circle of the ovoid, or of the first ellipse.
[b]Using for the ovoid $a = 1.52350$ and $e = 0.14115$, values differing only slightly from the elements of the first ellipse.

E_1, E_2 = observed elongations of Mars from the Sun,

C_1, C_2 = angles of commutation of Mars with the Earth.

The form of the planetary orbit, whether circle, ellipse, or ovoid, will be of little moment, just as long as it is nearly symmetrical about the line of apsides.

We are given:

$\angle M_1SM_2$, with rather high accuracy (from the times of observation, by the vicarious hypothesis),

SM_1, SM_2, approximately (from the ovoid theory),

everything about the Earth's orbit, with very high accuracy (from Kepler's solar theory),

E_1 , E_2, with high accuracy (from Tycho's work).

We wish to find SM_1 and SM_2 (with greatest possible accuracy). Procedure:

Step 1: Solve ΔST_1M_1 in which the given parts are ST_1, SM_1, E_1.
Solve ΔST_2M_2 in which the given parts are ST_2, SM_2, E_2.

Step 2: Check that $C_1 + C_2 + M_1SM_2 = T_1ST_2$, which is known from the Earth's orbit and the interval between the observations.

Step 3: If this test is satisfied, the distance is correct. If the sum in the left member exceeds the right member, the assumed distance is too large, and vice versa. The process is repeated with different values of the distances until a satisfactory check is obtained.

Using this method and the vicarious hypothesis Kepler computed geocentric longitudes from 28 directly determined distances in all parts of the orbit. This gave residuals ranging from −5.6′ to +5.3′. By a change of 1′ in the position of the line of apsides, the orbit could be improved to about ±4′. But the signs of the residuals were systematically negative in Cancer and positive in the opposite part of the sky. Furthermore, they were not due mainly to erroneous distances because such an error would have produced residuals of opposite signs on opposite sides of opposition.

Kepler now embarked upon a long and rigorous demonstration to confirm his earlier assumption that the plane of the orbit passes exactly through the Sun. (As he pointed out in the very beginning of his work at Prague, if the orbit plane passed through the "center of the ecliptic," i.e., through the mean position of the Sun instead of the Sun itself, an error of $1°1.6'$ would result in the longitude of the ascending node of Mars's orbit!) The position of the line of nodes was very closely confirmed. That it passed exactly through the Sun was a discovery of greater importance to the solution of the orbit problem than anything except the adoption of the Copernican viewpoint.

Kepler's Check of His Ovoid Theory by 40 of Tycho's Observations

Kepler next undertook to compare the distances obtained from other observations with those of his oval theory. There were now reliable distances known for about 40 different points of the orbit, found from the application to Tycho's data of Kepler's two methods of directly finding the distances from observation.

Kepler first used five of the best distances near the line of apsides to derive improved values of the semimajor axis and the eccentricity of the oval.[15] The result of the five determinations was as follows: aphelion distance $d_a = 1.66510$; perihelion distance $d_p = 1.38173$; eccentricity $= 0.14169$ (or 0.09301 on a scale of $a = 1.00000$). However, the previous enormous amount of work had convinced Kepler that $e = 0.09265$ was the best possible value of the eccentricity. Also, he believed his derived value for perihelion distance to be less reliable than that for aphelion. He therefore thought he could find a better value of the perihelion distance by using the aphelion distance in the following formula for an ellipse: $(1 + e)/(1 - e) = d_a/d_p$. Thus $(1 + 0.09265)/(1 - 0.09265) = 1.66510/d_p$, from which $d_p = 1.38274$. After some further dickering he adopted

$d_a = 1.66465$ (previously 1.66780),

$d_p = 1.38234$ (previously 1.38500),

$e = 0.14115$ (previously 0.14140),

$a = 1.52350$ (previously 1.52640).

It was with these elements that Kepler computed the residuals given in Table VII.3. The result was that the oval retires as far within the observed orbit as the circle goes beyond it. Continuing the computations for all 40 of the known distances using his ovoid orbit method, Kepler convincingly demonstrated directly from Tycho's observations that all the observed points fell everywhere halfway between the ovoid and its enveloping circle (i.e., almost exactly halfway between the first, or "auxiliary," ellipse and its major auxiliary circle).

Not content with this detailed comparison between theory and observation, Kepler further computed, for the seven oppositions used in his second method of finding observed distances, that the residuals in longitude from his ovoid orbit always showed the planet too

[15] At small mean anomalies (none of the five exceeded 12°) the correction necessary to derive a maximum or minimum distance from observed ones is practically the same for any smooth orbit and could be obtained closely enough from the circle of the vicarious hypothesis.

far behind in longitude before opposition and too far advanced after opposition, as would be expected if the actual distances were greater than those given by the ovoid. For an observation, i.e., in the intermediate part of the orbit (in 1595 at mean anomaly $\approx 90°$), the error amounted to $20'$! This he even checked with observations of his own. His contemporary D. Fabricius had also observed that the planet was far too advanced at the quadratures. He had complained about this in answer to a letter from Kepler explaining the oval theory, stating that the distances given by the oval were all too small in the intermediate part of the orbit. But, being a confirmed Ptolemaist, Fabricius had not pursued the subject further. Had he been more open-minded, he might have anticipated Kepler in proving the orbit an ellipse!

Kepler's Rejection of His Ovoid Theory

Kepler had thus disproved his cherished ovoid theory quite as effectively as he had earlier shattered his vicarious hypothesis (which, however, he kept for years as a reliable tool for supplying opposition longitudes). The abandonment of the ovoid theory must have been particularly depressing to Kepler, since he had based it on two principles that he thought were physical in nature. One was the propelling force of the Sun's motrix, varying in strength inversely as the distance; the other was the "magnetic" effect of the planet drawing itself toward and pushing itself away from the Sun. Both of these effects now had to be abandoned. His deepest overall objective, namely to give physical reasons for the motions of the planets, had now been shattered.

Kepler's Accidental Discovery of His Second, or Final, Ellipse

But within a few days out of these ruins arose by a sudden insight the clear and final solution to his great problem. From all his numerical labors he had preserved certain facts which unconsciously combined in his mind until an accidental discovery made him feel as if he had suddenly awaken from a long sleep! His thinking went something like this: He knew from his diligent numerical work that the maximum defect at 90° from the apsides of the observational orbit and its enveloping circle was 0.00660 on a scale of 1.52350 for the semimajor axis of the oval. This meant that on a scale of 1.00000 for the mean distance of Mars the maximum width of the lune between the circumscribing circle and the real orbit was 0.00432. In comparison the maximum width of the lune between the ovoid and its enveloping circle (occurring also at the quadratures) was 0.00858. Half of this, or 0.00429, was so near 0.00432 as to indicate that the real orbit at this point was halfway between the first ellipse and its major auxiliary circle, which was identically the circle enveloping the ovoid.

The next step was to derive the dimensions of the ellipse. By observation it was known that the equation of center (i.e., the maximum optical equation) was $5°18'$. He now remembered that $\sec 5°18' = 1.00429$, an excess of 0.00429 over unity. If in the oval theory, at eccentric anomaly $= 90°$, the distance of Mars from the Sun were made 1.00429 instead of 1.00000, the orbit would exactly match the observations at this point and would be halfway between the original oval and its enveloping circle. Further, if every ordinate (distance perpendicular to the line of apsides) of the first, or auxiliary, ellipse were multiplied by $\sec 5°18'$, the points would lie on a second ellipse, everywhere halfway between the first ellipse and the enveloping circle. This was the final solution to his problem. To test the observations, and later to draw the whole ellipse, Kepler computed distances and angles relative to the focus (the Sun) using the elegant method of *distantiae diametrales* (diametral distances), described below. From his study of the works of Euclid and Apollonius on the properties of ellipses, Kepler knew of course that the focal distance r to any point of an ellipse of semimajor axis a and eccentricity e, at an eccentric anomaly E, is given by $r = a + ae \cos E$, or $r = a - ae \cos E$, according as E is reckoned from the aphelion or the perihelion, respectively.

Construction of Kepler's Final Ellipse by Diametral Distances

In Figure VII.18 let P_1P_2 be two points on the ellipse $ApP_1B_1PeP_2B_2Ap$, with center C, focus S, semimajor axis a, and eccentricity e. Let $ApCD_1$ $(= PeCD_2) = E$ be the eccentric anomaly of points P_1 and P_2 on the ellipse. Let $CD_1 = CD_2 = a$ and $CS = ae$. Drop SD perpendicular to the diameter D_1D_2. By projection $CD = ae \cos E$. Thus $DD_1 = a + ae \cos E$ and $DD_2 = a - ae \cos E$. Drop D_1F_1 and D_2F_2 perpendicular to the major axis $ApPe$. With radius $r_1 = D_1D$ and center S, describe an arc locating P_1 at the intersection of the arc with the perpendicular D_1F_1. Similarly locate P_2 using $r_2 = D_2D$ on the perpendicular D_2F_2. P_1 and P_2 are then the points on the ellipse corresponding to the eccentric anomaly E. The results of testing Kepler's final orbit by 24 of Tycho's best observations (grouped to show 12 directions in space) are exhibited in Table VII.5. In this table the values for Mars determined directly from Tycho's observations are denoted by O, while C refers to the corresponding values in the final ellipse calculated with $a = 1.52350$ and $e = 0.09265$.

Table VII.5. Comparison between Observed and Computed Values of the Radius Vectors and Heliocentric Longitudes for Kepler's Final Ellipse

Average Mean Anomaly (from Aphelion)	Radius Vector (scale: $a = 1.52350$) $(O - C)$	Heliocentric Longitude (from Aphelion) $(O - C)$
11°37.0′	−0.00034	−1.6′
43 23.5	−0.00138	−1.6
70 55.0	+0.00027	−1.6
87 09.4	+0.00062	−0.3
113 00.0	−0.00158	−2.2
162 00.0	−0.00093	−0.8
−11 37.0	+0.00120	−0.7
−43 23.5	−0.00060	−0.2
−70 55.0	+0.00143	−0.0
−87 09.4	−0.00060	−0.3
−113 00.0	−0.00047	+1.8
−162 00.0	−0.00109	+0.4
Average Residual	±0.00088 (±0.00135 A.U.)	±0.96′

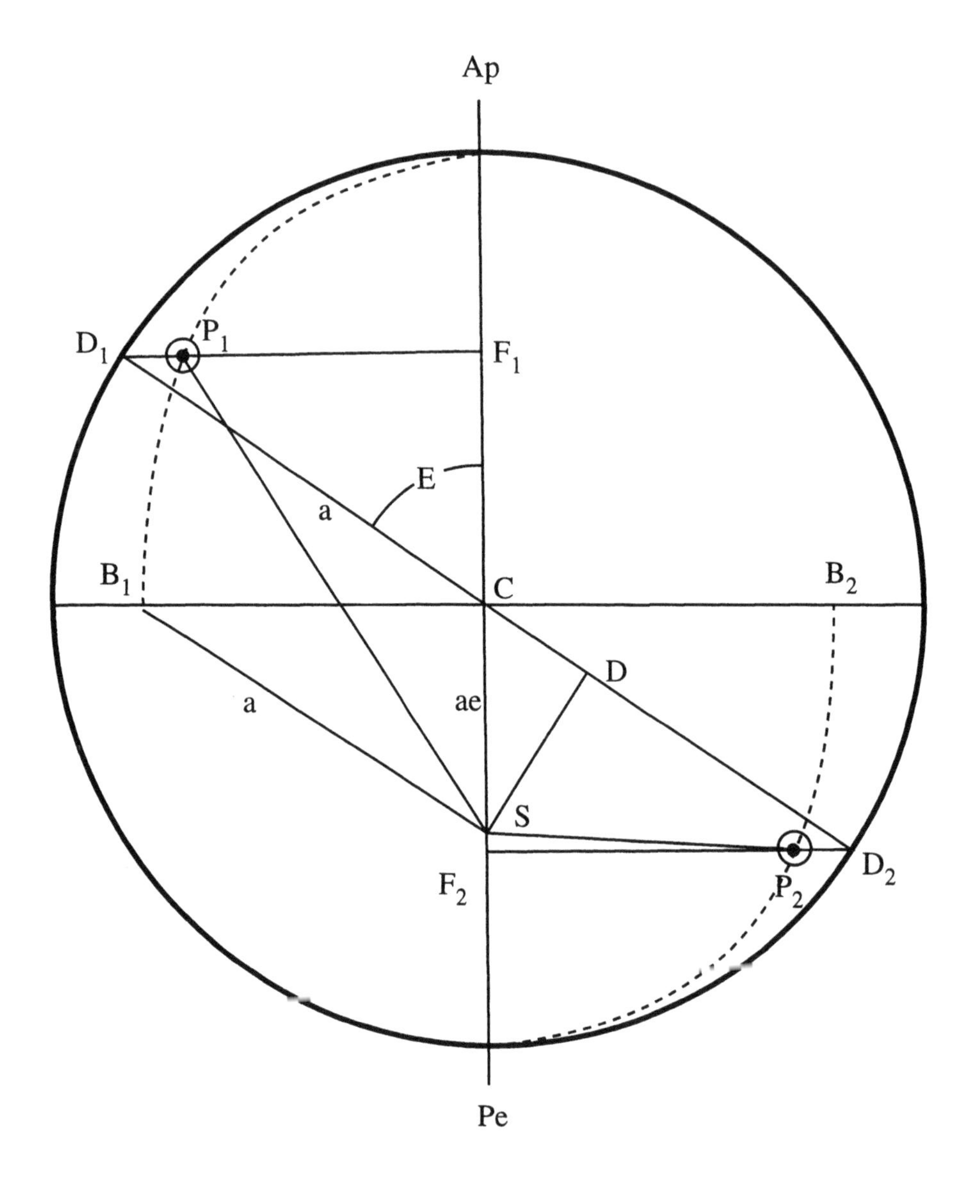

Figure VII.18 — Construction of Kepler's final ellipse by diametral distances.

These results indicate that Kepler had by 1605 reached one of his major goals, namely to represent the distances of Mars from the Sun to about 1/1000 part of the radius vector, and the heliocentric longitudes to about $1'$. Not yet satisfied that the orbit was everywhere an ellipse, he proceeded to compute a table of the distances in the ellipse for every whole degree of eccentric anomaly. Using this table to check the observed distances of all the remaining observations he proved in a most convincing manner that Mars's orbit was everywhere the ellipse halfway between the first, or auxiliary, ellipse and its enveloping circle.

Kepler's Search for a Physical Cause of Elliptic Motion: His First Magnetic Orbit

At this stage, since "Kepler's Equation" had not yet been proposed, the placing of a planet in an elliptic orbit at a given epoch was a long and laborious process. Perhaps this explains in part why he now, instead of forgetting all about circular uniform motions, made several attempts to retain them. He replaced the elliptic orbit by a construction assuming a retrograde uniform circular motion in an epicycle of radius e sliding with direct uniform motion on a circular deferent and of radius equal to the semimajor axis of the ellipse centered on the Sun F. This construction gave rise to a libration along the diameter of the epicycle (or of the orbit) with the same period as the planet. An enormous amount of time and energy was now spent on finding a law for this libration.

In order to derive a physical law for the motion of the planets, Kepler introduced a quasi-magnetic tendency whereby the body of the planet was permeated by a quasi-magnetic field, always directed at right angles to the line of apsides of its orbit. In his later writings, for example in *Epitome*, the cause of the constancy of the orientation in space of each planetary body was attributed, not to a magnetic field at right angles to the line of apsides (different for each planetary orbit), but to an "animal soul" inherent in each planet. This soul had the quality (through the quasi-magnetic fibers firmly embedded within the planet) of eternally tending to preserve the direction given to it at Creation, perpendicular to the line of apsides.

The motion of the planet was believed to be due to two simultaneous "forces," the solar motrix and the planetary motrix. Each planetary body was made up of three different kinds of "fibers." Although the fibers were everywhere finely intermixed or at least solidly connected, their effects are most easily visualized by considering them localized as follows: first, there were the "mass-giving" fibers, possibly the core of the planet, which in the first version of the planetary theory had the property of pinning down the planet at any given point of space. The planet was thus tethered in the space of the solar motrix "as by a hook" and would therefore rotate eastward with the speed of this space at the distance of the planet. Second, there were the "circular" fibers, possibly a mantle about the core of the planet, which had the property of rotating eastward eternally in or parallel to the plane of the planet's equator. This motion was also bequeathed to the planet at Creation. It would drag along the space about the planet, thus producing a secondary or planetary motrix, which would account not only for any direct axial rotation but also for the direct motion of satellites, if any. (Mars, for instance, had two satellites, as "proved" by Kepler in two ways!) Third, there were the "magnetic" fibers, not rotating with the planet and having two poles of opposite tendencies, the one solar seeking and the other solar repelling. These tended to be perpendicular to the line of apsides of the planet's orbit and either permeated the planetary body or, if restricted to a parallel bundle, were centrally placed

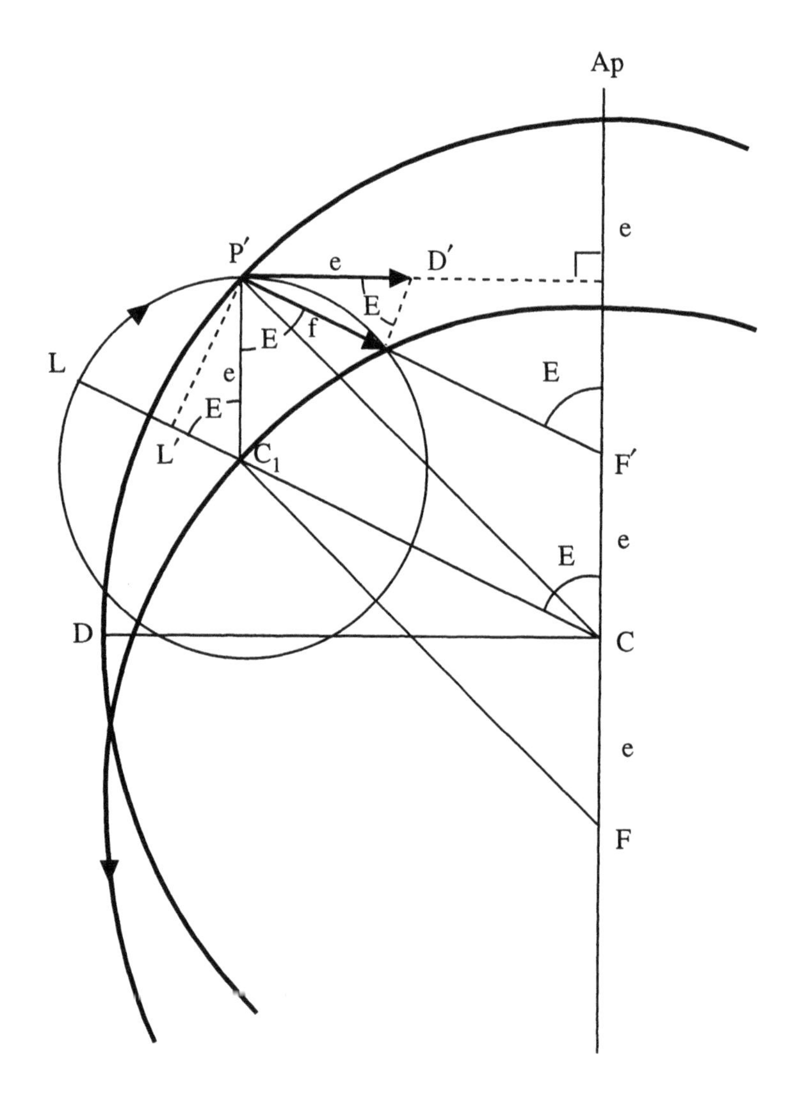

Figure VII.19 — Kepler's laws of libration and total libration resulting from magnetic forces.

and firmly interconnected with the core of the planetary body. The Sun-seeking pole was directed against the eastward motion at aphelion, as ordained at Creation.

The Sun was a strong magnet presumably built of the same kinds of fibers, except that its magnetic fibers were radially arranged so that one pole was at the center and the other uniformly spread over the surface. It was in constant eternal uniform rotation, with aesthetic and morphological reasons giving a period of about 3 days. Kepler states that the semidiameter of the Sun's body was to the semidiameter of Mercury's orbit as the semidiameter of the Earth's body was to the semidiameter of the Moon's orbit. Although these ratios were 1/42 and 1/54, Kepler considered that they were nearly enough equal to be significant. He therefore said (without any authority except perhaps a pretty Pythagorean notion) that there existed the following equality between the periods of the bodies involved: the rotation period of the Sun was to the orbital period of Mercury as the rotation period of the Earth was to the orbital period of the Moon. Hence, if x denotes the rotation period of the Sun, we have $(x/88) = (27/1)$, or $x = 3.2^{\mathrm{d}}$. He further considered that the rotation period in the solar motrix varied along the radius and was proportional to the inverse distance from the Sun.

The motion of a planet was influenced by the Sun's magnetic force as follows: Since at the aphelion of the orbit, the planetary magnetic poles were equidistant from the Sun, there was a balance of attraction and repulsion. The Sun therefore neither attracted nor repelled the planet, and the instantaneous motion was at right angles to the line of apsides with an orbital speed equal to that of the solar motrix at the aphelion distance. As the planet moved eastward away from the line of apsides, the solar-seeking pole moved nearer to the Sun than the solar-repelling pole. Therefore, the planet was slightly magnetically attracted by the Sun. Due to this attraction the planet now fell closer to the Sun and to the center of the orbit and increased its speed as it moved into a faster rotating region of the solar motrix. This effect continued until after perihelion, when the solar-repelling ends of the fibers were more turned toward the Sun (or toward the orbital center – at first Kepler did not distinguish between the two) than were the solar-seeking ends. At perihelion the magnetic fibers were once more perpendicular to the line of apsides and the orbital speed was a maximum. After perihelion the planet was magnetically repelled outward toward the circle enveloping the orbit, and its speed, being governed by the slower moving solar motrix at the greater distances, was now reduced until the original conditions were reached again at aphelion.

Kepler's Law of Libration for the Magnetic Orbit

Once the assumption was made that the directions of the magnetic fibers are always perpendicular to the line of apsides, the strength of the radial force on the planet due to the Sun's magnetic attraction on the fibers, and therefore on the body of the planet, follows a sine law, as illustrated in Figure VII.19 where e is the eccentricity and E the eccentric anomaly. Since, according to Aristotelian dynamics, the distance moved in any infinitesimal time interval is proportional to the strength of the moving force f during the interval, this distance, or "instantaneous libration" along the radius to the center of the orbit (later toward the Sun), is proportional to the sine of the eccentric angle E. Thus Kepler's law of libration was $f = e \sin E$. In *Astronomia Nova* Kepler explains the physical cause of the deviation of the planet from the enveloping circle of its orbit in this way.

Kepler's Law of Total Libration for the Magnetic Orbit

Having assumed the above law of libration, Kepler now investigated the resulting sum total of all the radial motion from the beginning up to any given position, i.e., from the aphelion to the point whose eccentric anomaly equals E. This "total libration," which is the radial distance LL', he said was equal to $G = e \operatorname{vers} E$. After having read the works of the famous Alexandrian mathematician Pappus, he made a proof using infinitesimals in a very daring manner, indiscriminately substituting lines for surfaces, radii for circles, heights for spherical segments, etc. The outcome of all this, however, was the correct result (we now know) that the total shortening of the radius of the enveloping circle (or later of the radius vector to the Sun) was proportional to the versed sine of the eccentric anomaly! This illustrates Kepler's remarkable knack for hitting upon correct results from erroneous, or at least non-rigorous, arguments.

The rigorous value of the libration G is shown by Caspar (1937) in a note to *Epitome* to be as follows: In modern notation, Kepler wants to find the value of $G = \int_0^E e \sin E \, dE$. This is $e\,(-\cos E)\,|_0^E = e[-\cos E - (-1)] = e(1 - \cos E) = e \operatorname{vers} E$.

Kepler's Proof That the Magnetic Orbit Is an Ellipse

Kepler immediately proved that the above law of total libration results in an exact elliptic orbit. According to Apollonius the ellipse has two quasi-centers, which Kepler had called the foci in his book on optics. Further, it was known that in an ellipse all ordinates perpendicular to the major axis have values equal to the values of the ordinates of corresponding points on the enveloping circle shortened in the ratio b/a of the minor to the major semiaxes. In Figure VII.20 let A = the Sun; B = the center of the ellipse PHE'; PR = the major axis; P = the aphelion; R = the perihelion; A, F = the foci of the ellipse; $AB = BF = e$ = the eccentricity of the ellipse; $BG = BD$ = the radius of the enveloping circle; $BP = AE' = a$ = the semimajor axis; $BE' = b$ = the semiminor axis; H = a point on the orbit; G = the point on the enveloping circle corresponding to H; $\angle PBG = E$ = the eccentric anomaly of H; ρ = AH = the radius vector or solar distance of H; F' = the foot of the ordinate of H and of G; $PF' = x$ = the total libration corresponding to the eccentric anomaly E; $GF' = y$ = the ordinate of G; $HF' = y'$ = the ordinate of H. Let $a = BR = BP = BD = BG = AE' = 1$. If H is a point on the ellipse, we must now prove that $e = AB/BP = AB = BF$ and that $HF'/GF' = y'/y = b/a = \sqrt{1 - e^2}$. From $\Delta AF'H$,

$$y'^2 = \rho^2 - (2e + FF')^2 = \rho^2 - (e + \cos E)^2.$$

The maximum length of ρ at aphelion is $AP = \rho_m = 1 + e$, and this is shortened by libration due to the orbital motion according to the formula $x = e \operatorname{vers} E = e(1 - \cos E)$. Hence the length of the radius vector H at any point with eccentric anomaly E is $\rho = AH = AP - x = (1 + e) - e(1 - \cos E) = 1 + e \cos E$.

Therefore, we have from $\Delta AF'H$:

$$\begin{aligned} y'^2 &= AH^2 - AF'^2 = (1 + e\cos E)^2 - (e + \cos E)^2 \\ &= 1 + 2e\cos e + e^2\cos^2 E - e^2 - 2e\cos E - \cos^2 E \\ &= 1 + \cos^2 E(e^2 - 1) - e^2 \\ &= 1 + (1 - \sin^2 e)\,(e^2 - 1) - e^2 \\ &= (1 - e^2)\sin^2 E, \qquad \text{or} \end{aligned}$$

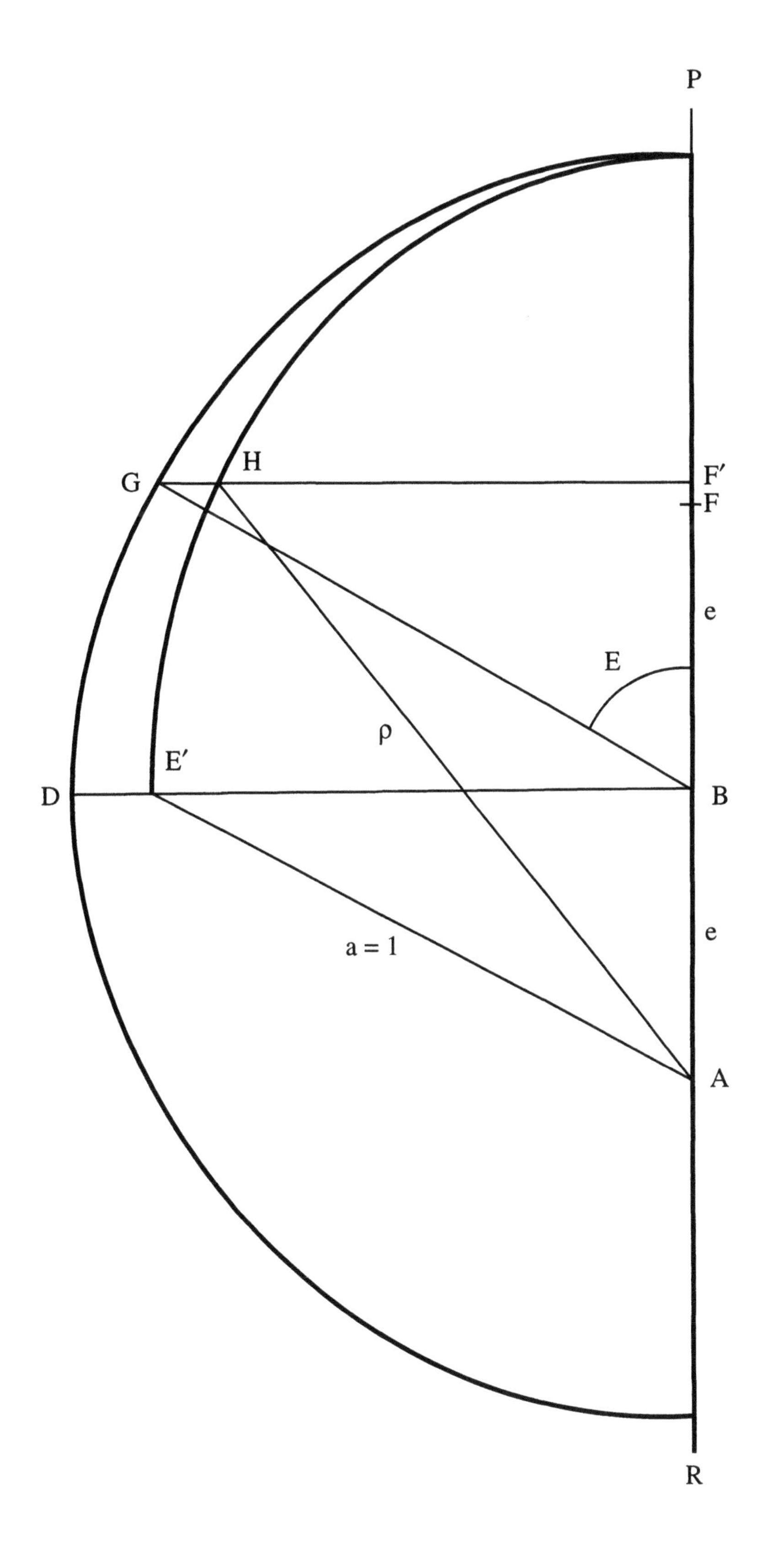

Figure VII.20 — Kepler's proof that his simple magnetic orbit is an exact ellipse.

$$y' = \sqrt{1-e^2}\sin E.$$

But $y = \sin E$, from $\Delta BF'$S. Therefore, $y'/y = \sqrt{1-e^2} = b/a$. Q.E.D. The locus of H is the orbital ellipse described under the influence of Kepler's total libration.

Kepler's Construction of an Ellipse from Its Total Libration on the Radius of Its Major Auxiliary Circle

Kepler now developed an elegant geometrical method to lay off points on an orbit. This entailed an exact libration on the diameter of the epicycle (or of the orbit, which is the same) defined by the versed sine of the eccentric anomaly. The construction adopted by Kepler, shown in Figure VII.21, was as follows: Choose two points C and F, distant e, on a straight line AA''. With radius $CA = a$, describe circles $AP'A''$ at center C and $A'BA'''$ at center F. Draw FC_1 and extend it to L. On C_1 as center, with $CF = e$ as radius, describe an epicycle $P'L$. This construction automatically makes $CF = P'C_1 = e$; $CP' = FC_1 = a$; $\angle P'C_1\,L = E$; $CD = BF = a$; and $CB = b$ when $E = 90°$. From P' drop perpendiculars $P'P'' = p$ and $P'P''' = d$. This automatically makes $LP'' = e \operatorname{vers} E$. Now locate the planet as follows: With center F and radius $FP'' = (a+e) - e\operatorname{vers} E$, which is the total libration, draw an arc of a circle intersecting $P'P'''$ in P, the position of the planet on an ellipse $APBA''$. (Note that the planet is neither at the point Y on the radius vector to P' nor at Z on the radius of the major auxiliary circle, but at P, which is on the perpendicular from P' to the line of apsides.) This construction gives an exact ellipse $APBA''$ for the orbit of the planet, while points such as Y and Z, both at the same solar distance as P, give various kinds of "ovals." The proof that the locus of P is an ellipse is as follows: In Figure VII.21, draw $FP = \ell$ and $FP' = \ell'$. Let $FP'' = \ell''$; $P'PP''' = d$; $PP''' = c$; and $CP''' = d'$. By inspection of Figure VII.21: $\ell = \ell''$; $\ell'^2 - \ell''^2 = \ell'^2 - \ell^2$. From $\Delta FP'P'''$: $\ell'^2 = d^2 + (d'+e)^2$. From $\Delta FPP'''$: $\ell^2 = c^2 + (d'+e)^2$. By subtraction and identification

$$\ell'^2 - \ell''^2 = d^2 - c^2. \tag{10}$$

Now, $\Delta P'P''C_1 \approx \Delta P'P'''C$. (Each has $\angle E$ and $\angle 90°$.) Therefore,

$$d^2 : a^2 : p^2 : e^2\,. \tag{11}$$

From $\Delta P'P''F$: $p^2 = \ell'^2 - \ell''^2 = d^2 - c^2$, by equation (10). From ΔFCB, $e^2 = a^2 - b^2$. Substituting in equation (11),

$$d^2 : a^2 = (d^2 - c^2) : (a^2 - b^2).$$

By composition,

$$d^2 : c^2 = a^2 : b^2 \qquad \text{or} \qquad d : c = a : b \qquad \text{or} \qquad c = (b/a)d.$$

This is the projection relation between ordinates on an ellipse and the corresponding ordinates on its major auxiliary circle. P thus lies on an ellipse $APBA''$ with center C, enveloping circle $AP'DA''$, semimajor axis $AC = a$, semiminor axis $CB = b$, and eccentricity e.

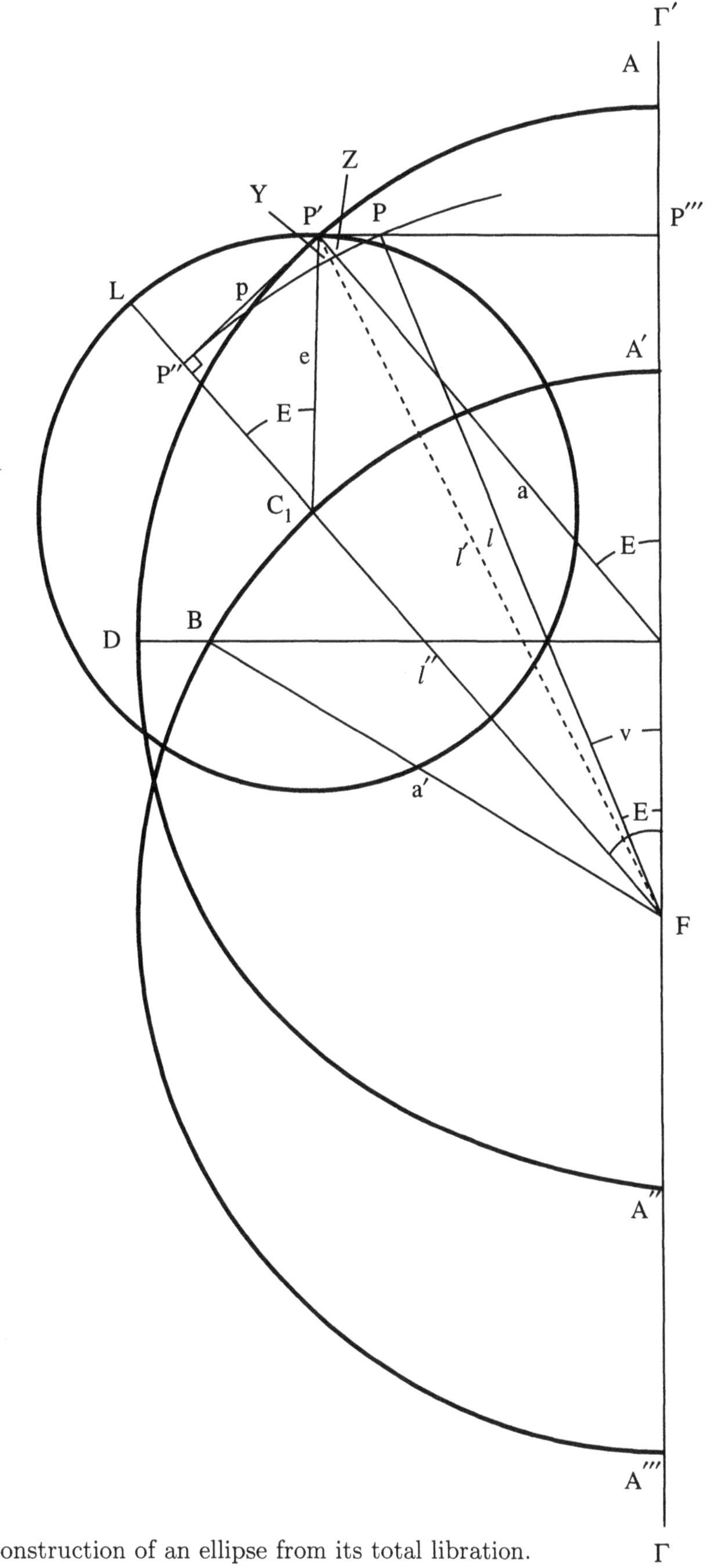

Figure VII.21 — Kepler's construction of an ellipse from its total libration.

Kepler's Abandonment of His Epicylic First Magnetic Orbit

Although this construction gave a perfect ellipse, Kepler seems to have used it only partially to find distances of Mars as affected by libration along the epicycle. To retain high accuracy in the results, he computed (rather than geometrically constructed) the relevant quantities. He soon found to his dismay that the longitudes in the first and fifth octants were $5\frac{1}{2}'$ greater than by the vicarious hypothesis, and in the third and seventh octants they were as much less. He therefore abandoned this theory, ascribing its failure to an error in the libration. In reality the error was due to his placing the planet on the radius of the major auxiliary circle at Z in Figure VII.21, instead of (correctly) placing it at P, on the perpendicular to the line of apsides. He was thus forced to return to the ellipse, which at this stage he considered to require a totally different process of computation involving his law of areas for locating a body in time.

First Application of the Areal Law to the Final Ellipse

Kepler now tentatively adopted the ellipse as final and applied himself to the difficult problem of finding a method of exactly, yet easily, placing a body at any given time along an elliptic orbit. In doing this he fully recognized the relations between the ellipse and its major auxiliary circle, and he continued to construct the ellipse using his elegant method of diametral distances. To place the planet at the proper point of the orbit he ran off the eccentric anomaly on the major auxiliary circle and then found the true anomaly by projection from circle to ellipse. This process would correctly place the planet on the perpendicular to the major axis from the terminal point for the eccentric anomaly. However, as a true product of his age, Kepler was obsessed with the idea that, as in his oval theories, the planet should be placed on the radius of the major auxiliary circle extending to the point measuring the eccentric anomaly. But the results were no better. Although the projected areas (referenced to the center of the ellipse) did grow uniformly with time, at first they did so at the wrong rate. We can see that this was because the mean anomaly M rather than E increases uniformly, and also because the uniformly growing areas should be referenced to the Sun, not to the center of the ellipse. This shows how Kepler temporarily abandoned his Second Law.

Another Attempt with the Areal Law on an Exact Ellipse

Kepler now tried to develop a more rational way to use his law of areas, applying it directly to the ellipse that he had already constructed. The law depended, he thought, upon the inverse relation of orbital speed and distance. This was simple in the case of a circular orbit, where the elements of arc are always at right angles to the radius. But in an ellipse they make a constantly varying angle. Undoubtedly he shrunk from the idea of directly connecting distances traveled along elements of arc with reciprocals of the radius vectors, when these elements were inclined by varying amounts to the radius vectors. Instead he calculated the sum of the diametral distances to every degree of eccentric anomaly, finding a value of 360.0. With this aid to placing the planet (equivalent to an areal law) he then tried an orbit with the planet positioned at a point on the radius of the major auxiliary circle, as in his oval theory, not at a point on the perpendicular to the major axis as in an ellipse! The result was the same as before, i.e., it gave maximum residuals of $\sim \pm 5'$ in longitude. Furthermore, the longitudes of the points where the diametral distances cut the perpendiculars to the major axis, i.e., of the points on the ellipse, were the same as found with the vicarious hypothesis and entirely agreed with observation. Kepler thus finally (in 1604) accepted this ellipse with the Sun at one focus as the actual physical orbit of Mars.

Astronomy with the Final Ellipse: Kepler's Equation

Having proved that Mars follows an elliptic orbit, Kepler now described a method for finding the true anomaly from the eccentric anomaly, as well as two cumbersome methods for finding the eccentric anomaly from the true anomaly. He had now solved the problem of finding the various anomalies, one from the other, as well as the radius vectors, for which he continued to use his diametral distances. But he found his methods, involving summations for every degree, too impractical. It was inelegant and cumbersome to construct a table for every degree, minute, and second of arc in the orbit in order to fix any past, present, or future planetary position. Yet he did not believe that there was any geometrically rigorous method of effecting a solution. He made a start, however, by posing what has ever since been referred to as Kepler's Problem. The problem was: From a given point in the diameter of a semicircle, draw a straight line dividing the area of the semicircle in any given ratio. In modern terms the problem is expressed by Kepler's Equation: $M = E - e \sin E$, where M is the mean anomaly, E the eccentric anomaly, and e the eccentricity. This problem, which ever since Kepler's time has exercised the ingenuity of the greatest mathematicians, is, as Kepler himself correctly stated, insoluble in rational terms. No rigorous solution can be obtained because, as Kepler remarks, a circular arc and the straight lines drawn about it, such as the sine, are quantities of quite different kinds, or, as we would say, the equation is transcendental and can be solved only by approximation. (During the last 350 years there have been over two hundred papers published on the solution of this equation!) For most purposes expansion into a series of a few terms gives the required accuracy.

Confirmation of Kepler's Final Elliptic Orbit by Celestial Latitudes of Mars

The celestial latitudes of the planets had throughout history given astronomers the most trouble. Even Kepler's otherwise admirable vicarious hypothesis had broken down completely with its failure to explain the latitudes of Mars. This is what finally and convincingly led Kepler to reject all perfectly circular orbits. The failure of the vicarious hypothesis proved that in no circle whatever would the distances, and therefore the latitudes, be correct (even though the longitudes were satisfactory). But when he determined the true forms, sizes, eccentricities, and orientations of the orbits of Mars and of the Earth, and when the Sun was placed at the focus of both orbits, the calculated distances (Table VII.6) were all in agreement with those determined from observations. The agreement is indeed remarkable for an orbit made from naked-eye observations without the availability of modern methods for the adjustment of the constants.

As a result of the good agreement between the observations and an assumed orbit of constant spatial orientation, the inclination of the plane of the orbit was established to be unvarying. This completely eliminated the oscillations in various directions of the epicycles (Ptolemy), as well as the periodic tilts of the planes of the orbits (Copernicus).

Table VII.6. Comparison between Observed (O) and Computed (C) Values of the Geocentric Celestial Latitudes of 12 of Tycho's Best Observations of Mars's Opposition[a]

Date	Constellation	Residual ($O - C$)
1580 Nov. 18	Ari	$-5.50'$
1582 Dec. 28	Gem	+2.67
1585 Jan. 31	Cnc	+0.67
1587 Mar. 6	Leo	−0.00
1589 Apr. 14	Vir	+2.00
1591 June 8	Sco	−2.00
1593 Aug. 26	Aqr	+1.25
1595 Oct. 31	Ari	+2.80
1597 Dec. 14	Tau	+3.00
1600 Jan. 19	Gem	+0.75
1602 Feb. 21	Cnc	+0.60
1604 Mar. 29	Vir	+2.90
Average residual		±1.98

[a]As quoted by Kepler in Chapter 62 of *Astronomia Nova*.

In addition, a clear explanation was now supplied for a phenomenon which had puzzled astronomers from Hipparchus to Tycho: celestial latitudes were not always greatest exactly at opposition. If the opposition occurred when the planet was near the perpendicular to the line of intersection of the terrestrial and planetary orbits, where for some days the change in the distance in the older systems is but little affected by any circular motions of both bodies, then the greatest geocentric latitude of the planet would indeed occur just at opposition. But if one or both bodies were moving in ellipses of even moderate eccentricities, then a distance change different from that caused by circular motions might entail a greater latitude some days before or after opposition. To quote Small's (1804) description: "If a planet be observed before an opposition at a certain distance from the Earth and at a certain heliocentric latitude, and if that distance due to the elliptic shape of either or both orbits should increase toward opposition in a greater ratio than the sine of the heliocentric latitude, then both the heliocentric and the geocentric latitudes would diminish toward opposition."

Table VII.6 indicates a preponderance of positive residuals, which may be a consequence of the unavailability of the method of least squares for the adjustment of the elements. In any extended series of observations a few large residuals may be expected. The first and largest residual, that of 1580, could be by chance, but might also represent an error in reduction or an effect of inexperience during the earliest observational work. (The first observation with the accurate 7-foot mural quadrant at Uraniborg was taken in 1581.)

The average residual of $1.98'$ seems to indicate that Tycho's planetary observations were on the average good to $2'$. This limit is reduced, however, as a consequence of the

Table VII.7. Comparison of Kepler's Orbital Elements for Mars with Those Obtained by Modern Methods from the Same Observations

	a	e	Γ	i	Ω	M_0
a)	1.5237	0.0931	328°36.4′	1°50.8′	46°36.5′	307°10.8′
b)	1.5237	0.0931	328 37.1	1 50.8	46 36.5	307 10.2
c)	1.5235	0.0926	328 39.8	1 51.0	46 33	307 27
d)	1.5237	0.0930	328 32.4	1 51.1	46 32.7	307 13.2

a) Values from Tycho's observations as used by Kepler.
b) Modern values from Tycho's observations, corrected for all relevant perturbations.
c) Kepler's best values as given in the *Rudolphine Tables*.
d) Values from Newcomb's *Planetary Tables* (1898), for the epoch 1591.0.

following considerations presented in Preuss (1971). In that volume, F. Schmeidler shows that Kepler's omission of orbital perturbations known to us today had no effect on the results within the accuracy at his disposal. From modern celestial mechanics it appears that there actually were perturbations of several arcminutes present in the positions of Mars during the period 1580–1600, but that they were so distributed as to cause only a small change in the derived elliptical orbital elements. Schmeidler also shows that the average error of a planet observation by Tycho was 1′ or at most 2′, as Kepler had always maintained.

A different way of elucidating the statement that not accounting for the perturbations in Tycho's opposition longitudes did not vitiate Kepler's work is to compare the values of orbital elements resulting from Kepler's work with the best modern estimate of their values in 1591, the epoch chosen by him as being central to his data. Such a comparison, shown in Table VII.7, reveals a remarkably close agreement. In spite of the presence of sizable perturbations their net effects were apparently minimal on the values of the final elements. The final orbit is illustrated in Figure VII.22.

Kepler's Second Magnetic Orbit

As we have seen, in his first theory of a magnetic orbit of a planet Kepler had assumed that the planetary magnetic fibers were kept constantly perpendicular to the line of apsides of the planet orbit, a direction ordained by the Creator and obeyed by the planet through the agency of an "animal soul" (probably unconscious, but always obedient). With the further assumption that the total radial drop-off from the enveloping circle (the so-called libration toward the center) was proportional to the versed sine of the eccentric anomaly, he proved rigorously that the resulting orbit is an exact ellipse. However, he soon decided that the libration, to be physical, would have to be directed toward the Sun rather than toward the center of the orbit. The question then arose whether the orbit would still be an ellipse if the same amount of drop-off were directed along a radius vector to the Sun rather than toward the center of the orbit. The answer to this question is "No!" for only a radial drop-off of amount e vers E will produce an ellipse.

In order to solve this dilemma Kepler developed a second magnetic orbit. For this he permitted the direction of the magnetic fibers of the planet to vary slightly during the revolution of the planet, producing just such a tilt that the net result of the Sun's

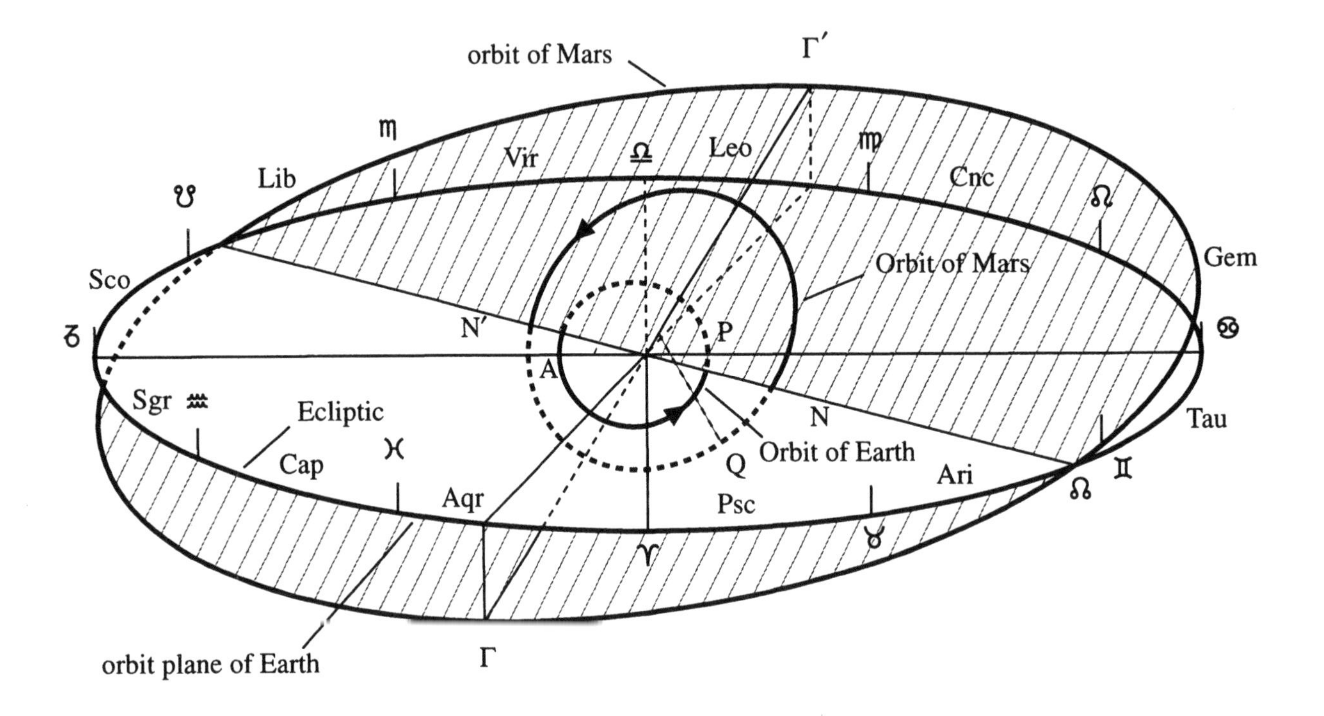

Figure VII.22 — The orbits of Earth and Mars (to approximate scale) as seen from the north side of the ecliptic plane. Both planets move in a direct sense in elliptic orbits of small eccentricity.

attraction on the tilted fibers very closely compensated for the fact that the drop-off from the enveloping circle was not radial, but along a solar radius vector. This deviation of the fibers from the original direction was called the *conversion* (or change of the original direction of the planetary magnetic fibers) and was taken to be due to the Sun's unequal "grip" upon the two hemispheres of the planet, the Sun-seeking and the Sun-repelling ones. The amount of conversion was guided by the planet's animal soul, which could "sense" the direction of the Sun.

To appreciate the result of this ingenious modification, let us retrace a few elementary developments. Let us assume (with Kepler) that at any point in the orbit the tilt of the magnetic fibers toward their original direction is equal to the optical equation at that point. In Figure VII.23 let P, I, N, E_1, R be positions of a planet on a circle of unit radius and center B. Let $\mathsf{m} = PP' = IH = NQ = E_1G' = RR'$ be the positions of a "converted magnetic" force originally at right angles to the line of apsides, making angles x, ϕ, y with the perpendiculars to the line of apsides. These angles are equal to the optical equation of the Sun as seen from the points I, N, E_1, respectively. Let A be the position of the Sun on the line of apsides at a distance e from the center B. Draw the perpendicular IS to the line of apsides. Draw IB, IA, NA. By definition:

$\angle PBI = E =$ the eccentric anomaly,

$\angle PAI = v =$ the true anomaly,

$\angle BIA = x =$ the optical equation at $I =$ the assumed "conversion" at I,

$\angle BNA = \phi =$ the equation of center $=$ the maximum optical equation.

Drop $p = BF$ perpendicularly on AI, and AH' perpendicularly on IH$'$. Let $f =$ the component of m along the radius vector AI, producing the libration along the radius vector from the planet to the Sun. To find x, the optical equation:

From ΔBIF, $\sin x = \dfrac{p}{BI} = \dfrac{p}{1} = p.$

From ΔBNA, $\tan\phi = \dfrac{AB}{BN} = \dfrac{e}{1} = e.$

By division, $\dfrac{\sin x}{\tan\phi} = \dfrac{p}{e}.$

From ΔBAF, $\dfrac{p}{e} = \sin v$, and

$$\sin\ x = \tan\phi \sin v \tag{12}$$

If $PINE_1R$ had been an ellipse, N would be the end of the minor axis, NA would equal 1, and equation (12) would read

$$\sin\ x = \sin\phi \sin v = e \sin v\,, \tag{12$'$}$$

where $\sin\phi = e$. In the case of Mars the difference in the formulas is inappreciable for $v = 90°$, since $\arcsin(\sin e) = 5°19'$, whereas $\arctan e = 5°18'$. From direct distance measurements based on observations Kepler proved in *Astronomia Nova* that Mars is at mean distance when $E = 90°$. Hence equation (12$'$) is correct for finding x.

The value of f can be found in various ways. The following method requires only elementary plane geometry:

Draw HG perpendicular to IA. From ΔIHG,

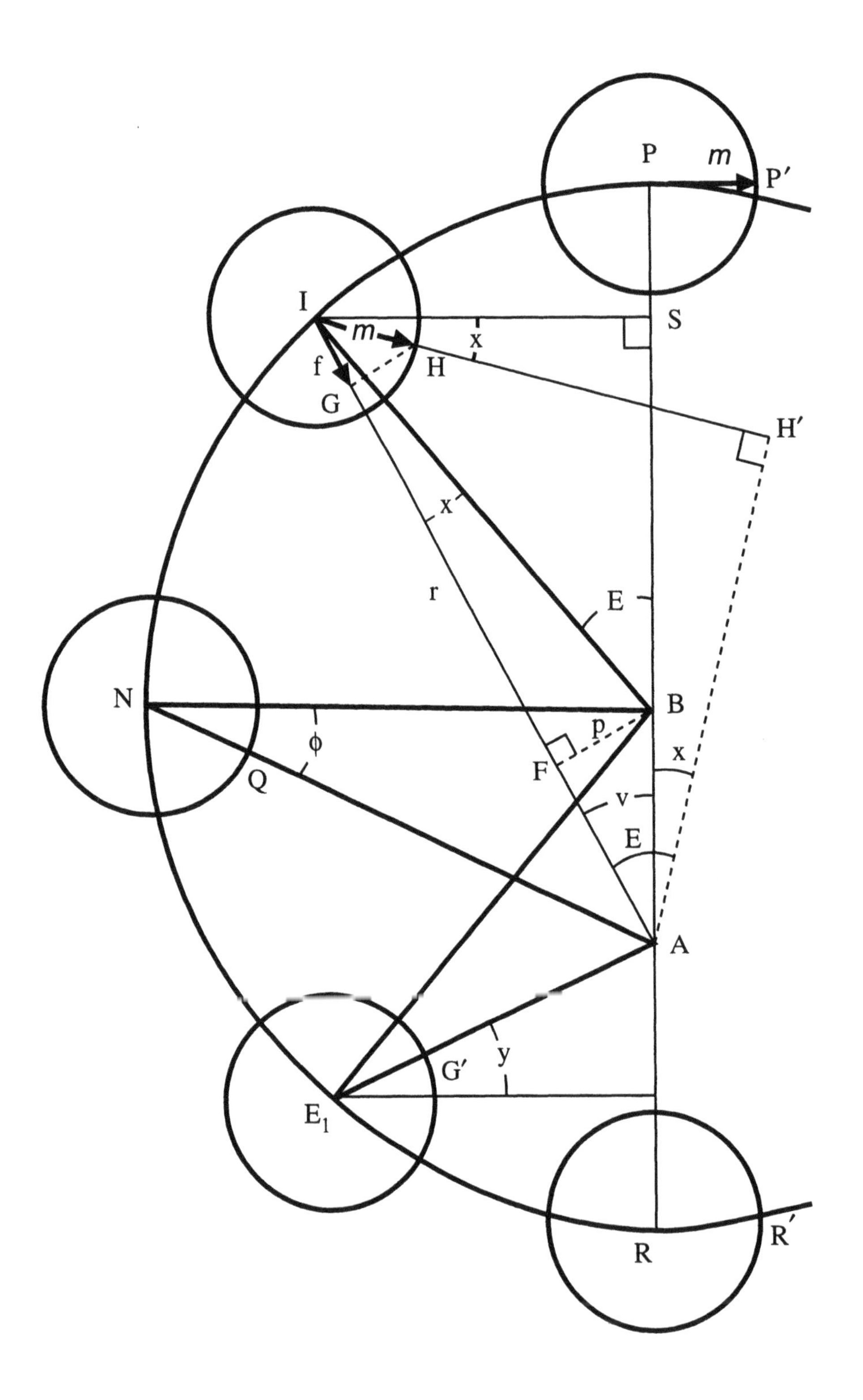

Figure VII.23 — Kepler's proof that "magnetic conversion" leaves the orbit an exact ellipse.

$$f = \mathsf{m} \sin \angle IHG. \tag{13}$$

To find $\angle IHG$, by hypothesis, the "conversion" at I is

$$\angle SIH' = \angle BIA = x = \text{the optical equation at } I.$$

By definition, $\angle SAI = v$. From ΔBIA, $E = x + v$. By construction

$$\angle SAH' = x \; (= \angle SIH) .$$

From Figure VII.23,

$$\angle IAH' = \angle SAI + \angle SAH' = v + x = E,$$

$$\angle GIH = 90^\circ - (v + x) = 90^\circ - E.$$

From ΔIHG,

$$\angle IHG = 90^\circ - \angle GIH = 90^\circ - (90^\circ - E) = E.$$

Using this in equation (13),

$$f = \mathsf{m} \sin E . \tag{14}$$

According to Kepler the meaning of equation (14) is that due to the tilt of the fibers the force f, although now active along the radius vector, nevertheless can be with you and varies as the sine of the eccentric anomaly only – not as the sine of the true anomaly. According to Aristotelian dynamics this force also measures the distance moved along its line of action in an infinitely short interval. If the drop-off were radial, i.e., along IB (Fig. VII.23), the orbit would be an exact ellipse centered on B, as proved earlier. But Kepler at first thought he could neglect the slight difference in direction between a radial drop-off and a drop-off along the radius vector. (This difference is much smaller than that between the direction of the radius and the perpendicular to the major axis.)

Thus we see that the shape of the second magnetic orbit (or, as Kepler thought, the "physical" orbit) is not exactly elliptical. Yet, like his oval orbits it passes through the ends of both axes of the true ellipse. At the octants it differs in longitude from the ellipse by about $\frac{1}{4}e^2$ radians, as we now know, and would have given residuals of about this size. But Kepler never subjected his magnetic orbits to the searching observational tests to which he had subjected his circular and oval orbits. Throughout the *Epitome* (1618) he made the tacit assumption that all real planet orbits were elliptical. As far as he was concerned, the Astronomical Revolution had been completed in 1604!

Comparison between the Shapes of Some of Kepler's Ovals and the True Ellipse

It was mentioned earlier that Kepler's first oval differed slightly from his first (auxiliary) ellipse by being too broad around the aphelion and too narrow around the perihelion. This statement may perhaps be best appreciated by looking back at Figure VII.11, in which a case near the first octant is exhibited. In this figure, let

M = the mean anomaly about the equant = $\angle HAM'$ (taken near the first octant),

E = the eccentric anomaly = $\angle HCM'$ (in the vicarious hypothesis),

v = the true anomaly = $\angle HSM'$ (in the vicarious hypothesis),

r = the true distance = SM'' (using $\angle HSM'' = E$),

SM' = the true direction of the planet (using v),

M''' = the true position of the planet (using v of the vicarious hypothesis and r of the modified bisection of eccentricity hypothesis).

Lay off the circular arc through M'', of center S and radius r. Then the planet is at M''' on the radius vector SM' to the Sun.

Compare this position M''' with a point on the ellipse corresponding to the eccentric anomaly E. On an elliptic orbit this point is at F on $M'F'$, which is perpendicular to the line of apsides HI at F′. It is obvious from this construction that M''' falls inside the ellipse HFI, and that the anomaly v of M''' is too large by $\angle M'''SF$. Hence, as computed by the oval, the perpendicular distance of M''' from HI is too large and the longitude too advanced; i.e., the oval is too broad around the aphelion. A similar construction with the angle near the third quadrant shows the distance too small and the longitude retarded relative to what the elliptic orbit would yield.

At a later stage Kepler constructed several other incorrect curves having true elliptic distances (found by his method of diametral distances) but incorrect anomalies. All had the same major axis and were ovals of slightly different shapes depending on whether the planet was placed on the radius vector to the Sun or on the radius to the center of the enveloping circle. The construction in Figure VII.24 shows how to locate points on two of the ovals. Let

M = the mean anomaly,

$\angle ACR' = 45°$, the eccentric anomaly,

$CS = CF = e$.

Then $r = SE = C'R'$, using the method of diametral distances. Drop the perpendicular $R'F'$ to the line of apsides AP. Intersect $R'F'$ at E by an arc of radius $r = R'C'$ and center S. Then E is the planet's place in an ellipse $AEBP$ of eccentricity e and center C.

Kepler was for a time so convinced that the planet must be located on the radius of the auxiliary circle to R', or at D on the radius vector from R' to the Sun, that he constructed two erroneous curves: (1) through R, the intersection of $R'C$ and the arc of radius r and center S, and (2) through the intersection D of the said arc with the radius vector from R' to the Sun. In both cases these constructions gave ovals, not ellipses. They were too broad around the aphelion and too narrow around the perihelion.[16] These representations obviously gave too long distances and too large true anomalies in the first octant, and too short distances with too small longitudes in the third octant.

Kepler's Correction of His Law of Linear Orbital Velocity

We will conclude this chapter by noting attempts by Kepler to rationalize two of his laws governing the motions of the planets. As has been shown, Kepler proved to his satisfaction

[16] A similar construction in the second octant, using r', E', R'', D' instead of r, E, R, D, places R'', D' inside the ellipse at the second octant.

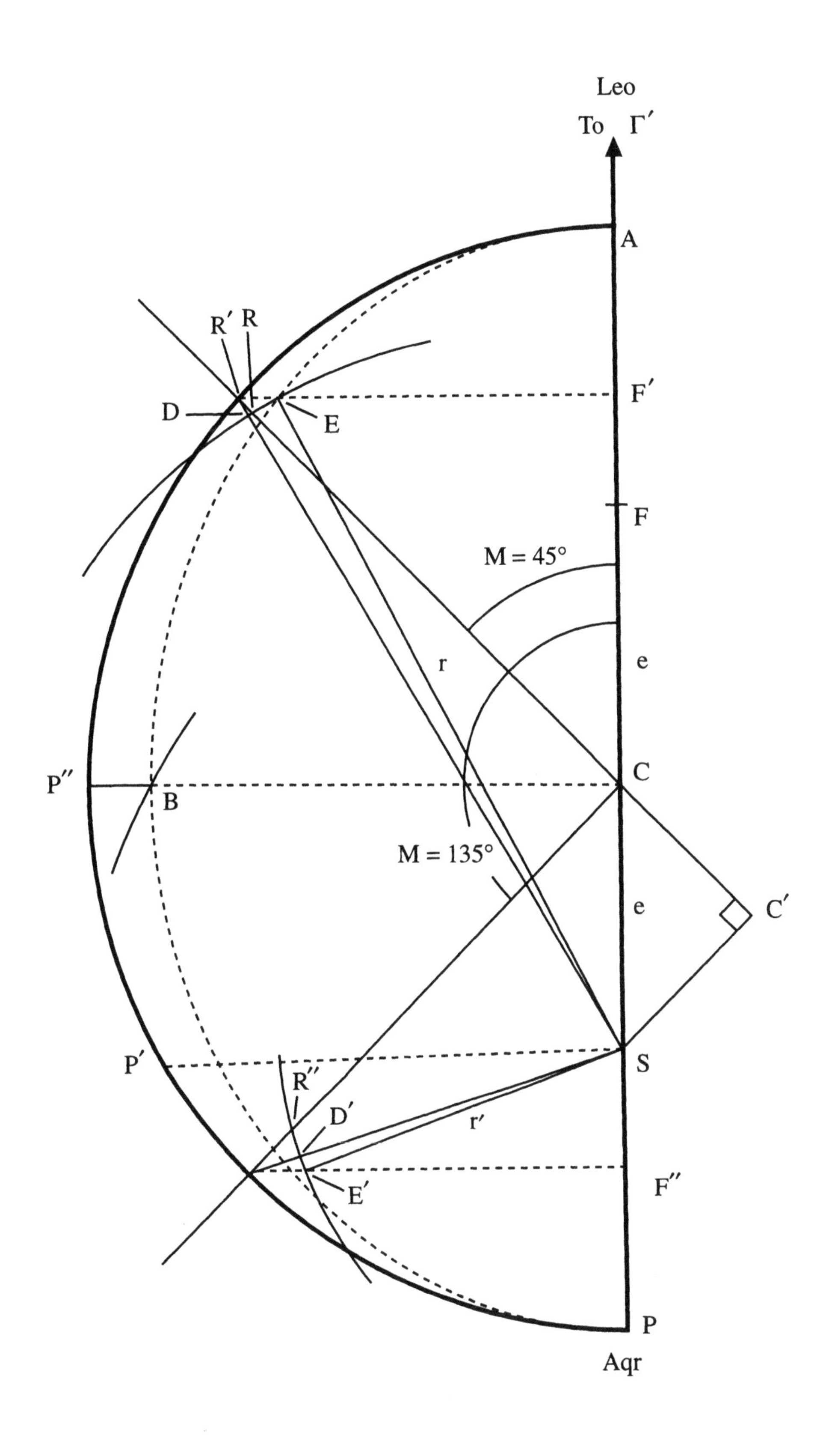

Figure VII.24 — Comparison between the ellipse and two of Kepler's ovals.

the First Law, that of the exact elliptical shape of the orbit, by assuming a total libration on the radius vector proportional to the sine of the eccentric anomaly. This proof is rigorously true for a libration of the specified amount along the radius of the major auxiliary circle and not along the radius vector of the ellipse, but the discrepancy for Mars's orbit is small. Throughout his major work he used an erroneous law of linear planetary orbital velocity, namely $v \sim 1/r$, where r equals the length of the radius vector from the Sun to a point of the orbit. But this law is rigorously true only for two points of the ellipse, the aphelion and the perihelion. From Newtonian theory the law of linear velocity valid for all points of the orbit is $v \sim 1/p$, where p is the length of the perpendicular from the occupied focus upon the tangent to the ellipse at the point considered. In the case of Mars, however, only a small discrepancy arises from assuming the wrong law for short intervals of time.[17] But when used to compute areas bounded by ovals or ellipses and summed for macroscopic time intervals, the accumulated errors are appreciable.

In *Epitome* (1618) Kepler took up again the "proof" of the law of areas that he had originally advanced in *Astronomia Nova*. His proof was based on the summation of the times taken to pass opposite pairs of small arcs of the orbit (with his still erroneous law of linear velocity). Although this proof was only approximate, it did lead him to modify his law of linear velocity. He now (correctly) made it apply only to velocities perpendicular to the radius vector.

In Figure VII.25, let

$A =$ the Sun,

$P =$ a point on the orbit OPT,

$PD =$ the tangent to OPT at P,

$r =$ the radius vector of P,

$p =$ the perpendicular on PD from A,

$ds =$ the linear orbital velocity at P,

$d\sigma =$ the component of linear orbital velocity perpendicular to r,

$\phi =$ the angle between the radius vector and the perpendicular upon the tangent.

Kepler now says: "The velocity component perpendicular to the radius vector r is inversely proportional to the distance r." Calling ds the tangential (i.e., the total orbital) velocity and $d\sigma$ its component perpendicular to r, we have by his new law in ΔPdc: $d\sigma = k/r$, where k is a constant of proportionality. Caspar (1937) points out in a note to his edition of the *Epitome* that, since (in ΔPdc) $d\sigma = \cos\phi\, ds$, or $ds = \sec\phi\, d\sigma$, Kepler's new law becomes $ds = k\sec\phi/r$. But ΔPdc and ΔADP are similar. Hence, $r = p\sec\phi$. We thus find $ds = k/p$, the true law of linear orbital velocity, which was first clearly stated and geometrically proved by Newton.

[17]For example, it would hardly have been conspicuous, in Magini's 1619 tables of hourly motions of the planets, which were used to extrapolate Tycho's observations for short intervals of time (in order to refer various observed quantities to a standard time).

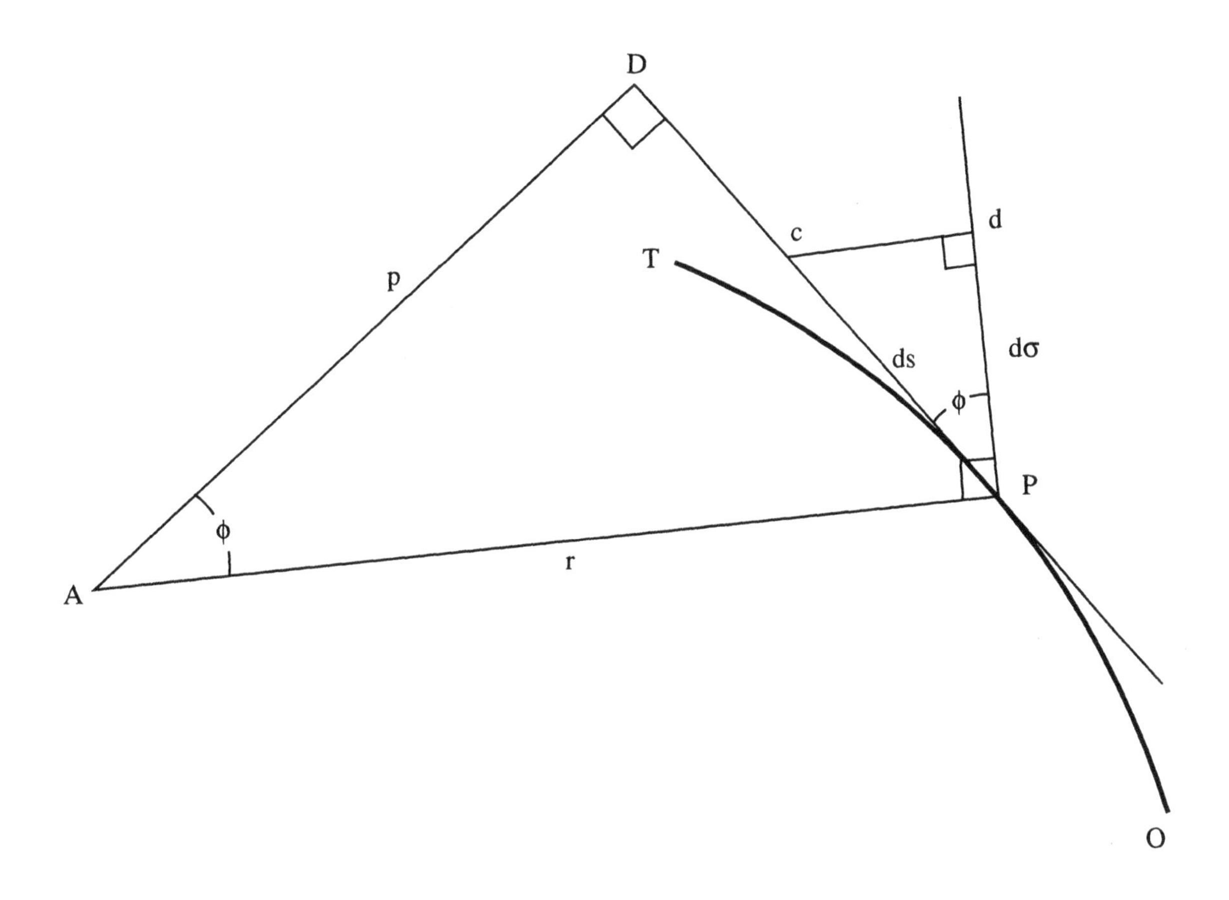

Figure VII.25 — Kepler's correction of his law of linear orbital velocity.

Some Curiosities Found in Kepler's Works

Kepler's "Proof" of His Third Law

In the "Harmony of the Universe" section of the *Epitome* Kepler says: "In order to dispel an unbearable state of monotony in the Universe, the Creator made the following choices of the relations between distances and velocities, masses, and densities of the planets:

$$\frac{v_1}{v_2} = \frac{a_2}{a_1}, \qquad \frac{m_1}{m_2} = \frac{\sqrt{a_1}}{\sqrt{a_2}}, \qquad \frac{d_1}{d_2} = \frac{\sqrt{a_2}}{\sqrt{a_1}}."$$

Consider any two planets moving in circular orbits of radii a_1 and a_2. Assume that the "mass" of any planet is proportional to the square root of the radius of its orbit, i.e., let $m \sim a^{\frac{1}{2}}$. According to Aristotelian dynamics, resistance to motion ("sluggishness") is proportional to "mass," and speed was instantly conferred to a body acted upon by a "projectile" force. Its magnitude would instantly attain and preserve a value inversely proportional to the resistance to motion. Thus $v \sim 1/m$, or equivalently $v \sim 1/a^{\frac{1}{2}}$.

Now consider two planets m_1 and m_2. For these we have accordingly $v_1/v_2 = a_2^{\frac{1}{2}}/a_1^{\frac{1}{2}}$. Of course s, the distance covered, is proportional to a, and the period will be $P = 2\pi a/v$. Combining these statements, we have

$$P_2/P_1 = (2\pi a_2/v_2)(v_1/2\pi a_1) = (a_2/a_1)(v_1/v_2) = (a_2/a_1)(a_2^{\frac{1}{2}}/a_1^{\frac{1}{2}}) = a_2^{3/2}/a_1^{3/2}.$$

Or $(P_2/P_1)^2 = (a_2/a_1)^3$, which "proves" Kepler's Third Law for circular orbits.

Kepler's Views on Stellar Distances

Although Kepler subscribed to the Copernican idea as a model of the Universe, he supplied his own refinements and estimates of numerical quantities describing its various parts. On the whole he revised values in the right direction, but in the cases where his values were most in error they came from his fanatic belief that mathematical harmonies from the workings of the Divine Mind permeated all of nature and could be ascertained by mental deduction and diligent observation. Thus, having assumed that the solar parallax was certainly less than $1'$, he found no conflict in adopting a value of 14 million miles for the Sun's distance and therefore 65,000 miles for its radius (a substantial improvement over the Copernican values of 6 million miles and 27,000 miles). For the fixed stars, having assumed (with Copernicus) that they showed no measurable annual parallax, he resorted to the following philosophical speculation. First he assumed them all to be in a spherical shell of thickness about 8 miles, concentric with the Sun. Now, since the Divine Mind could build only by proportion or progression, he assumed without any further argument the following proportion to hold true: $R_* : R_S :: R_S : r_\odot$, where

R_* = radius of fixed-star sphere,

R_S = radius of Saturn's orbit,

$r_\odot$ = radius of Sun.

Hence,

$$R_* = \frac{R_S^2}{r_\odot} = \frac{(9.5)^2(1.4 \times 10^7)^2}{6.5 \times 10^4 \times 5.88 \times 10^{12}} \text{ lt-yr} \approx 0.045 \text{ lt-yr}.$$

Kepler's "Proof" That Mars Has Just Two Moons

When Galileo discovered four satellites of Jupiter in 1610, Kepler immediately jumped to the conclusion that Mars had two, and that it could have no more nor less! His argument was as follows: The function of satellites is to furnish the planetary inhabitants (whose existence was taken for granted) with night illumination. Mercury, being so close to the Sun, required no extra illumination and therefore had no satellites. Venus could use a satellite, but the Venusians were so extremely sinful that God had withheld from them the favor of possessing a moon. The inhabitants of Earth were also sinful, but acting from Grace (the Protestant meaning of this concept) God had given them one moon! Jupiter had four moons as revealed by the telescope. Mars, lying between the Earth and Jupiter, must have two moons and only two because the second term of either *(a)* an arithmetic progression starting at the Earth with first term 0 and difference 2 or *(b)* a geometric progression starting at the Earth with first term 1 and ratio 2 gives Jupiter (corresponding to the third term) the number 4 and Mars (the second term) the number 2.

	Mercury	Venus	Earth	Mars	Jupiter	Saturn
Either	0	0	0	2	4	6
Or	0	0	1	2	4	8

In either case Mars must have two satellites. It follows equally that Saturn must have six or eight satellites. Kepler deplored the fact that he did not have a giant telescope with which he might anticipate Galileo in discovering two satellites of Mars, or six or eight of Saturn!

Concluding Remarks

One could continue to review a wealth of examples of Kepler's viewpoints regarding observational, astronomical, and optical phenomena (as has been done to some extent in the notes by Rosen [1965]). One finds him in possession of a remarkably modern viewpoint on many subjects. He was first of all an astronomer, although mystic contemplation (Neoplatonism) played a great part in his mental processes. But the examples of his purely astronomical work given here have shown that not only did he absolutely prove observationally that Mars's orbit is an ellipse, but he also supplied rigorous mathematical proofs of his results, although these were based upon what must today be considered entirely erroneous premises! This may have been inevitable prior to Newton's great dynamical clarification.

When one contemplates the extreme good fortune of Kepler to be able to work with Tycho (or sometimes perhaps in spite of Tycho), interesting possibilities arise. Everyone agrees, of course, that both the accuracy and completeness of Tycho's observations of Mars and Kepler's unswerving adherence to his own standards of accuracy of the agreement between theory and observation were necessary for the convincing replacement of the circles and epicycles of Ptolemy, Copernicus, and Tycho by one simple curve, the ellipse. The final proof (Table VII.5) was overwhelming and at one stroke ended nearly two thousand years of astronomical labor. Yet it demanded an entirely foreign and difficult viewpoint (that of nonuniform, elliptic motion) as a replacement for uniform circular motion. But suppose now that (1) Kepler had replaced his self-adopted standard of accuracy of 2′ by a less stringent one and been satisfied with residuals of 4′ or 5′ in size. At several places in his labyrinth of investigations he could have refined the work to meet this standard with the result that he would never have been forced to give up his cherished oval orbit, which could be closely reproduced by a system containing a deferent and one or two epicycles. Or suppose that (2) Kepler had relaxed his Copernican conviction and refined one of the elliptiform ovals studied by the Alfonsine astronomers. One of these, Azarquiel's oval, can in fact be made to represent the elliptical deferent of Mercury's orbit to an accuracy of 3′ in longitude. This curve could presumably be made to yield a still better representation of Mars's orbit, one requiring a very close scrutiny not only of the size but also of the run of the residuals in order to check its suitability. Or suppose that (3) Tycho's observations had been still more accurate, say good to 10″ or 20″, and Kepler had adopted a correspondingly high standard of accuracy. In this case the perturbations on Mars's orbit by the other planets, notably Jupiter and the Earth, would have become apparent and would have complicated the primitive empirical derivation of an elliptical orbit to such an extent that it is very doubtful that the correct solution could have been found. Or suppose finally that (4) Tycho had not assigned to Kepler the task of "defeating" his "most powerful enemy," Mars, but instead put him to work on some relatively easy project, such as improving the orbits of Jupiter and Saturn.

In closing we may note that the present account, though designed to describe some of Kepler's purely astronomical activities, has of necessity included some of his mystical opinions. These were necessary to a full understanding of the developments in *Astronomia Nova.* A recomputation of Kepler's complete work was undertaken by the French astronomer J. Delambre and also checked in several places by Small (1804). Both studies indicated a few numerical errors in the original 900 large, completely filled worksheets of *Astronomia Nova.* It is pointed out that these errors happened to cancel each other in a remarkable sequence of good luck. Thus there is no doubt that luck helped Kepler's progress in a number of instances, but a study of his work as a whole reveals unequivocally that the greatest characteristic of his remarkable genius, the one that carried the project to its conclusion, was perseverance. He was first of all the rigorous astronomer who designed, computed, and wrote *Astronomia Nova*, as well as the one who constructed and published the *Rudolphine Tables.* The seemingly scatterbrained, mystical, and medieval fantasies should be considered as inspirational aids to drive the completion of his great astronomical revolution.

Bibliography

Small, Robert (1804). *An Account of the Astronomical Discoveries of Kepler.* London: T. Gillet. Reprint, with Foreword by W. D. Stahlman, Madison: University of Wisconsin Press, 1963.

Herz, Norbert (1887). *Geschichte der Bahnbestimmung von Planeten und Kometen.* I Theil. *Die Theorien des Alterthums.* Leipzig: B. G. Teubner.

Dreyer, J. L. E. (1890). *Tycho Brahe: A Picture of Scientific Life and Work in the Sixteenth Century.* London: Adam and Charles Black. Reprints, New York: Dover, 1963; and Gloucester, Mass.: Peter Smith, 1977.

Herz, Norbert (1897). "Algemeine Einleitung in die Astronomie." In *Handwortenbuch der Astronomie,* ed. W. Valentiner. Breslau: Eduard Trewendt.

Dreyer, J. L. E. (1905). *History of the Planetary Systems from Thales to Kepler.* 2nd ed., rev. by W. H. Stahl. Cambridge: Cambridge University Press. Also available as *A History of Astronomy from Thales to Kepler.* New York: Dover, 1953.

Berry, Arthur (1910). *A Short History of Astronomy.* New York: Scribner's.

Petersen, Arthur (1924). *Tyge Brahe ved Udvalget for Folkeoplysnings Fremme.* Copenhagen: G. E. C. Gad.

Duncan, J. C. (1926). *Astronomy: A Textbook.* New York: Harper's.

Russell, H. H., Dugan, R. S., and Stewart, J. Q. (1926). *Astronomy.* Vol. I. *The Solar System.* Boston: Ginn.

Smart, W. M. (1931). *Textbook on Spherical Astronomy.* Cambridge: Cambridge University Press. 6th ed., rev. by R. M. Green, 1977.

Antoniadi, Eugène M. (1934). *L'Astronomie Egyptienne depuis les Temps les Plus Reculés Jusqu'à la Fin de l'Epoque Alexandrine.* Paris: Gauthier-Villars.

Caspar, Max, trans. (1937). *Epitome Astronomiae Copernicanae.* Bd. 7 of *Johannes Kepler: Gesammelte Werke.* Munich: C. H. Beck.

Zeller, Karl, trans. (1943). *Erster Bericht über die 6 Bücher des Kopernikus von den Kreisbewegungen der Himmelsbahnen.* Munich: R. Oldenbourg.

Rosen, E., trans. (1965). *Kepler's Conversation with Galileo's Sidereal Messenger.* New York: Johnson Reprint Co.

Ahnert, Paul (1971). *Astronomische-chronologische Tafeln für Sonne, Mond und Planeten.* Leipzig: J. A. Barth.

Preuss, Ekkehard, ed. (1971). *Kepler-Festschrift.* Mittelbayerische Druckerei.

Thoren, Victor (1990). *The Lord of Uraniborg.* Cambridge: Cambridge University Press.

Evans, James (1998). *The History and Practice of Ancient Astronomy.* Oxford: Oxford University Press.

Milton Keynes UK
Ingram Content Group UK Ltd.
UKHW052207280924
448807UK00002B/13